IET ENERGY ENGINEERING SERIES 98

Large Scale Grid Integration of Renewable Energy Sources

Other volumes in this series:

Volume 1 **Power circuit breaker theory and design** C.H. Flurscheim (Editor)
Volume 4 **Industrial microwave heating** A.C. Metaxas and R.J. Meredith
Volume 7 **Insulators for high voltages** J.S.T. Looms
Volume 8 **Variable frequency AC motor drive systems** D. Finney
Volume 10 **SF_6 switchgear** H.M. Ryan and G.R. Jones
Volume 11 **Conduction and induction heating** E.J. Davies
Volume 13 **Statistical techniques for high voltage engineering** W. Hauschild and W. Mosch
Volume 14 **Uninterruptible power supplies** J. Platts and J.D. St Aubyn (Editors)
Volume 15 **Digital protection for power systems** A.T. Johns and S.K. Salman
Volume 16 **Electricity economics and planning** T.W. Berrie
Volume 18 **Vacuum switchgear** A. Greenwood
Volume 19 **Electrical safety: a guide to causes and prevention of hazards** J. Maxwell Adams
Volume 21 **Electricity distribution network design, 2nd Edition** E. Lakervi and E.J. Holmes
Volume 22 **Artificial intelligence techniques in power systems** K. Warwick, A.O. Ekwue and R. Aggarwal (Editors)
Volume 24 **Power system commissioning and maintenance practice** K. Harker
Volume 25 **Engineers' handbook of industrial microwave heating** R.J. Meredith
Volume 26 **Small electric motors** H. Moczala *et al.*
Volume 27 **AC–DC power system analysis** J. Arrillaga and B.C. Smith
Volume 29 **High voltage direct current transmission, 2nd Edition** J. Arrillaga
Volume 30 **Flexible AC transmission systems (FACTS)** Y.-H. Song (Editor)
Volume 31 **Embedded generation** N. Jenkins *et al.*
Volume 32 **High voltage engineering and testing, 2nd Edition** H.M. Ryan (Editor)
Volume 33 **Overvoltage protection of low-voltage systems, Revised Edition** P. Hasse
Volume 36 **Voltage quality in electrical power systems** J. Schlabbach *et al.*
Volume 37 **Electrical steels for rotating machines** P. Beckley
Volume 38 **The electric car: development and future of battery, hybrid and fuel-cell cars** M. Westbrook
Volume 39 **Power systems electromagnetic transients simulation** J. Arrillaga and N. Watson
Volume 40 **Advances in high voltage engineering** M. Haddad and D. Warne
Volume 41 **Electrical operation of electrostatic precipitators** K. Parker
Volume 43 **Thermal power plant simulation and control** D. Flynn
Volume 44 **Economic evaluation of projects in the electricity supply industry** H. Khatib
Volume 45 **Propulsion systems for hybrid vehicles** J. Miller
Volume 46 **Distribution switchgear** S. Stewart
Volume 47 **Protection of electricity distribution networks, 2nd Edition** J. Gers and E. Holmes
Volume 48 **Wood pole overhead lines** B. Wareing
Volume 49 **Electric fuses, 3rd Edition** A. Wright and G. Newbery
Volume 50 **Wind power integration: connection and system operational aspects** B. Fox *et al.*
Volume 51 **Short circuit currents** J. Schlabbach
Volume 52 **Nuclear power** J. Wood
Volume 53 **Condition assessment of high voltage insulation in power system equipment** R.E. James and Q. Su
Volume 55 **Local energy: distributed generation of heat and power** J. Wood
Volume 56 **Condition monitoring of rotating electrical machines** P. Tavner, L. Ran, J. Penman and H. Sedding
Volume 57 **The control techniques drives and controls handbook, 2nd Edition** B. Drury
Volume 58 **Lightning protection** V. Cooray (Editor)
Volume 59 **Ultracapacitor applications** J.M. Miller
Volume 62 **Lightning electromagnetics** V. Cooray
Volume 63 **Energy storage for power systems, 2nd Edition** A. Ter-Gazarian
Volume 65 **Protection of electricity distribution networks, 3rd Edition** J. Gers
Volume 66 **High voltage engineering testing, 3rd Edition** H. Ryan (Editor)
Volume 67 **Multicore simulation of power system transients** F.M. Uriate
Volume 68 **Distribution system analysis and automation** J. Gers
Volume 69 **The lightening flash, 2nd Edition** V. Cooray (Editor)
Volume 70 **Economic evaluation of projects in the electricity supply industry, 3rd Edition** H. Khatib
Volume 72 **Control circuits in power electronics: practical issues in design and implementation** M. Castilla (Editor)
Volume 73 **Wide area monitoring, protection and control systems: the enabler for smarter grids** A. Vaccaro and A. Zobaa (Editors)
Volume 74 **Power electronic converters and systems: frontiers and applications** A.M. Trzynadlowski (Editor)
Volume 75 **Power distribution automation** B. Das (Editor)
Volume 76 **Power system stability: modelling, analysis and control** B. Om P. Malik
Volume 78 **Numerical analysis of power system transients and dynamics** A. Ametani (Editor)
Volume 79 **Vehicle-to-Grid: linking electric vehicles to the smart grid** J. Lu and J. Hossain (Editors)
Volume 81 **Cyber-physical-social systems and constructs in electric power engineering** Siddharth Suryanarayanan, Robin Roche and Timothy M. Hansen (Editors)
Volume 82 **Periodic control of power electronic converters** F. Blaabjerg, K. Zhou, D. Wang and Y. Yang
Volume 86 **Advances in power system modelling, control and stability analysis** F. Milano (Editor)
Volume 88 **Smarter energy: from smart metering to the smart grid** H. Sun, N. Hatziargyriou, H.V. Poor, L. Carpanini and M.A. Sánchez Fornié (Editors)
Volume 89 **Hydrogen production, separation and purification for energy** A. Basile, F. Dalena, J. Tong, and T.N. Veziroğlu (Editors)
Volume 90 **Clean energy microgrids** S. Obara and J. Morel (Editors)
Volume 92 **Power quality in future electrical power systems** A.F. Zobaa and S.H.E.A. Aleem (Editors)
Volume 93 **Cogeneration and district energy systems: modelling, analysis and optimization** M.A. Rosen and S. Koohi-Fayegh
Volume 95 **Communication, control and security challenges for the smart grid** S.M. Muyeen and S. Rahman (Editors)
Volume 97 **Synchronized phasor measurements for smart grids** M.J.B. Reddy and D.K. Mohanta (Editors)
Volume 100 **Modeling and dynamic behaviour of hydropower plants** N. Kishor and J. Fraile-Ardanuy (Editors)
Volume 101 **Methane and hydrogen for energy storage** R. Carriveau and David S.-K. Ting
Volume 905 **Power system protection, 4 volumes**

Large Scale Grid Integration of Renewable Energy Sources

Edited by
Antonio Moreno-Munoz

The Institution of Engineering and Technology

Published by The Institution of Engineering and Technology, London, United Kingdom

The Institution of Engineering and Technology is registered as a Charity in England & Wales (no. 211014) and Scotland (no. SC038698).

First published 2017

The Institution of Engineering and Technology
Michael Faraday House
Six Hills Way, Stevenage
Herts, SG1 2AY, United Kingdom

www.theiet.org

British Library Cataloguing in Publication Data
A catalogue record for this product is available from the British Library

ISBN 978-1-78561-162-9 (hardback)
ISBN 978-1-78561-163-6 (PDF)

Typeset in India by MPS Limited

Contents

Chapter 1

The power grid as part of a 100% renewable energy system

Math H.J. Bollen[1] *and Antonio Moreno-Munoz*[2]

1.1 Introduction

Worldwide, energy demand is constantly increasing. It is expected that global energy consumption will increase more than a third in the period to 2035. Among all the energy demand, the residential sector represents an average 30% of the total energy consumption in most of the developed countries, showing an upward trend in the last 10-year period with an average growth of about 5% in the European countries. Currently, most of this demand is being met by fossil fuels, which contributes to increased emissions of greenhouse gases and the warming global. It is true that last years' falling prices increased demand for fossil fuels, but prices apparently cannot fall much further. OECD countries have been reducing demand for almost a decade; in China, fossil fuel consumption has also flattened. Only developing countries show rapidly expanding fuel demands, though at a slower rate.

Renewable energy sources (RES) can be divided into two categories: dispatchable (e.g., hydro, geothermal and biomass power) and non-dispatchable or variable renewable energy or VRE (e.g., solar and wind) due to the dependence of its production on weather conditions. Of late, VRE played only a minor role in most countries' energy portfolio. While public institution and non-governmental organisations desired an increase in the share of RES in their energy market, the economic issues for new sources (wind and solar) were challenging. Up until recently, VRE has remained a more expensive alternative because of the upfront costs of installing these technologies – requiring public subsidies to get a look-in. However, according to [1], a major change of trend in the growth of low carbon technologies is observed. Prices for inputs – particularly for PV panels and wind turbines – have come down so far that VRE is now cost-competitive with retail electricity in an increasing number of markets globally. The results of the analysis show that the main driver of grid parity is the decrease in VRE system components price. Public subsidies have been helping wind and solar get a foothold in the global energy

[1]Luleå University of Technology, Skellefteå, Sweden
[2]Universidad de Córdoba, Córdoba, Spain

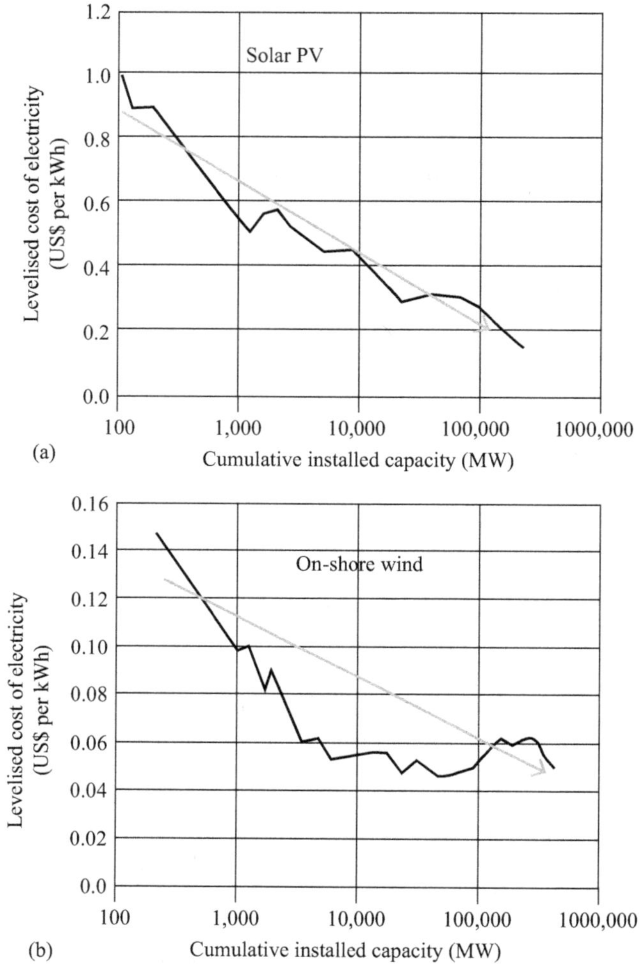

Figure 1.1 *(a) Solar PV costs have decreased as installed capacity has increased. The logarithmic trend line is close to linear and can be explained by the learning curve in manufacture of PV modules and by reduction of soft costs. (b) On-shore wind costs have decreased as installed capacity has increased. Cost increases at the end of the last century were from higher site costs and larger turbine designs, but the trend appears to be continuing downward. Courtesy of the Pembina Institute. https://www.pembina.org/pub/true-price-of-wind-and-solar*

markets, but the real driver of lower prices has been economies of scale. As seen in Figure 1.1, installed capacity increase and competition among module manufacturers in established solar markets led to significant reductions in module costs [2]. The reason is that it is a technology, so over time the improvements increase efficiency and lower prices. Likewise, the price of batteries is falling between 40% and 60%, so that there is already talk of a boom in the energy storage market. Distributed energy resources (DER) include distributed generation, or on-site small-scale power generation sources, but any distributed energy storage system (ESS) can also be classified as a DER system. Recent reports indicate that the global ESS market for smart grid is expected to grow 73% through to 2020, also forecasting that solar energy combined with batteries and electric vehicles will soon be cheaper than grid electricity and fuel in key markets such as Germany, Spain and Italy.

As stated in a recent study for solar electricity, roughly 30 countries are currently at grid parity. And it is expected that electricity prices worldwide will double over the next 10–15 years making the case for solar grid parity even stronger. In the increasing energy production of recent years is observed a clear decrement of the energy produced from solid fuels (18.8%) and a higher penetration of new renewable sources, whose annual average growth rate is currently 7% [3]. Although grid parity by itself is no guarantee of market creation, the outlook for VRE looks promising, and equally optimistic prospects are expressed in the commercial and utility scale segments. Thereby, in a 'business as usual' scenario, RES could reach 25% of the global energy mix already by 2018. But this industry is nowadays speeding up towards the goal of the cent per cent renewables.

1.2 Status and trends in the voluntary use of renewable energy

Voluntary green power markets are those where consumers and institutions voluntarily purchase RES and other DER to meet their electricity needs. In the mid-1990s, there were few options for commercial and institutional customers who wanted to employ RES. An end user could install a solar photovoltaic system in place, or he/she could select a retail provider that combines renewable energy certificates with normal power, in a deregulated market. By the middle of the last decade, the solar power purchase agreements (PPAs) were born. They are allowed more frequently in locations with deregulated electricity markets. Since then, a number of companies, universities and others have negotiated long-term purchases of renewable energy through them. Consequently, the most popular method of solar installations transitioned from ownership to PPAs, and the upward trend continues as more countries approve third-party financing. Purchasing options in the voluntary markets for renewable energy have continued to evolve and diversify until today, as has been thoroughly reported [4]. Some of the new purchase options are providing a hedge against future increases in electricity prices or other benefits as discussed below.

1.2.1 World's leading companies pledge to use 100% renewable

Clearly, we are at a turning point today how corporate and institutional organisations seek to use renewable energy. Renewable energy markets have fundamentally changed in profound ways. Commercial prosumers may be even able to reach self-consumption ratios of close to 100%, covering a fair amount of their base load by DER. Nearly two-thirds of the Fortune 100 and nearly half of the Fortune 500 companies are committing to 100% renewable power, by joining the RE100 [5] initiative. Worth to mention the leadership in solar power generation carried out by many of them, operating over 445 MW of solar capacity at more than 950 facilities in the United States. These non-energy companies are becoming major DER players, with a view to diversifying their energy supply, hedging against fossil-fuel price and supply volatility, reducing their long-term operating costs and earning market-based returns on investment, as well as cutting emissions that contribute to global warming. The scale of their energy savings is likely to grow substantially. They are tapping solar and wind and creating new markets, no longer only just responding to them. So, as companies in adaptation to being more dynamic participants in the energy marketplace, they are developing DER in emplacements where policies conditions are most acceptable. Regulatory inconsistency could dissuade companies from meeting ambitious commitments because of uncertainty around the price, supply, and deployment of VRE.

1.2.2 The islands leading towards 100% renewable

Particularly, the most ambitious targets are being raised on small isolated islands. This is not by coincidence; most islands rely on fuel imports to run their standalone grid. These shipments of diesel, oil or natural gas are very expensive. It is not only the savings, it also means less reliance on imports, increasing energy security. Hawaii announced its plans to transition its electric grid to 100% renewables by 2045. El Hierro, one of the Spanish Canary Islands off the coast of Africa, inaugurated year ago a hybrid wind-hydro power plant to serve its population of 11,000 and run power-hungry desalination plants. The hybrid plant has just completed one year of successful operation [6].

Many other islands are expanding the amount of electricity from renewables, as has been addressed elsewhere [7]. Obviously, an island seems like a hard place to VRE like wind and solar or even DERs. Island grids are usually isolated and weak island grids generally have to pay more attention to power quality and reliability (PQR), raising the overall costs of DER.

1.3 The 100% renewable energy system

A vision for a 100% renewable energy system is shown in Figure 1.2. The renewable energy can be in the form of wind or solar power, but also as small- or large-scale hydropower installations. Also certain types of biomass are considered as renewable energy. For this chapter, the discussion about what exactly constitutes renewable energy is not relevant. Important is only that there will be a mix of energy sources of different sizes and at different locations, a lot of which will be constrained in time and in location. Solar and wind power will only be available when the sun is shining or when the wind is blowing. A lot has been written about this already, where the somewhat inappropriate and confusing term 'intermittency' is typically used. But the actual situation is much more complicated. Next to the

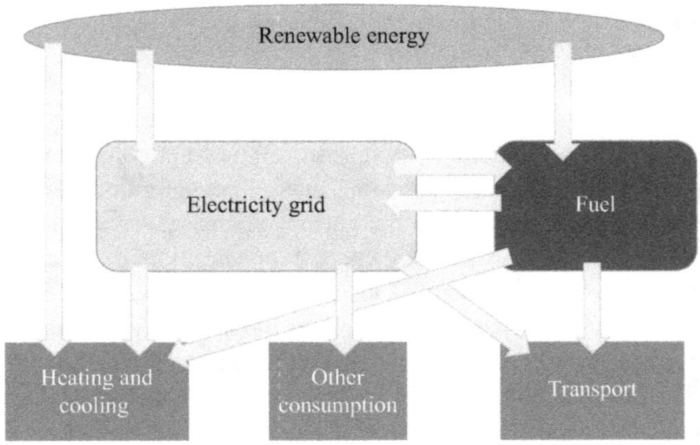

Figure 1.2 A vision of a 100% renewable energy system; the arrows indicate the different energy conversions

variations in availability, the availability of sun and wind is also difficult to predict. This holds especially for regions with fast-changing weather patterns, like north-western Europe. The availability of hydropower is easier to predict, at least up to a few months in advance, and at least where it concerns large-scale reservoir-based hydropower. Unfortunately, this is also the type that is most difficult to build in many parts of the world, not only technically but also politically. At timescales longer than a few months, also hydropower becomes less predictable. The amount of water in the reservoirs depends on the amount of rain or snow in the preceding weeks or months. Countries with large amounts of hydropower (like Spain and Sweden) are very much aware of the existence of 'wet years' and 'dry years'. To which extent there is a correlation between dry years, years with little sun, and years with little wind are not known at the moment. But the overall impression from various studies is that the probability of lack of energy will reduce by mixing different sources of renewable energy, for example, combining wind power, solar power and hydropower.

The renewable energy will be used for a range of applications. An important part of the transition to a 100% renewable energy system will be an overall shift from energy-intensive activities to non-energy-intensive processes as well as an overall increase of efficiency of the energy-intensive processes. Some of the on-going or heavily discussed changes are:

- Insulation of buildings to reduce the amount of energy needed for space heating or cooling;
- A shift from gas heating to electric heating and from direct electric heating (resistive heating) to heat pumps;
- A shift to more energy-efficient (electric) heating, ventilation and air-conditioning;
- The shift from incandescent lamps to LED lamps;
- The shift from gasoline-powered cars to electric cars and the continued electrification of the railways.

But a certain energy demand will remain, which is the one indicated in Figure 1.2. Energy efficiency is beyond the scope of this book; it is assumed that the energy demand is the one that is necessary and that is the most efficient one.

For simplicity, a distinction is made in Figure 1.2 between three types of energy consumption: 'heating and cooling'; 'transport'; and 'other consumption'. Note that we use the physically incorrect expression 'energy consumption' when referring to an energy conversion process where the entropy increases. In almost all cases it is obvious to the reader what is meant by energy consumption, despite the term being physically incorrect. Note also that the vision in the figure does not include a 100% electrical transport system. Instead, some of the transport will use fuel. But this fuel will not come from fossil fuel but from renewable sources, like biogas. Part of the fuel will also be produced from electricity through processes that go under the terms 'Power2Gas' and 'Power2Liquid'. (We will use the general term 'Power2Fuel' where needed.) During periods with a surplus of fuel and/or a shortage of electricity, either locally or globally, the fuel can be turned back into electricity.

Heating and cooling will be partly using renewable energy directly, e.g., solar panels for water heating, partly based on electricity, and partly using the fuel from renewable sources or from Power2Fuel. In the latter case, this will most likely be in the form of combined heat and power (CHP, co-generation). Burning fuel just for space heating is rather inefficient compared to modern heat pumps and CHP installations.

Energy conversions, the arrows in Figure 1.2, numbered in Figure 1.3, are an important part in the energy system. It is these transitions that allow electricity to become the heart of the energy system. But these conversions are also, in many cases, where the losses are. The lower the losses, the sooner a 100% renewable energy system with sufficient reliability can be reached. The amount of conversions thus has to be limited to get an efficient energy system. Increasing the efficiency of

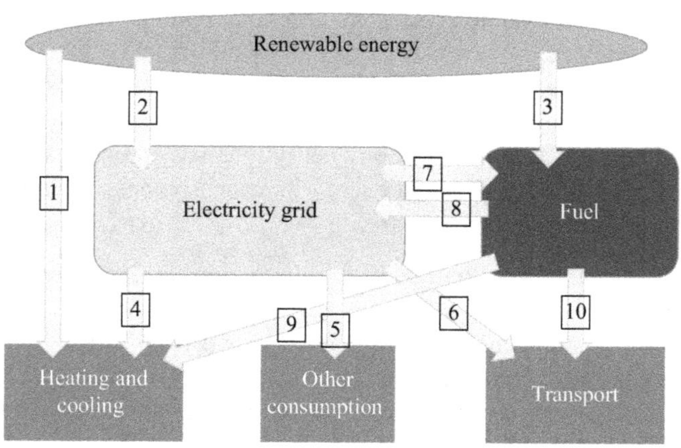

Figure 1.3 Different energy conversion processes in a 100% renewable energy system

the various conversions is an important development as well, but also this is beyond the scope of this book. It should be noted however that the conversion technology to and from electricity can impact the electric power system as well. Examples are harmonics from adjustable-speed drives and short-term voltage collapse due to air-conditioners based on direct-driven induction motors.

The following conversion processes can be identified from Figure 1.3:

1. Using renewable energy directly for heating and cooling; for example, solar panels for water heating;
2. Renewable electricity production; hydropower; wind power; solar power and other future sources like wave or tidal power. The increase in efficiency of, for example, solar cells is important here;
3. Fuel from renewable sources, like biomass;
4. Using electricity for heating and cooling; in an efficient energy system this is accomplished using heat pumps. In cold climates resistive heating or fuel-based heating may be used as an additional source during very cold periods.
5. Other use of electricity, like lighting, washing, computing or industrial processes;
6. Use of electricity for transport. This can be direct use of the electricity like with electric trains, trams and subways. For electric cars the use will more likely be in the form of batteries that will be charged from the electric power grid. The transport system will be hybrid but most or all vehicles may be based on one single source.
7. Power2Fuel: using electricity to produce gasses or liquids with high-energy contents. This could become an important form of storage ensuring the stability of the electric power grid.
8. Burning fuel to produce electricity. This can be in the form of the conventional Carnot cycle, as part of CHP installations or by means of fuel cells. There are high expectations for the latter, but for the time being the first and second one remain the ones with the highest efficiency.
9. Burning of fuel for space heating or for other heating processing like in industrial production;
10. The use of fuel for powering vehicles like cars, boats or planes.

1.4 Flexibility

Many textbooks on electric power system and many texts on the integration of renewable electricity production in the electric power system, contain one or both of the following two statements: 'The operation of the electric power system requires that production and consumption of electricity have to be always equal to each other'; 'The reason for this is that electricity cannot be stored'. Both statements are commonly made and are strictly speaking incorrect. When looking at a timescale of milliseconds, there is a continuous conversion between electrical and magnetic energy in the electric power system, both of which are stored for a few milliseconds. This is related to the concept of 'reactive power'. At timescales of seconds, there are subtle differences between production and consumption that are visible in the form in frequency variations. They are these small differences, and the associated frequency variations, that allow the balance between production and consumption to be kept at longer timescales, through what is referred to as 'power-frequency control'.

Electricity or electric energy (electromagnetic energy would be the correct physical term) is a kind of energy and as such it can be stored. However, there are no known processes to store electricity in a way useful for large-scale use. For the purpose of most of this book, we can therefore assume that electricity cannot be stored and that production and consumption of electricity have to be equal to each other at all times. We will return to energy storage (note the difference between 'energy storage' and 'electricity storage') in Section 1.6.

In an electric power system, the consumption varies with time over a range of timescales. As production and consumption have to be equal, the production has to vary in exactly the same way as the consumption. In the existing power system, the variations are set by the consumption and various control systems ensure that the production follows the consumption. From a consumer viewpoint, this means that electricity is available whenever wanted. With conventional production units this can be achieved and actually it has for many years been the aim of the power system planning and operation to ensure that this is always the case. An alternative way of looking at this is that the power system is dimensioned in such a way that it can cope with the worst case; i.e., the highest amount of consumption. It is obvious that this can easily result in overinvestment, but electricity is such an important commodity for the society that this was considered a price worth paying.

This general planning approach is on the way out, however, for a number of reasons. One reason is that overinvestment has become less acceptable in the society. Another reason is the introduction of free competition between companies producing electricity. The transport part of the electric power system (the wires) remains in the hands of a monopoly company. But even in the transport part of the power system, there is less willingness to overinvest. An important driving force here is the availability of technology to shift some of the flexibility from production to consumption.

Flexibility on the consumption side is not completely new and some applications have been around for many years. For example, many countries used to have time-of-use tariffs were electricity was more expensive during peak hours. Also did some countries, France being one example, have a tariff option where electricity was more expensive on days with high expected demand in relation to the available capacity. Several of those schemes disappeared with the restructuring of the electricity market in the 1990s but have attained renewed attention rather recently.

In a 100% renewable energy system, many of the sources of electricity production cannot be easily controlled so as to follow the consumption. This will either require a very large installed capacity of renewable electricity production or

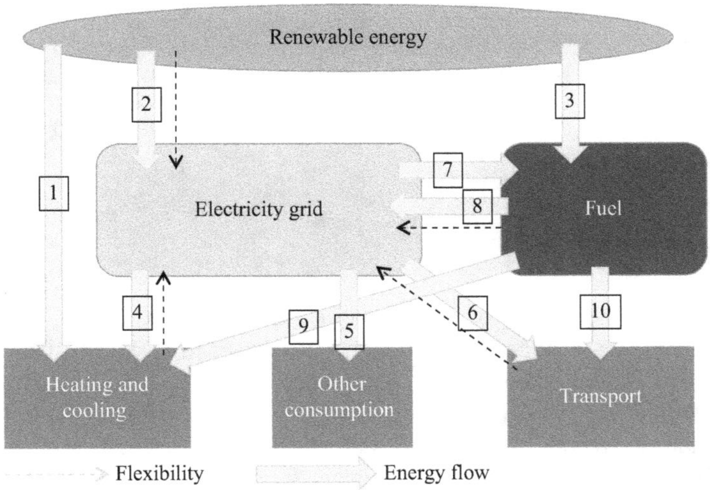

Figure 1.4 Sources of flexibility for the electricity grid in the 100% renewable energy system

alternative sources of flexibility. The latter is the most likely solution and the following sources of flexibility are indicated in Figure 1.4.

- Conversion 2: curtailing the sources of renewable energy. The main source of flexibility is formed by reservoir-based hydropower. Already in the conventional power system, in many countries, hydropower is an important source of flexibility. But also solar and wind power can contribute to flexibility by occasionally being curtailed.
- Conversion 4: electricity for heating and cooling. The long thermal time constant of most buildings (several hours to days) makes that it is possible to adjust the electricity consumption to the availability of RES.
- Conversion 6: electricity for use in transport. This concerns especially the use of electricity for charging batteries for electric vehicles. With direct-driven transport like electrified railways, the flexibility is much less.
- Conversion 8: burning fuel for producing electricity. This is equivalent to the use of conventional fossil fuel-based production units.

Production and consumption should be equal to each other at timescales from a few seconds and upwards. At the shortest timescale, the power-frequency control takes care of the balancing. Differences between production and consumption result in frequency changes that are detected by the power-frequency control. As the frequency is the same throughout the power system, each device (production or consumption) can contribute to the balancing at this shortest timescale. Production units can increase their production when the frequency drops; heating, cooling and battery charging for electric vehicles can reduce their consumption when the frequency drops. This so-called primary control takes place at a timescale of seconds.

In the conventional power system, an unbalance between production P_{prod} and consumption P_{cons} does not immediately result in instability, because of the availability of energy in the rotating mass of the generators. Instead the generators change speed, which is visible throughout the system as a change in frequency, $\frac{dt}{dt}$, around a frequency f_0:

$$\frac{df}{dt} = \frac{f_0}{2} \cdot \frac{P_{prod} - P_{cons}}{E_{kin}}$$

The kinetic energy E_{kin} of the rotating mass in the generators is important to limit the frequency change and therewith give the control system time to react. Modern types of renewable energy are connected to the grid through a power-electronics interface that do not contribute to the kinetic energy available for system stability. Keeping the system stable at these timescales is a major power-system challenge for the 100% renewable energy system.

At longer timescales, beyond the power-frequency control, the issue is not so much the actual balancing but ensuring that there is sufficient margin available with production and/or consumption to allow them to be brought in balance with each other. The term 'operating reserve' is used to describe this in the conventional electricity grid. Maintaining sufficient operating reserve is an important task of the system operator in the conventional grid and it is likely to be an important task in the 100% renewable energy system as well. In conventional power systems, there are strict rules for the minimum amount of operating reserve that shall be available up to several hours ahead of time. The further ahead of time, the more reserves should be available as the uncertainty in both production and consumption increases. Those rules do however only consider availability of power production capacity. Those rules do not consider the energy aspect of the reserves. In the 100% renewable energy system, the limitations may be set by the available energy instead of by the available power. Several of the storage options, to be discussed in Section 1.5, are limited in time up to a few hours. When planning more than a few hours ahead, those cannot always be relied upon. Alternative ways of scheduling reserves are needed, where the time element becomes very important.

At longer timescales, weeks through years, we enter the realm of 'operational planning' and 'planning'. For these timescales, the expected maximum consumption is compared with the amount of production capacity that is available with a high reliability. In a system with a large amount of hydropower, somewhat different long-term planning methods would be needed. In the 100% renewable energy system, such alternative methods are likely needed throughout.

1.5 The role of the electricity transport

At a local level, the need for balancing between electricity production and consumption becomes less strict. The reason for this is that the rest of electricity grid can serve as a buffer as long as the amount of electric power exchanged with the rest of the grid is less than the transport capacity.

For the electricity grid as a whole, production and consumption should be equal to each other:

$$\sum_i P_i = \sum_j C_j$$

where P_i is the electricity production at location i and C_j the electricity consumption at location j. For each location in the grid, the exchange T_i with the rest of the grid is equal to the difference between local production and consumption:

$$P_i - C_i = T_i$$

The exchange with the rest of the grid is limited by the total transport capacity $T_{\max}$ of all connections to the rest of the grid, so that the following inequality should hold:

$$|P_i - C_i| \leq T_{\max}$$

Both net consumption, $C_i - P_i$, and net production, $P_i - C_i$, are limited by the transport capacity. As mentioned before, the requirements locally are not as strict as for the electricity grid as a whole. Local balancing is only needed when otherwise the transport capacity is exceeded, i.e., when there is a large difference between local production and local consumption. All the different forms of flexibility (see Section 1.4) and of energy storage (see Section 1.6) can also be used locally to avoid that the transport capacity is exceeded. In that way the local consumption and/or the local production can be increased without the need for investment in additional transport capacity. The limits locally are not set by the variation of the production with time and typically not even by the unpredictability of the production capacity. The limits are set instead mainly by the increase of net production and/or net consumption locally.

Note that 'transport capacity' not only refers to the thermal capacity but also includes voltage drops or rises, harmonic distortion, stability, etc. The 'hosting capacity approach' has been introduced for a systematic and transparent study of the amount of additional production or consumption that can be connected locally without the need for additional flexibility or transport capacity.

1.6 The role of energy storage

The flexibility offered by consumption and production, as discussed in Section 1.4, is to a large extent related to energy storage. Part of the flexibility is offered by changes in behaviour, which would be an interesting and important social study field. But the main source of flexibility, and an important technical study field, is the use of energy storage so as to decouple production and consumption of energy. The main storage options are shown in Figure 1.5; these storage options contribute to the stability of the system. The duration of the storage is important in ensuring the reliability of the energy system in the longer term. The longer the energy can be stored, the more useful the storage option is. Also important are the energy losses

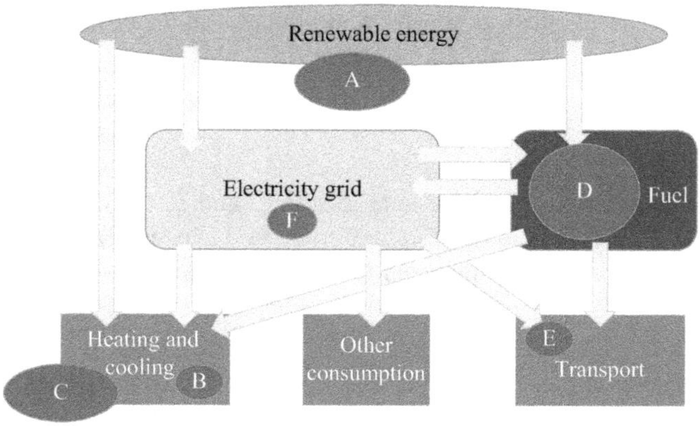

Figure 1.5 Different forms of energy storage in the 100% renewable energy system

during the storage and the energy needed to allow for the actual storage. For several of the storage options energy conversion is needed, which is also associated with losses.

The following energy storage options are present in the 100% renewable energy system (see Figure 1.5):

(A) Storage of the renewable energy, for example, with reservoir-based hydropower. The energy can be stored in this way for several months, with losses in the form of evaporation of the water in the reservoir. Also some solar-power installations have the option of storing energy in the form of hot molten salt. The energy loss is in the form of cooling down of the salt. This limits the storage time to a few hours, but plans for demonstration projects mention 24-h storage times.

(B) Space heating and cooling are associated with natural storage in the thermal capacity of the building that is being heated or cooled. Thermal time constants range from a few hours to more than 24 h. The storage time depends on the temperature difference and on the comfort zone in temperature, but storage times of up to about 1 hr seem possible. This natural storage has the additional advantage that there is no conversion loss and no standby-loss associated with it, which makes it very energy efficient.

(C) Heat and cold can also be stored, typically in the form of hot or cold water. By using good thermal insulation, the thermal losses can be limited and storage is possible for several hours. An extreme case is to store energy in an aquifer underground where storage up to several months is possible. Stockholm Arlanda airport uses such a system for seasonal storage of both heat and cold.

(D) Fuel is a natural form of energy storage and storage can take place almost indefinitely. With some fuel types there is some energy use associated with the storage, for example, cooling of the storage installations. The long storage possibility makes that fuel is likely to become an important buffer in ensuring the reliability of the energy system. However, the conversion losses are high both from electricity to fuel and from fuel to electricity.

(E) Part of the future transport will be powered from batteries that are charged from the electricity grid. Also this is a form of natural storage and the energy can be stored under long time without high losses. However, as the main purpose of the stored energy is to power the electric vehicle, its storage duration is in practice limited to a few hours. Assuming that a large part of the vehicle fleet will be electric and powered from batteries, there is likely to be always a substantial amount of energy available in those batteries for balancing the electricity grid.

(F) A final option, that is very much studied, is to maintain global or local balance between production and consumption of electricity by temporary storing a surplus in batteries located at strategic locations.

An alternative view of energy storage, from the viewpoint of the electricity grid, is shown in Figure 1.5. Three types of storage are visible here: storage of primary energy before it is converter to electricity; storage of electricity in another form; and storage of energy in another form before being utilised. There are advantages and disadvantages with all three types. Storage of electricity (Type II) has as the main disadvantage that two conversions are needed. However, with modern batteries (lithium ion being the most appropriate one at the moment), the conversion losses are relatively small. The advantage of storing electricity is that it can take place almost anywhere in the grid, not restricted to production or consumption sites. The location does not matter for the overall energy balance in the electricity grid, but an appropriate choice of the location makes it possible to limit the need for transport capacity in the grid.

An example of Type I energy storage in Figure 1.6 is the storage of water behind the dam and the storage of fuel for thermal power generation. Examples of Type II are battery storage and pumped storage. An example of Type III is the storage of hot water in a tank, for heating purposes a few hours later. In the last two cases the conversion losses are small, but there are storage losses and especially pumped-storage is limited to certain locations.

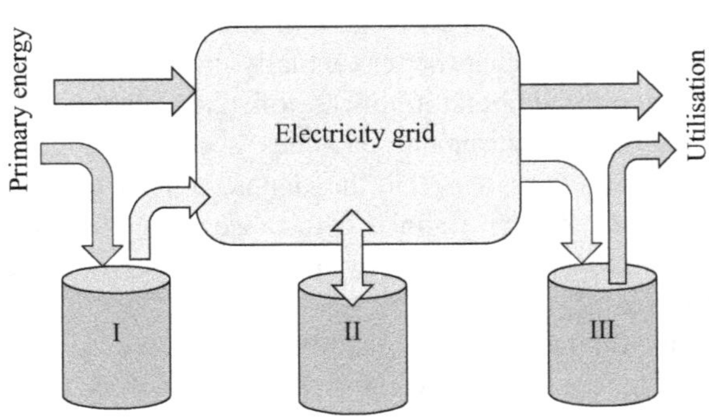

Figure 1.6 Energy storage in relation to the electric power system

1.7 Reliability in the 100% renewable energy system

Reliability of supply has been rather well studied for the conventional electricity grid. A distinction is thereby made between the reliability of the electricity production, the reliability of the electricity transmission (transport over long distances), and the reliability of the distribution (transport over shorter distances). All three will be different in the 100% renewable energy system as envisaged in Figure 1.2 and it may become more difficult to distinguish between the three hierarchical levels (production, transmission and distribution). This will strongly impact the way in which reliability is calculated and the way in which it is included in operation and planning.

Some things will however most likely not change. The consumer, the producer and the system operator will look at reliability in different ways. Each of these viewpoints and the way in which this will likely change is discussed in the forthcoming sections.

1.7.1 The view of the electricity consumer

Like in the conventional system, the consumer wants the electricity to be available when needed. Reliability for the consumer has to do with the availability of electricity on demand. It can be quantified in terms of number of interruptions per year and duration of interruptions. The difference with the conventional energy system will be in when the electricity is needed. There will however still be applications that require instant access to electricity, like lighting or making the perfect cup of tea. Note again that we do not consider in this book any change in habits that adjust our activities or demands to the availability of energy. This does not imply that such changes would not be important, it only means that this is beyond the engineering realm and instead a social science issue.

Even for applications with natural energy storage, there will be a certain demand for electricity, to recharge the energy storage. But the energy does not have to be available any moment in time. Instead there is the need for a certain amount of average capacity over a certain period of time. In other words, the reliability requirements are in terms of energy capacity over a certain period of time instead of power capacity at all times. An alternative and more fundamental way of looking at it, and one that matters actually more for the customer, is to set requirements on the room temperature in case of space heating or cooling. In case of battery charging, the requirement becomes a certain minimum amount of energy in the battery at the moment that the vehicle is needed. In the conventional electricity grid, an interruption is defined as the absence of electricity. In the future grid alternative definitions are needed, for example, the temperature being outside of a certain comfort zone.

With electricity for heating, the reliability becomes even more complicated when also other energy sources are available. The same absence of energy capacity from the electricity grid may be an interruption one day and not an interruption another day, depending on the availability of other energy sources.

The overall conclusion is that reliability becomes different for different consumers are it is no longer possible to come up with an overall definition that covers all consumers.

1.7.2 The view of the electricity producer

The electricity producer needs the electricity grid to transport the produced energy. An interruption, in the producer view, would be a situation where the produced energy cannot be transported to the consumers (which should include storage). This may be due to the absence of a connection (e.g., a cable outage), a local surplus of production or a global surplus of production. The former is an issue for conventional production units as well; the other ones are a concern only when there are no or insufficient storage facilities at the production side.

In the conventional power system, the electricity price varies with time. Therefore the price that electricity producers get paid for the electricity they produce varies with time. When a power producer is not able to deliver the potentially produced energy, it results in a loss of income. In the conventional system, surplus of production is more likely to occur during periods of high production. As production follows the consumption, this corresponds with periods of high consumption which are also typically periods of high electricity price. In the 100% renewable energy system, the marginal costs will be low or even zero with low electricity prices as a result. The inability to deliver electricity to the grid may not be a concern for the electricity producer. It may however be a concern for the system operator as it could lead to a shortage of electricity.

1.7.3 The view of the system operator

The role of the system operator is to ensure that electricity is available when needed. The system operator thus has to limit the number of events in which there is a shortage of electricity, both locally and globally. As we saw before, electricity demand comes in different forms: in the form of instantaneous demand where a certain amount of power is demanded at a certain moment in time; and as a more diffuse demand where a certain amount of energy is demanded over a certain period of time but without specific demands on when in that period the energy should be available. The latter type of demand is more flexible, but when there is a shortage of supply capacity the diffuse demand will at a certain moment turn into instantaneous demand. With heating or cooling, this is when the room temperature comes outside of the comfort zone. With charging of electric vehicles, this moment is more difficult to define, but also here the energy demand will turn into power demand at a certain moment. If the system operator is not able to supply that instantaneous demand, the system will have failed in its primary requirement.

Seen from the system operator, there are three types of events to be avoided, where production and consumption are treated equally:

- There is no galvanic connection between one or more consumers and producers and the rest of the grid.

- Demand cannot be supplied locally or local surplus production cannot be transported to the rest of the grid.
- There is a global shortage or surplus of production.

The primary aim, or should we say the engineering aim, of the energy system is to ensure a supply of energy to the different processes. In terms of Figure 1.2, the bottom row is what matters. In that context, the situations listed below are those that should be avoided. It is those that are referred to as interruptions.

- There is no galvanic connection between a consumer and the rest of the grid. This is covered by the conventional 'distribution reliability', which used to be the easier form of reliability calculations. However also this will become more complicated in the 100% renewable energy system because of the presence of production units and storage close to the consumers as well.
- Demand cannot be supplied locally, due to the net local demand (consumption minus production) exceeding the transport capacity from the rest of the grid.
- Demand cannot be supplied globally, due to the total demand exceeding the total production.

The system operator must avoid all these situations. Avoiding the first one requires reliable components and/or the presence of backup supply paths; it is what conventional distribution companies are doing all the time. Avoiding the second and the third situation is a system operator issue where demand and supply of electrical energy should be used in an optimal way. On the demand side is the consumption of electricity (conversion 4, 5 and 6), the conversion of power to fuel (conversion 8) and charging of any storage in the grid. For the consumption, a further distinction should be made between power demand and energy demand. On the supply side is the production directly from renewable sources (conversion 2), the conversion from fuel to electricity (conversion 9) and the discharging of any storage in the grid. If we consider fuel as a form of storage, there are four different parts in the balance between demand and supply of electricity:

To illustrate how this impacts the operation of the electricity grid, consider five different levels of available electricity production from RES, as shown in Figure 1.7. Note that there is not only a capacity of power towards the storage (charging the storage) but also a capacity of power from the storage to the grid (discharging the storage). It is part of the role of the system operator to decide which of the two should be chosen at any moment in time. Some thoughts are given below, based on the supply capacity in relation to the demand.

Supply level 1. The total supply capacity, directly from renewable sources plus by discharging the storage, is not enough to cover the power demand. The result is that not all the power demand can be fulfilled. All the available storage discharging capacity will be used to limit the amount of demand that is not fulfilled.

Supply level 2. The amount of supply capacity directly from renewables is not sufficient to cover the power demand, but by using part of the discharging capacity of the storage the power demand can be supplied. The remaining

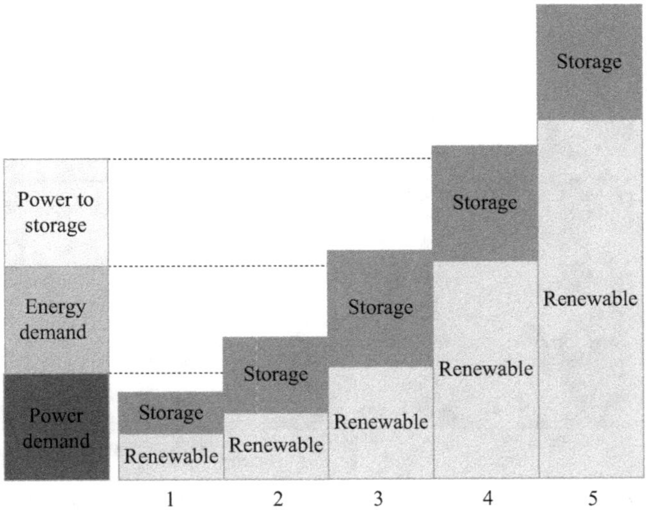

Figure 1.7 Different levels of supply (five columns on the right) in relation to the electricity demand (column on the left)

storage capacity can either be saved for later use or be used to cover some of the energy demand. This will be an optimisation issue, where the state of charge of the storage, the expected future demand and the expected future production from renewables will have to be considered.

Supply level 3. The amount of supply capacity directly from renewables is sufficient to cover the total power demand. The remainder can be used to supply part of the energy demand and/or to charge the storage. When there is sufficient energy in the storage, the stored energy can even be used to supply the total energy demand. The optimisation of the charging/discharging of the storage versus supplying the energy demand is one of the tasks of the system operator.

Supply level 4. The amount of supply capacity directly from renewables exceeds the sum of power demand and energy demand. In that case the total power demand will be supplied and the remainder will be used to charge the storage.

Supply level 5. The amount of supply capacity directly from renewables exceeds to sum of power demand, energy demand, and charging capacity of the storage. In that case all demand should be fulfilled and the remaining amount of renewable energy will be curtailed.

A number of assumptions have been made here to simplify the description. The aim of the list above is not to give a detailed set of instructions, those would be far more complicated, but to illustrate how the flow between charging and discharging depends on the amount of available production and consumption. For example, transport is not considered here. When including batteries for transport, both

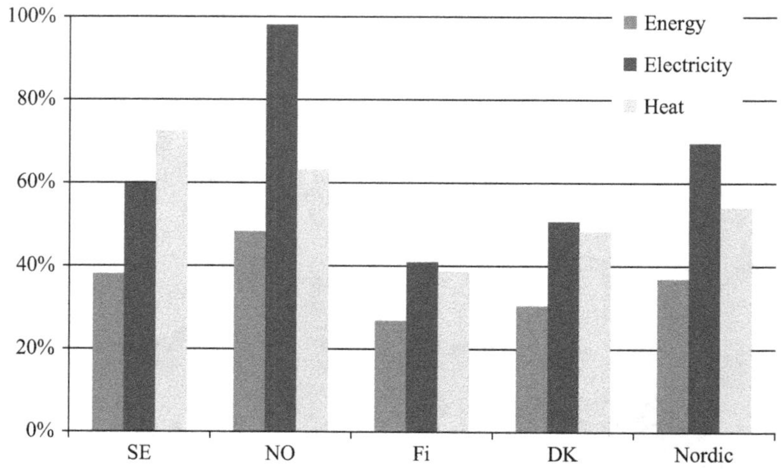

Figure 1.8 Current share of renewables in energy, electricity and heat in the Nordic countries (Own elaboration on data from IEA)

surplus of production and surplus of consumption will have to be avoided, which makes that fully charged is not always the preferred state of the storage. Also are alternative energy sources, for example, renewables directly used for heating instead of via electricity, not included in the scheme here.

1.8 The transition stage: towards 100% renewable energy

The existing energy system has not yet reached to goal of 100% renewable energy. In fact, the energy system in many countries is far from 100% renewable. The situation in the Nordic countries (Sweden, Norway, Finland and Denmark) is shown in Figure 1.8 as an example. A substantial part of the electricity is already produced from renewable sources: mainly hydropower in Sweden and Norway but a growing amount of wind power especially in Denmark but since a few years also in Sweden. Norway's electricity is produced almost exclusively from hydropower. However, when looking at the energy system as a whole, including transport, less than half of the sources are renewable.

The transition stage, with less than 100% RES, is shown in Figure 1.9. Non-renewable energy forms a large part of the fuel mix, used for transport and for generating electricity. The result is that several of the energy conversions become partly or completely non-renewable. These are indicated as grey in the figure.

This transition period is very important as it is impossible to build a completely renewable energy system that will simply replace the conventional one. Instead the 100% renewable energy system has to grow out of the existing energy system. During the transition, the energy system has to fulfil all its requirements on reliability and voltage quality. This places certain requirements on renewable energy production, as are discussed further in Section 1.9.

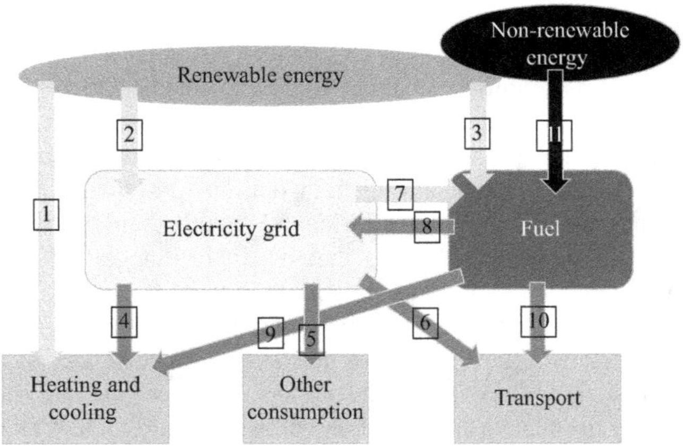

Figure 1.9 Energy flows in the not-yet-100% renewable energy system

During the transition period, the system should be such that the use of non-renewable sources is minimised. Also should market mechanism for setting of electricity prices be such that they encourage the building of more renewable generation units.

1.9 Renewable energy integration issues

While VRE is no longer a marginal business, its rapid growth is already presenting technical issues in some parts of the world. In some markets, on certain times of the year, this may have a particular effect on the generation-demand balance at peak times. This is particularly the case in areas where peak demand for air conditioning often coincides with warm summer days, such as in California or southern Europe. Wind too often is well suited to cover part of the energy demand in the early evening, and in the UK is at its strongest in the afternoon and evening. This means that the production of solar and wind energy has the potential to significantly erode one of the most important markets for peaking generation.

What is more, as lot of solar PV is fed into the low-voltage distribution system, in certain locations, for example, has led to power back-feeding into the circuit, causing over-voltage conditions, congestion on certain nodes and other PQR issues. A growing body of literature exists regarding the impact of these slow voltage variations, or anomalous behaviour of the facilities under fault conditions and their incidence into the grid protection systems, and other technical issues of integrating VRE at large scale [8].

The consensus of the scientific community is that, given current grid configurations, there are limits to the ability of distribution networks to accommodate VRE. So, there remain serious concerns about the impact on the distribution grid of VRE at large scale. Accordingly, some utilities are putting a halt to the integration of new installations, mainly rooftop solar, in certain locations, freezing permit, while solutions are sought.

Meanwhile, some countries are already moving to address technical challenges posed by VRE at large scale. Opportunities for improvement in the electric sector should be based on the necessity of embracing the challenges of technological improvements that allow producers to both optimise production to reduce grid impact, thus rendering VRE more 'grid friendly', enable them to make this type of electricity production dispatchable, so that it can be integrated into the future context of smart grids without compromising PQR.

VRE originally focused more on tapping the maximum power from the source. It omitted to make any contribution to power system PQR neither to operate in a coordinated mode with the rest of the system. Pioneer international standards [9–11] have been deemed to be conservative, as they force DER to trip off-line quickly when an abnormal condition is reached, partly in order to isolate the DER quickly, and to limit the risks of unplanned islanding. In conventional power grids, however, the simultaneous trip off-line of a large number of DERs due to short-term voltage or frequency fluctuations (e.g., as a result from a fault in the transmission network) results in a loss of supply to the network, which can threaten the stability in the grid. There can be a cascading effect at distribution level in areas with high VRE concentration due to simultaneous tripping. The capability of DER to remain grid-connected during such short-term disturbances – also named Fault Ride-Through (FRT) capability – is decisive for the large-scale renewables integration.

Consequently, it is increasingly seen as important that VRE plays a role in supporting network stability. Some of the functions commonly included are static voltage control to limit slow voltage variations; dynamic operational support in case of voltage dips; active power limitation in case of rising frequency or operational safety risks; and provision of reactive power for network re-phasing. And this is being incrementally required by interconnection standards in a number of countries, such as Germany [12] and Spain [13] (Figure 1.10). These new standards work in the opposite direction, mandating DER to remain connected for a certain period of time in case of faults and support the grid voltage by injections of reactive power.

Recent developments of power electronics technologies, such as advanced inverters, have enabled wind generating units to possess performance comparable or even superior to those of conventional generating units. Beyond maximum power-point tracking, islanding detection and basic energy feeding and power conversion functions, smart or multi-functional PV inverters can participate in frequency control through active power regulation (with proper control algorithms) and can provide additional FRT capabilities, voltage control, reactive power compensation, and local power quality and stability enhancement, since they are already based on the same advanced power electronics converter technology. Consequently, the drawbacks associated with simultaneous loss of a large amount of generation capacity have been virtually solved.

Seeing the trend, the only possible alternative would be the 'grid defection'. Having a significant number of citizen independent of the grid is being seen by several agencies as inevitable [14,15]. Increasingly, we will see more adoption of

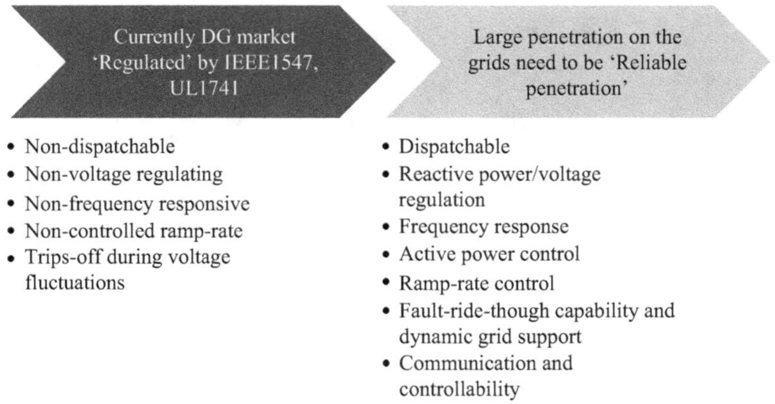

Figure 1.10 Requirements in grid codes

residential ESS, which is the key thing that allows people to choose whether to maintain grid connection or go off-grid. Although solar power plus battery ESS is still not economic for most consumers, when system prices come down enough that the payback period drop under the psychological barrier of ten years, they will become more financially appealing in several countries. Current trends suggest that in many countries – such as Spain, Italy, Australia and Germany – this combination will be cheaper than electricity purchased from the grid in just a few years, and in some cases it already is. While the report predicts that most customers will not defect from the grid, nevertheless, grid use might be reduced.

Therefore, given the challenges posed by VRE, the need persists for increased flexible operation and faster dispatch paradigms to cover fluctuations in demand. The essence of this market is evolving to times where the required peak generation will be less predictable, and the number of hours that are employed to operate the amount of capacity required regularly will reduce. Much effort has been already done to better understand the impact of those approaches in the grid. Some technical solutions to accommodate increased power system flexibility are already underway and must be adopted as soon as possible to cope with this new reality [16].

1.10 The prosumer role

Other efforts, next to the ones mentioned in the previous section, are essential to integrate DER technology. Among them, probably the most important is the political effort. Long-term policies are needed that provide a predictable and reliable market and regulatory framework compatible with the objectives of today's society.

A new paradigm is emerging worldwide. The reasons vary, but include a desire to improve the energy efficiency of the community, protecting the environment, and the desire to participate actively in decision-making in the energy

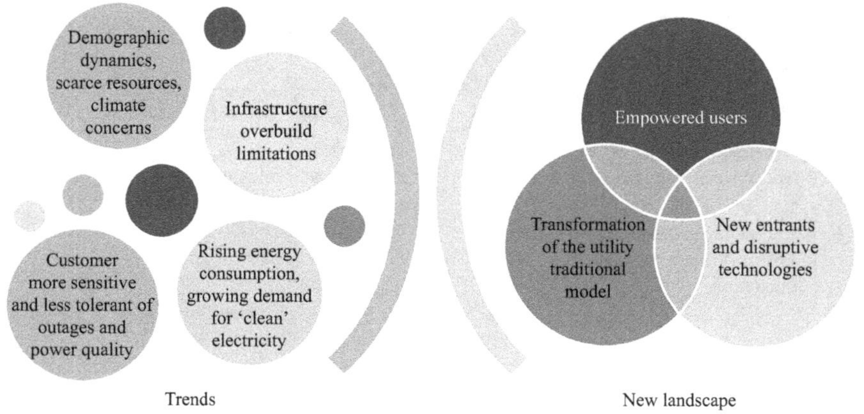

Figure 1.11 Trends affecting the power sector

sector (Figure 1.11). These motivations are coming together under the name of 'Sustainable energy communities' (SECs) movement. In some countries, over the last decade, citizens' ownership of RES has become a reality: in Germany, for example, 50% of all plants are owned by households, farmers and cooperatives, while only less than 10% are owned by large energy companies. The rest are owned by other private investors or companies that have emerged as rivals to traditional utilities.

The desire to regain sovereignty over the way they produce and consume energy has been echoed by most relevant governments and organisations at the European Union level (European Commission through the Intelligent Energy Europe program) and in North America (U.S. Department of Energy, Natural Resources Canada): they consider that to increase the social acceptance of renewable energy, it is necessary to adopt a holistic approach around the figure of the so-called 'Prosumer' or producer–consumer, initially a grid-connected consumer that also produces energy (partially for self-consumption); but would be willing to manage the complete cycle of energy in the near future, by himself/herself or by the support of a third party. Within a DER paradigm, SECs are a particular case of local production units in which the DER plant is owned, at least in part, by members who are quite involved in the achievement of the SEC objectives.

The third player enters the picture as an entity other than the community member or the utility that rule the DER on the customer side, or even beyond the meter. They represent new opportunities, but also threats to consumer base of a monopoly. In a physical sense customer-owned and third-party DER systems are often indistinguishable. These new participants, including consumer goods manufacturers and telecommunications companies, are already making decided moves towards residential, providing: energy audits and personalised energy efficiency guidelines, integrated solutions for building monitoring and even electric load aggregation packages. As more opportunities beyond-the-meter arise, competition for value with energy-related products and services will increase.

However, even the most ambitious programmes of renewable energy often lack specific policies for the prosumer. Instead, policies are often focused on large VRE facilities, for example, wind farms or solar parks, for many reasons: project costs per kW are usually lower because of economies of scale, fewer stakeholders are involved, local utilities are more familiar with their grid integration and operation, etc. In addition, development banks tend to prefer funding larger, centralised plants over DERs. Yet the above-mentioned ownership agreement could be critical in providing financing flexibility to customers interested in DER, allowing substantial increases in its penetration.

1.11 Key technologies

While it is clear that a system with large-scale VRE integration needs mechanisms to balance supply and demand, there are plenty of ways this can be done besides paying for new fossil fuel back-up plant. Making the transition requires substantial shifts in ideas: new operating procedures, adoption of the newest information and communication technology, evolved business models for electric utility, and updated regulatory frameworks. This new book looks at several new approaches to ease the various shifts identified above and ensure the success of the transition to a sustainable energy system. These approaches must include increasing interconnection capacity among geographical areas. Higher interconnection and transmission capacity at regional, national and international levels allow greater flexibility in power transmission between the regions with surplus of renewable generation and other regions with demand to reduce variability. Another option is the hybridisation of the contribution of different DER to complement each other at different times (e.g., solar power during the day, wind power overnight). This means including concentrated solar power, biomass and biogas plants to provide base-load power. Generation curtailments need to be avoided in order not to miss any kWh produced. In general, the larger the geographic dispersion between plants, the lower the combined variability will be. A mix of technologies within a balancing area enhances congestion management and reduces the need for new (and back-up) generation. They can provide important advantages in terms of increasing generation flexibility and reducing the variability of the overall VRE contribution. Additionally, this can be improved by local ESS, and building up complementary demand response (DR) approaches to meet peak demand.

Demand-side management, which includes DR and investments in energy efficiency, is seen as an effective remedy to limit the required investments in transmission and operating reserves. With DR algorithms, the demand can be modified in response to the request of the system operator in order to accomplish a specific goal such as peak shaving, contingency control, or frequency control. DR can be activated via market price signals (time-based rates or other forms of financial incentives) to drive certain kinds of behaviour or responses from prosumers. This includes reducing load during peak times, and shifting it earlier or later in the day to off-peak times, based on the wholesale market price. This has

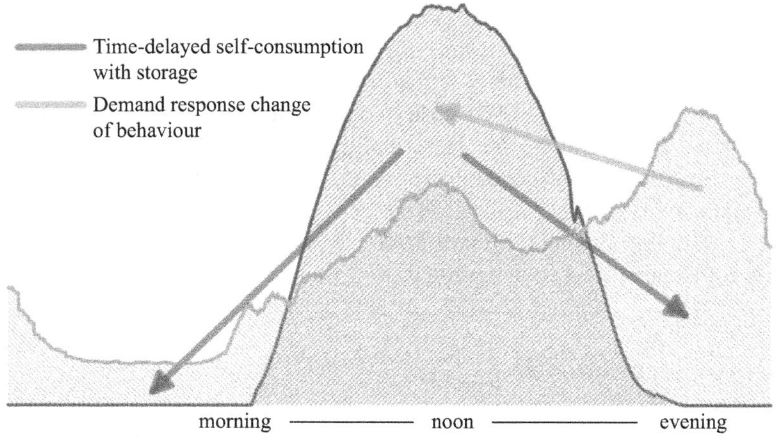

Figure 1.12 Integration of onsite PV generation, battery storage and load shifting to optimise residential energy consumption

the potential to change the demand profile to accommodate the variability of the VRE.

But most interesting here is to activate via local price signals: Some kind of local price signals can be provided by the energy management system based on grid hosting capacity (HC) information and generation-demand forecasts. These price signals can be used to establish thresholds or triggers beyond which specific DR measures are activated, smoothing demand peaks and/or meeting peak demand with VRE when available. In PV plants, for example, where maximum production takes place around midday, this DR program may eventually include 'Direct load control' of end-use appliances to shift this evening electricity demand towards off-peak periods to central hours of the day [17], as shown in Figure 1.12.

In both cases the use of the advanced metering infrastructure (AMI) as the baseline for this deployment is needed. Utilities have long collected customer usage data for billing purposes. However, monthly bills provide little or no information to customers about their actual use of electricity. Despite their initial promises, measuring the consumption of homes and businesses at 15-min or hourly intervals and take advantage of this information is still challenging and costly, even in territories with AMI or automated meter reading systems.

If usage data could be consistently obtained from the retail electric utilities, new software applications could be widely applied to a huge profit. Untapped DR resources or incentives combined with ESS may be useful for supplying the flexibility and reserve capacity desired and reducing the impact on the power system. With the adequate forecasting tools included (with a time horizon from minutes to 48 h), DR programs entail an important opportunity for the optimisation of consumption profiles to smooth the output profile and mitigate large injection peaks of their energy system before the net production reaches the grid.

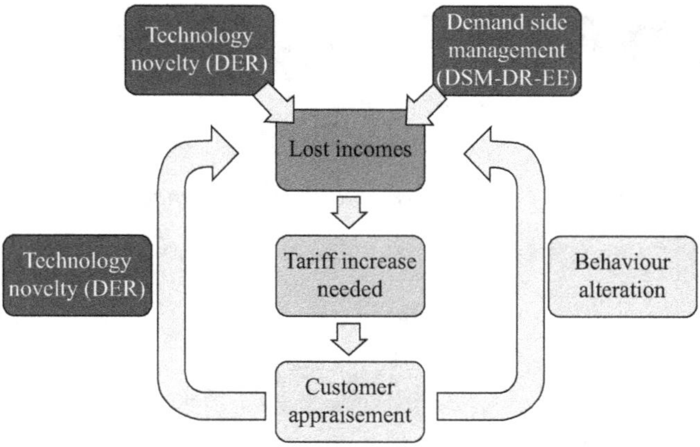

Figure 1.13 Representation of the vicious cycle of disruptive forces

Finally, an additional approach is to include on-demand dispatchable loads. Such 'demand sinks' could include thermal energy storage solutions (water heating, refrigerated warehouses, ice production, etc.), water desalination, as well as pumping and irrigation-related loads, which may be shifted to off-peak times, based on grid needs.

Integration costs are due not only to the specific characteristics of VRE technologies but also to the intrinsic power system flexibility to deal with greater variability. With the potential to become technically and economically feasible alternatives, many are growing very fast. Because of this confluence of forces, industry and stakeholders should proactively assess the impacts and possible solutions to address the disturbing challenges in a timely fashion. Even though their application still depends on specific states and markets, it is very likely they would disrupt many aspects of the current energy framework, especially with regard to electric utilities. Figure 1.13 illustrates the forces currently oppressing the traditional utility business model [18]. As DER reduces revenues, utilities try to recoup their costs by raising fares of traditional customers, creating an implicit cross-subsidy of DER users. And because the higher rates, in turn, create incentives for more customers to install their own DER, a vicious circle is formed.

To preserve long-term customer fidelity, to fully accommodate an increasing number of clients that are in the forefront of being 'prosumer', utilities should transform themselves from energy suppliers to energy service advisor (Figure 1.14).

In conclusion, the large-scale integration of VRE into power grids requires substantial transformations to increase the power system flexibility: (1) to allow bi-directional electricity flow, which is aimed to ensure grid PQR when including DER; (2) to establish DR mechanisms aimed at reducing peak-loads in order to deal with increased variability; (3) to expand grid interconnection at the regional and international level aimed at increasing balancing capabilities and stability; (4) include technologies and procedures to ensure adequate stability and control

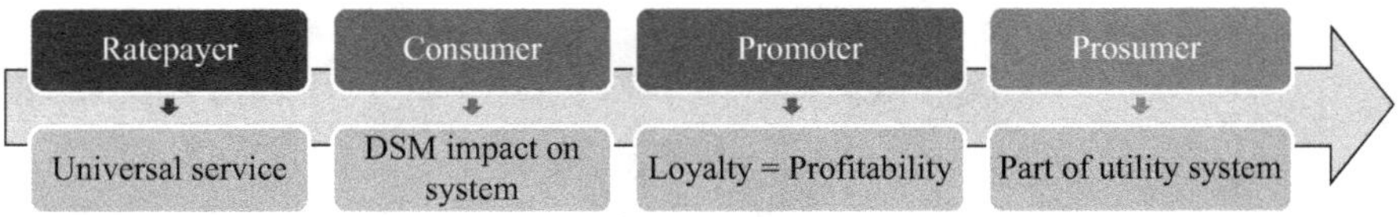

Figure 1.14 The journey to consumer empowerment

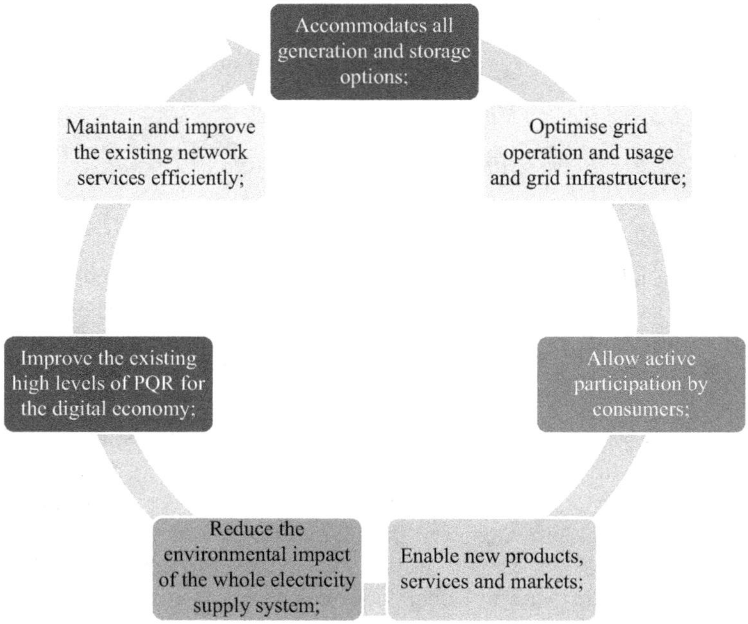

Figure 1.15 Benefits of the smart grids

(e.g., frequency, voltage, power balance) in the presence of VRE and (5) to exploit the ESS potential to store electricity surplus from VRE. Smart grids have the potential to benefit the whole value chain so that system operators can get the required power security and quality of supply (Figure 1.15).

References

[1] Deutsche Bank report: Solar grid parity in a low oil price era – Deutsche Bank Responsibility n.d. https://www.db.com/cr/en/concrete-deutsche-bank-report-solar-grid-parity-in-a-low-oil-price-era.htm (accessed March 10, 2015).

[2] Trancik JE, Jean J, Kavlak G, *et al.* Technology improvement and emissions reductions as mutually reinforcing efforts: Observations from the Global Development of Solar and Wind Energy. 2015.

[3] Eurostat – Statistical Office of the European Communities. Energy, transport and environment indicators. 2014. doi:10.2785/56625.
[4] Heeter J, Nicholas T. Status and Trends in the U.S. Voluntary Green Power Market (2014 Data). 2015.
[5] RE100 n.d. http://there100.org/ (accessed October 26, 2015).
[6] Gorona del Viento. El Hierro n.d. http://www.goronadelviento.es/ (accessed October 26, 2015).
[7] Renewable islands: Settings for success n.d. http://www.irena.org/DocumentDownloads/Publications/GREIN_Settings_for_Success.pdf, 2014 (accessed October 26, 2015).
[8] Commission IE. Grid integration of large-capacity renewable energy sources and use of large-capacity electrical energy storage. White Pap Oct 2012.
[9] IEC. Photovoltaic (PV) systems – Characteristics of the utility interface. IEC 61727-2004. 2004.
[10] IEEE. IEEE application guide for IEEE Std 1547™, IEEE standard for interconnecting distributed resources with electric power systems. 2009. doi:10.1109/IEEESTD.2008.4816078.
[11] V 0126-1-1. Automatic disconnection device between a generator and the public low-voltage grid. VDE. 2013.
[12] Elektrotechniker V. Technical minimum requirements for the connection to and parallel operation with low-voltage distribution networks. 2011.
[13] Ministerio de Industria de España. PO 12.3. Requisitos de respuesta frente a huecos de tensión de las instalaciones eólicas. BOE 2006;254:37017–9.
[14] Institute RM. The economics of load defection. 2015.
[15] Greentech Media. Australia Prepares for "Inevitable" Grid Defection n.d. https://www.greentechmedia.com/articles/read/australia-prepares-for-mass-grid-defection (accessed October 16, 2015).
[16] Martinez SR and Hughes W. Bringing variable renewable energy up to scale: options for grid integration using natural gas and energy storage. Washington, DC: World Bank Group; 2015. http://documents.worldbank.org/curated/en/185451468124481422/Bringing-variable-renewable-energy-up-to-scale-options-for-grid-integration-using-natural-gas-and-energy-storage.
[17] Palacios-Garcia EJ, Moreno-Munoz A, Santiago I, Moreno-Garcia IM, Milanes-Montero MI. Smart community load matching using stochastic demand modeling and historical production data. 2016 IEEE 16th Int. Conf. Environ. Electr. Eng., IEEE; 2016, p. 1–6. doi:10.1109/EEEIC.2016.7555885.
[18] Kind P. Disruptive challenges: Financial implications and strategic responses to a changing retail electric business. 2013.

Chapter 2

International requirements for large integration of renewable energy sources

A. Molina-García[1], *A.D. Hansen*[2], *E. Muljadi*[3], *V. Gevorgian*[3], *J. Fortmann*[4] *and E. Gómez-Lázaro*[5]

2.1 General overview

Most European countries have concerns about the integration of large amounts of renewable energy sources (RES) into electric power systems, and this is currently a topic of growing interest. In January 2008, the European Commission published the 2020 package, which proposes committing the European Union to a 20% reduction in greenhouse gas emissions, to achieve a target of deriving 20% of the European Union's final energy consumption from renewable sources, and to achieve 20% improvement in energy efficiency both by the year 2020 [1]. Member states have different individual goals to meet these overall objectives, and they each need to provide a detailed roadmap describing how they will meet these legally binding targets [2]. At this time, RES are an indispensable part of the global energy mix, which has been partially motivated by the continuous increases in hydropower as well as the rapid expansion of wind and solar photovoltaic (PV). The International Energy Agency's 2012 edition of the *World Energy Outlook* stated that the rapid increases in RES integration are underpinned by falling technology costs as well as rising fossil-fuel prices and carbon pricing, but RES integration is also encouraged by continued subsidies: from $88 billion globally in 2011 (compared to $523 billion in fossil-fuel subsidies in 2012 [3], with a share of $131 billion for electricity generation) to an estimated $240 billion in 2035 [4]. According to [3], in 2015 RES accounted for 22% of electricity generation, which was approximately the

[1]Universidad Politécnica de Cartagena, Cartagena, Spain
[2]Department of Wind Energy, Technical University of Denmark, Copenhagen, Denmark
[3]National Renewable Energy Laboratory, Golden, CO, USA
[4]HTW Berlin, University of Applied Sciences, Berlin, Germany
[5]Renewable Energy Research Institute, Universidad de Castilla-La Mancha, Albacete, Spain

same level as gas and about one-half the level of coal. Projecting into the future, it is estimated that by 2035 RES will account for almost one-third of total electricity generation, becoming the world's largest source of electrical power generation [3]. Nevertheless, to avoid excessive burdens on governments and consumers, subsidy measures to support new renewable energy projects need to be adjusted over time as power capacity increases and as the costs of renewable technologies fall [4]. Therefore, during the last decade, concerns about global climate change, finite conventional resources, and decreasing costs of renewables have been major arguments to justify the inclusion of RES and energy-efficiency policies in a significant number of countries [5]. In fact, most developed countries have been promoting policy principles and incentives to turn efficient and renewable scenarios into reality, and several are pushing for major renewable energy integration [6,9]. In this context, and aiming to reduce dependence on fossil fuels and foreign energy sources, electricity emerges as a sector wherein renewable energy policies and energy-efficient initiatives are most likely to be implemented because it can be generated from a large variety of fuels and technologies.

Moreover, 75% of the energy consumed by the residential/commercial sector is in the form of electricity, and approximately 35% of industrial energy consumption is in the form of electricity [10]. Current climate policy agreements in Europe require a reduction of at least 80% of all carbon dioxide (CO_2) emissions by 2050, eventually leading to the full decarbonisation of the power sector [11]. Consequently, in European countries the ratio of conventional power plants has been gradually decreasing while RES are becoming more relevant [12]. In this respect, nuclear energy could be considered an alternative and potential solution as a carbon-free energy source; however, as discussed in [13], nuclear technologies currently present some relevant drawbacks. For example, the Fukushima Daiichi incident has raised the public's concerns about nuclear energy, emphasising the fundamental safety weaknesses and potential inspiration for terrorism activities. Thus, despite significant growth in low-carbon sources of energy – shifts away from oil and coal toward natural gas and renewable sources – fossil fuels remain dominant in the global energy mix [14]. Displacing fossil fuel-based generation with renewable generation presents many desirable outcomes within the power sector, such as the reduction in pollution and CO_2 emissions. Also, resolving significant challenges can be met through these actions – for example, the reliable delivery of electrical power of acceptable quality nearly 100% of the time [15].

In the electrical sector, global installed capacity of RES was approximately 3.6 GW in 2000, representing 22.4% of new power capacity installations [16]. This capacity increased up to 47.4 GW in 2013, accounting for more than 72% of new installations. In 2014, the share of renewables in total electricity production exceeded 22%; hydropower accounted for 16.4%, and variable renewable energy (VRE) (e.g., PV and wind) for 3.6%. A number of countries already have a significantly higher individual share of VRE than the global figure. Denmark, Spain, Ireland, and Germany, for example, each have a share of more

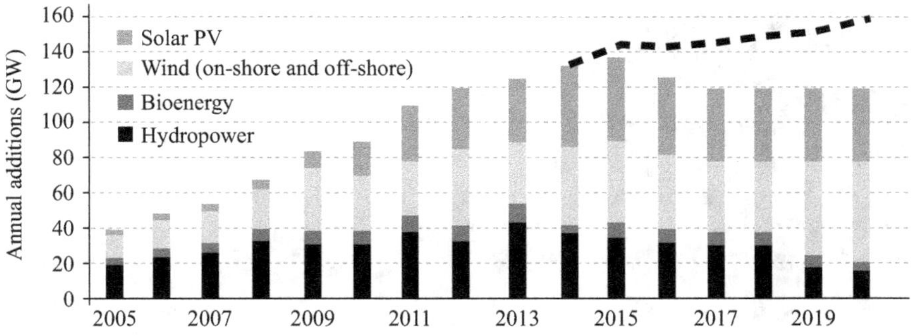

Figure 2.1 Global net additions to renewable power capacity – historical and forecast [8]

than 15%. Actually, the increased penetration of renewable energy to attain the goal of the Danish government to convert the present energy system into an entire renewable energy integrated system by the end of 2050 [16], will be accomplished by large scale of WP and PV plants. For example, the WP generation capacity in Denmark of 4,792 MW, supplying 33.2% of total electricity consumption in 2013 [17], being increased to 6,700 MW already in 2020 [18]. According to [19], the PV generation capacity in Denmark of 610 MW in 2014 will be increased at 1,000 MW in 2020 [20]. The International Renewable Energy Agency (IRENA) analysis indicates that the global share of VRE can be expected to increase by as much as 20% by 2030 [21]. Wind and PV resources are among the top technologies, with more than 115.4 GW and 80 GW installed, respectively [22]. According to [23], no other technology has reached or will reach enough maturity to challenge PV, wind, or natural gas in the next five years. In fact, wind power had a record-breaking year in 2014 after passing 50 GW for the first time, and it reached another milestone in 2015 as annual installations topped 63 GW, a 22% increase. China led the way, with a record 30.8 GW of new installed capacity, breaking the previous record it had set (in 2014) for installations in a single year [24]. Nevertheless, other renewable technologies (hydropower, biomass, concentrating solar power (CSP), geothermal, and ocean energies) have also increased in installed capacity during the past decade but to a lesser degree [25]. Figure 2.1 shows the global net additions to renewable power capacity, both historical and as forecasted. During the past three years, PV power plants have been the largest generation source in terms of installed capacity in Europe. Indeed, in 2013 the installation of PV power plants was more than 11 GW (31% of total capacity). This was only slightly behind wind power plants, with approximately 11.2 GW (31%); and natural gas, with 10.5 GW (23%). The installations of the remaining resources, including renewables as conventional types of generation, are far from these values. Figure 2.2 shows the estimated share of RES in global final energy consumption.

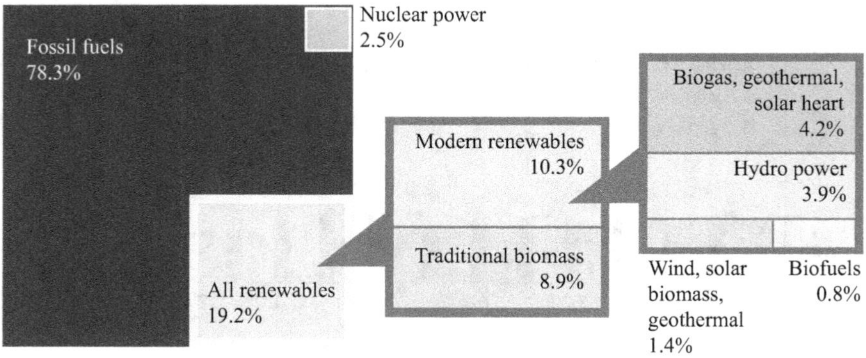

Figure 2.2 Estimated share of RES in global energy consumption [9]

2.2 Ancillary services in RES: comparisons among different countries

The forecasted integrations of high penetrations of RES, such as wind and PV, into the electricity supply around the world impose the requirement that they not be detrimental to the overall stability of the electric power systems. One way of ensuring this is to require RES to play a role not only in energy production but also in the delivery of ancillary services that are needed to ensure stability at both the transmission and distribution levels. Definitions of ancillary services can differ significantly [26]. Although some definitions emphasise the importance of ancillary services for system security and reliability, others mention the use of ancillary services to support electricity transfers from generation to load and to maintain power quality. Further, some definitions limit the contribution of ancillary services to the transmission network; others include distribution purposes as well. According to F. V. Hulle *et al.* [27], ancillary services can be defined as all grid-support services required by a transmission or distribution system operator to maintain the integrity and stability of the transmission or distribution system as well as power quality. Further, the amount of ancillary services needed for power systems in the future will increase with high shares of RES.

As grid-connected RES technologies are achieving significant penetration levels, interest in analysing the potential impacts of RES on the electric distribution grid is also increasing [28]. The primary focus of VRE, such as wind and PV, had been on the provision of energy. Early on, because most types of renewable generation were connected to the medium-voltage system, the corresponding distribution grid codes that applied treated these renewables comparable to loads (in this case 'negative loads') that had to be disconnected from the grid as fast as possible in the case of disturbances [29]. A gradual change in thinking began in 2001 when the German transmission system operator (TSO) E.ON Netz published a grid code [30] that required wind power plant operators to ride through disturbances and to provide additional functionality (in addition to feeding in power)

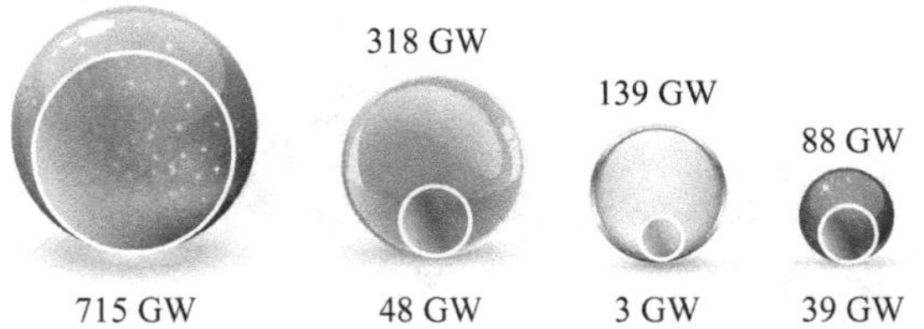

Figure 2.3 Net generation capacity added in 2000–2013 [9]

to support the operation of the conventional power stations. These functionalities included the provision of reactive power, the remote controllability of active and reactive power, the capability to keep operating after grid faults, and the capability to perform frequency down-regulation (e.g., to reduce the active power output in the case of an increase in frequency).

Today, such capabilities are usually referred to as ancillary services, and examples can be found in the literature – e.g., the authors of [31] focused on PV production for residential applications, and the authors of [32] analysed intermittent wind resources. Indeed, with an aim towards maintaining continuity and security in electricity supply, most countries with increasing penetration levels of RES have developed specific requirements to connect them to the grid (also known as a grid code). Grid codes aim to avoid undesired disconnection under the presence of disturbances – mainly voltage dips – and they establish rules and limits for the active and reactive power as well as expected performance under frequency and voltage oscillations. These requirements have mainly focused on wind power plants [33] due to their major presence in current power systems. In fact, annual wind power installations in the European Union have increased steadily during the past 12 years – from 3.5 GW in 2000 to 11.2 GW in 2013, with an average annual growth rate of 11%. PV generation has become the third most important renewable source, after hydropower and wind power. Its growth has been considerably high during the last few years. For example, at the end of 2009, cumulative installed PV capacity in the European Union was approaching 17 GW, and one year later it accounted for approximately 30 GW. In 2013, more than 136.7 GW of PV generation was installed globally – an amount capable of supplying at least 77 TWh of electricity every year [34] (Figure 2.3). Figure 2.4 shows the new installed and decommissioned power capacity at the end of 2015 in the European Union, and Figure 2.5 shows the power capacity at the end of 2015 in the world. Nevertheless, and despite the relevant presence of PV power plants in current power systems, few countries have developed specific technical requirements for these installations and their performance under disturbances. Usually these requirements have been quite similar to those for wind power plants, with the exception of the German Grid Code, which refers to technical guidelines and requirements specifically for the interconnection of PV [35].

Note that despite the increasing RES installations, the conceptual priority of RES in current grid codes is not to replace conventional generation but primarily to

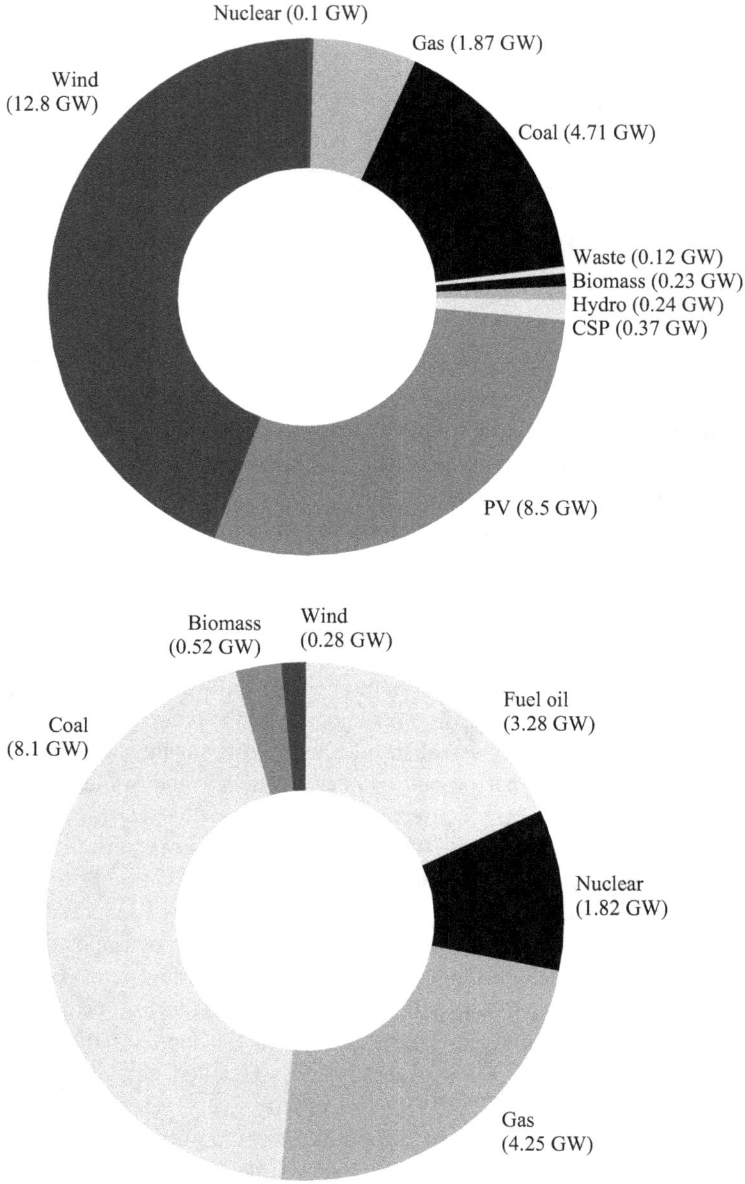

Figure 2.4 New installed and decommissioned power capacity at the end of 2015 in the European Union [16]

reduce consumption. This may be seen as a question of wording only, but with respect to the requirements this has a significant impact. For a long period of time, 'negative loads' such as RES were considered to behave like typical loads connected to a distribution system – i.e., they should limit the impact on the grid by

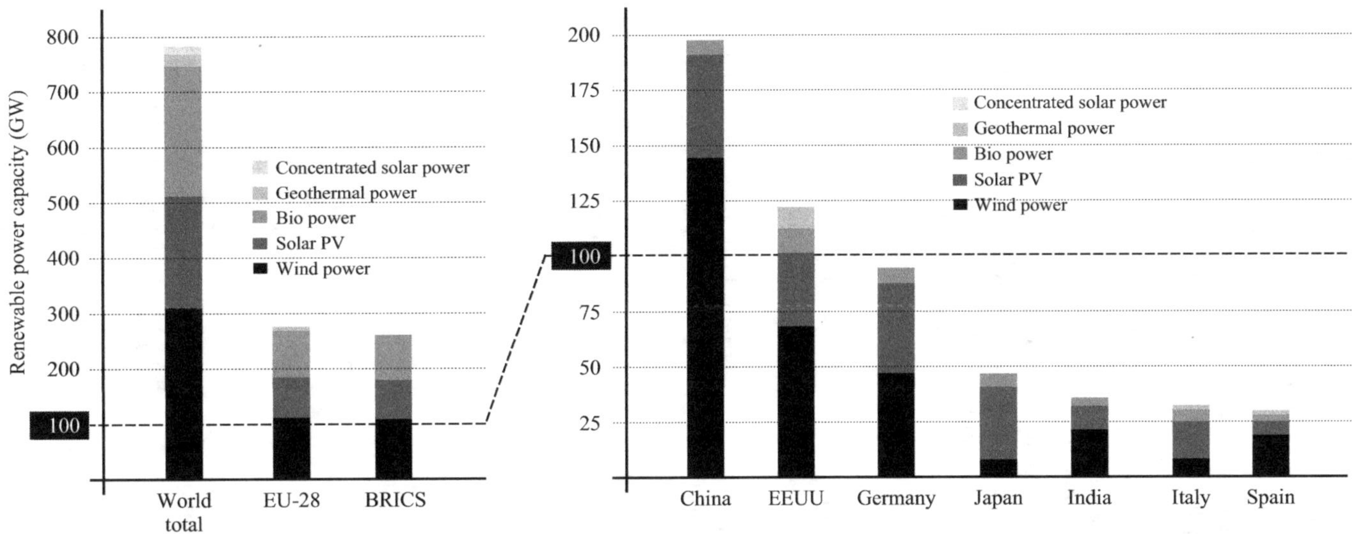

Figure 2.5 Renewable power capacity at the end of 2015 in the world, the 28 member states of the European Union (EU-28), the five major emerging national economies (Brazil, Russia, India, China, and South Africa; BRICS), and the top seven countries [13]

having a predefined power factor range and disconnect in the case of grid events such as relevant changes in voltage magnitude, voltage phase angle, or frequency.

Largely resulting from pressure by TSOs, requirements for RES – even those connected to a distribution system – usually try to limit their impact on the grid. Features such as the capability to support the grid are often referred to as RES being 'grid friendly,' and they are necessary to reduce possible impacts on the grid in the case of disturbances; however, note that these features always assume a stable grid consisting of a sufficient number of synchronous generators in operation that ensure basic services such as voltage stability (by providing voltage control and sufficient short-circuit power) and frequency stability (by providing primary control and system inertia). Today's grid code requirements describing ancillary services are not intended to replace the basic services provided by conventional power stations. Also, today's RES are neither allowed (by grid codes) nor capable (due to existing control and hardware design) to provide these services. The development of a next generation of RES that may provide such features on a larger scale (not only for a small, islanded system) is only beginning.

2.2.1 Active power reserves and frequency control

Frequency control services are related to the short-term balance of energy and frequency of a power system. According to the definition by the Union for the Coordination of the Transmission of Electricity [36], frequency control includes automatic (primary/secondary) and manual (tertiary) frequency regulation and operating reserves, as illustrated in Figure 2.6. This is the main service provided by generators (online for automatic services and online or offline for longer-term activated services). It can also be provided from flexible loads and storage units.

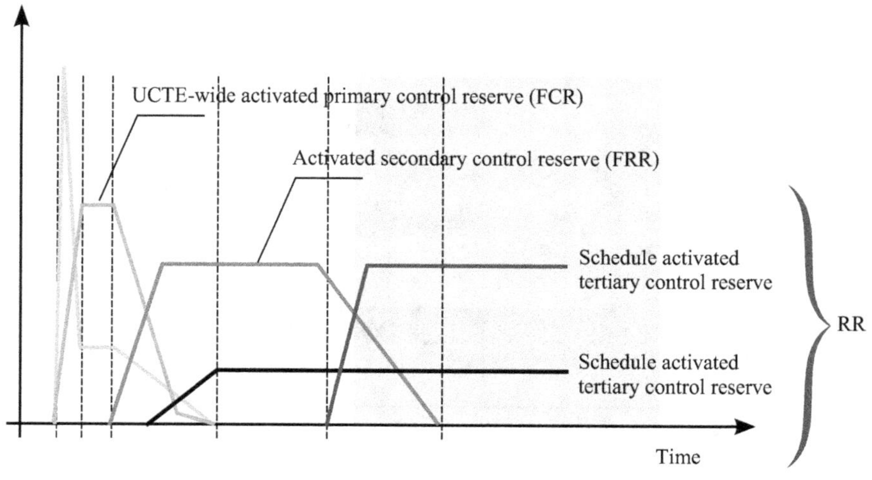

Figure 2.6 Principal frequency deviation and subsequent activation of reserves [37]

According to the ENTSO-E classification [37], frequency control reserves are defined as frequency containment reserves, frequency restoration reserves, and replacement reserves; see Figure 2.6. In Denmark, the TSO (Energinet.dk) published an ancillary services strategy for 2011–2015 [38] that classifies the ancillary services into frequency-controlled reserves, secondary reserves, manual reserves, and regulating power and properties required to maintain power system stability (e.g., short-circuit power, continuous voltage control, voltage support during faults, and inertia).

Recently, a demonstration project was conducted by the National Renewable Energy Laboratory in collaboration with Puerto Rico Electricity Power Authority (PREPA) to demonstrate the ability of a PV power plant to operate in spinning reserve mode by curtailment. Puerto Rico's transmission system consists of 230-kV and 115-kV lines, 38-kV subtransmission lines, and 334 substations. PREPA's typical summer daytime peak load is approximately 2.8 GW. Total installed generation in the region has a capacity of 6 GW, with 173 MW of wind and PV generation; the rest is based on petroleum and natural gas. AES's 20-MW Ilumina PV power plant, located in Guayama, Puerto Rico, consists of 40 inverters rated at 500 kWac each. Several solar irradiance sensors are placed strategically within the PV plant, and they are used to forecast the actual output of the PV plant. Based on the real-time forecast of the plant output, supervisory control is used to curtail the output of the PV plant partially by 20% (and subsequently 40%) to operate the PV plant with a spinning reserve of 20% from the maximum available power generation. The frequency control is demonstrated by the implementation of both the droop control and the automatic generation control. The demonstration projects were implemented successfully, as reported in [39].

2.2.2 Reactive power control/voltage control

From a system point of view, voltage is the prerequisite for any kind of power transport. By changing the amount of reactive power provided to a grid (by changing the reactive power reference of power stations and active or passive sources), the power flow in the grid can be controlled. In most countries, RES play only a very limited role in controlling the voltage of transmission systems.

At the level of the distribution system, voltage control services focus on maintaining power system voltage within the prescribed bounds during normal operation and during – and especially following – disturbances by keeping the balance between generation and consumption of reactive power. Voltage control includes reactive power supply (injection or absorption), and it can be provided by dynamic sources (generators, synchronous compensators) and static sources (capacitor banks, static voltage controllers, and FACTS[1] devices), including network equipment such as tap-changing transformers in the substations and loads. Voltage control has two targets (Figure 2.7):

- Steady-state reactive power/voltage control: The aim is to keep the voltage profile close to the desired profile and within the tolerance band margins

[1]Flexible alternating current transmission system.

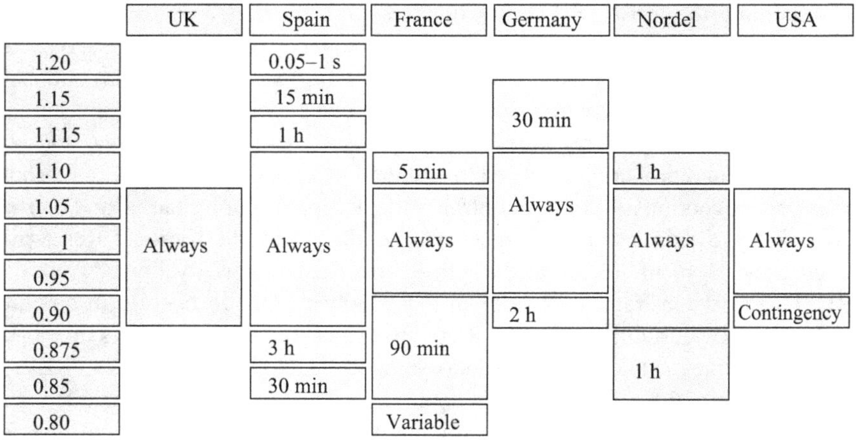

Figure 2.7 Voltage deviation and subsequent activation of reserves [40]

within the time frame of hours. This control is commonly achieved by injecting or absorbing reactive power at a voltage-controlled node. The TSO dispatches the reactive power using the active and passive reactive power sources that belong to different levels – generation, transmission, and distribution – using optimal power flow methods. Steady-state requirements are usually defined as power factor or voltage requirements at the level of a wind power plant, with typical response times in the range of seconds up to 1 min. The integration of RES power stations into a power system has usually been facilitated with economic incentives [41]. This is the case of reactive power generation. For example, in Spain in 2004, the Royal Decree 436/2004 introduced an incentive as a percentage of the average reference tariff. In 2007, a new incentive for reactive power was also introduced in Spain, IV.02, which had a scheme similar to that of the previous incentive. It is based on a reference value that is updated based on the consumer price index minus a correction value. In this new scheme, hourly values are used instead of quarter-hourly ones, and dispatches are introduced. This change was because dispatchers seek results at the wind power plant's point of interconnection, but measurements are performed at the level of the wind power plant's substation. The values (cent€/kWh) from these two schemes were 7.659 and 7.8441 in 2006 and 2007, respectively. This power factor incentive was suppressed in Spain in 2013.

- Dynamic voltage stability: The aim is to keep the network voltages in a dynamic time frame (seconds to minutes), thus preventing a slow voltage collapse event or limiting the depth and extension of an incident (e.g., loss of a main line, loss of generation unit). Dynamic voltage requirements are usually defined as reactive current response at the turbine level, with typical response times in the range of 30–50 ms.

In traditional power systems, VRE proportion has been considered very small relative to conventional generating units. As the penetrations of RES increase – especially wind and solar – grid codes compel renewable generation to contribute more significantly to power system voltage and reactive regulation [42]. Starting in 2006 [43], the capability to provide ancillary services, such as (limited) frequency control and voltage control, has become a requirement in grids that have a high penetration level of renewables. Nevertheless, knowledge gaps – especially at the level of distribution system operators – and thus the need for further research have been identified in the area of the provision of ancillary services from RES. For example, according to [42], in the context of ancillary services delivered from wind power plants, further investigations strengthening system reliability are necessary regarding faster and reliable communication (i.e., among wind power plants and system operator control rooms), dedicated tuning of the control strategies, estimation of available power, and coordination of offshore wind power plants to provide reactive power control or voltage control at their land-based point of coupling. In the context of ancillary services delivered from PV, further investigations are needed on available (regional) power estimation, faster and reliable communication and control within the plants, and improved control strategies [44].

Grid code design depends a lot on the knowledge of the system operator and the size of the wind power plant. In the United States, large wind power plants interconnected to the transmission system are common, and as a result they are treated as power stations based on synchronous generators using classic (PI) voltage control. Power factor control is usually not allowed. In case the short-circuit power of the grid is considerably higher than the reactive power capability of the wind power plant, pure P-control of the voltage is used ('voltage static') instead of a PI control. This is necessary because the reactive power capability of the wind power plant is too small to really control the voltage, and the controller of the wind power plant would be saturated. (See also [45] as an application in Europe.) In case a wind power plant is too far from the high-voltage grid, voltage excursions can be significantly limited if voltage control is used. In the case of short-circuit ratios (i.e., the grid's short-circuit power divided by the wind power plant's apparent power) of 5 and lower, usually voltage control is needed to keep the voltage within the normal operating range (see [46–48]). In the United Kingdom, a proportional voltage control ('voltage static') is commonly used. In Germany, power factor control had been common, but due to the improved voltage-stabilising capability of voltage control, voltage static at the level of the wind power plant is becoming the preferred method of control for newer, interconnected, medium-voltage wind power plants.

2.3 RES under disturbances: fault ride-through capability

With increasing penetration levels of renewable energy in power systems, mainly due to the integration of wind power plants, more responsibilities must be assumed by the wind power plants to support power system reliability [49]. In fact, before the fast increase of such distributed generation units at the low-voltage and medium-voltage levels of a grid, system operators traditionally controlled the grid through conventional power plants. This practice changed when significant amounts of wind and PV were integrated into the systems, and consequently renewables needed to be considered in planning and operating the grid [50]. For this reason, grid code requirements have been extended to low-voltage ride-through (LVRT) capability, which demands that generation units, including renewable power plants, stay online and supply a specified amount of reactive current to stabilise the voltage during faults.

Generally, LVRT defines steady and dynamic performances of the renewables in three continuous periods after a grid fault occurs: fault transient, fault continuous, and fault recovery [51,52]. According to the ENTSO-E [53], fault ride-through requirements should be extended to smaller generation units mainly due to the following aspects:

- Future generation systems will be based on a vast number of distributed, power electronics-based generators and RES that are variable and only partly dispatchable.
- To a significant extent, RES are connected to the distribution network.
- The objective is to proportionally allocate the requirements for future generators throughout Europe on a level playing field.
- It is of increasing importance that a single system event should not result in a large-scale shutdown of generation. RES generators need to be resilient to system faults by staying connected (and generating) during the initial voltage transients (as conventional generators do today).

Grid codes, in the widest set of the term, involve a set of rules for power system and energy market operation. They enable network operators, generators, suppliers, and consumers to function more effectively across the market. Different types of grid codes are depicted in Figure 2.8. Grid code requirements under voltage dips have typically been expressed as a function of positive-sequence voltage against time.

Connection codes	∘ Generator connection code ∘ Demand connection code ∘ HVDC connection code
Operating codes	∘ Operational security code ∘ Operational planning and scheduling code ∘ Load frequency control and reserve code ∘ Emergency procedure code
Planning codes	∘ Generator planning code ∘ Network planning code
Market codes	∘ Market rules code ∘ Network capacity allocations and congestion management code ∘ HVDC connection code

Figure 2.8 Different types of grid codes based on ENTSO-E [22]

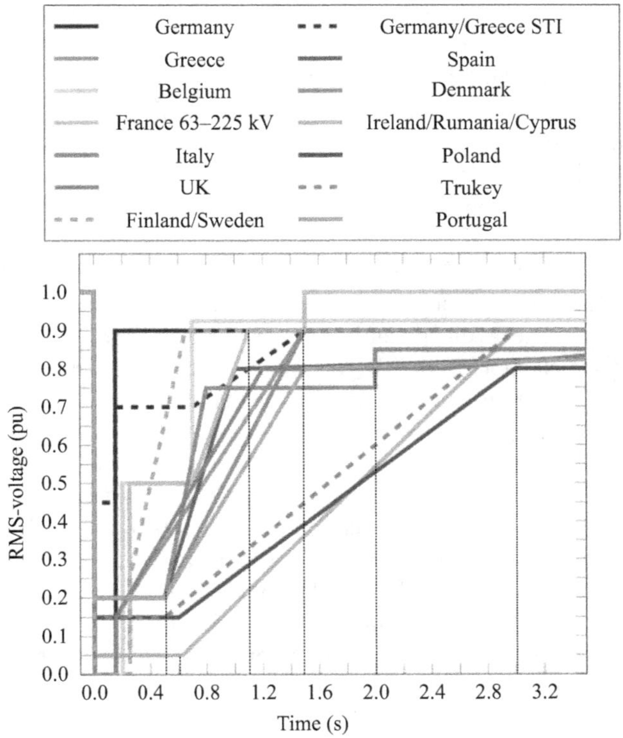

Figure 2.9 Grid code requirements of wind power plants: comparative of rms-voltage-limiting curves at the point of interconnection [54]

This is the case for current requirements for wind power plants submitted to voltage dips, which vary significantly from one country to the next, depending on the penetration level of the renewable technology as well as on the robustness of the national or regional power system. There is considerable variation even within the European Union [55], although standardisation is being implemented [37]. In Europe, countries that have high renewable power capacity, such as Germany and Spain, first developed specified requirements to ensure the continuity of the supply side in the presence of disturbances, mainly voltage dips, and connection and continuity requirements for RES are becoming more demanding. Figure 2.9 summarises the current grid code requirements provided by most European countries for wind installations connected to the grid. In all cases, if the rms voltage trajectories remain above the minimum voltage limits, the generating units shall remain transiently stable and connected to the system, without tripping a close-up, solid, three-phase, short-circuit fault or any unbalanced short-circuit fault on the transmission system [56]. In [57], the problem of testing power electronics and control systems through a finite set of conceivable fault voltage waveforms is addressed by setting forth a powerful computational technique for the design verification of wind turbine power electronic drives, and computational technique for design verification has been proposed by [58].

The industry standard for wind turbine testing is the International Electrotechnical Commission (IEC) Standard 61400-21 [59], which describes the test procedures for active power, reactive power, and fault ride-through testing. IEC 61400-27 [60] describes the corresponding generic simulation models and verification procedures for mode validation.

As variable generation units, PV power plants may lead to possible stability problems in power systems. Moreover, new power flow patterns should be studied in detail to evaluate the impact of high PV penetration on current electric grids [61]. In this context, and to avoid potential drawbacks associated with undesired PV power plant disconnections, TSOs have promoted strict technical requirements for RES connected to the grid and submitted to disturbances. Recent reports have focused on providing preliminary overviews about the risks for system stability caused by the disconnection of dispersed generation in case of over- or under-frequency events [62]. Different examples of grid code requirements can be found in various countries in Europe, such as Spain, Germany, and Italy. These countries have the biggest capacities of PV panels installed in Europe. The Spanish TSO (REE), as system operator of the national grid, developed the operation procedure *Requirements for Response to Voltage Dips of Production Facilities under the Special Regime (P.O.12.3)* [63]. These requirements were approved and issued in October 2006, and only wind power plants were called on to fulfil the specific requisites under the presence of such disturbances. In November 2010, the RD 1565/2010 was proposed and issued [64], supposing an extension of the previous continuity requirements to PV power plants in response to voltage dips. In Germany, in 2008 BDEW published the *Technical Guideline: Generating Plants Connected to the Medium-Voltage Network—Guideline for Generating Plants Connection to and Parallel Operation with the Medium-Voltage Network* [65]. This revised version of the German grid code proposes two types of generation power plants – type-1 and type-2 – depending on their performance under disturbances: type-1 includes generating plants wherein synchronous generators are either directly connected to the grid or only through the generator transformer; whereas the rest of the power plants – including RES – are classified as type-2. The general rules in Italy are summarised in the following reports: CEI 0-16 (which references technical rules for the connection of active and passive consumers to the high-voltage and medium-voltage electrical networks of distribution companies), CEI 11-20 (which discusses electrical energy production systems and uninterruptible power systems connected to the low-voltage and medium-voltage networks), and CEI 11-32 (which discusses electrical energy production systems connected to the high-voltage network). CEI 0-21 discusses the technical requirements for the interconnection of PV power plants. European grid codes thus present specific criteria for the interconnection of PV power plants and their performance under disturbances. In an attempt to compare these different requirements, Figure 2.6 shows the voltage-rms limitations for PV power plants at the point of interconnection. Both magnitude and the duration of the voltage dips according to the different European grid codes are depicted, characterising the severity of their

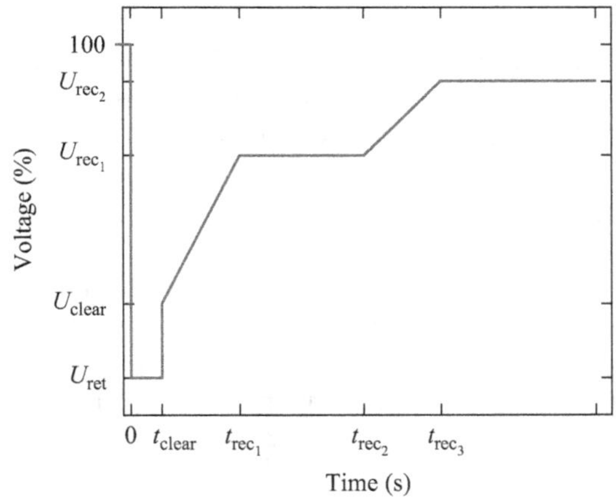

Figure 2.10 Guidelines for grid code requirements: voltage-limiting curves for generating units at the point of interconnection [37]

Table 2.1 Guidelines for grid code requirements: parameters for times and voltages within ranges defined by the network code [37]

Synchronous power-generating modules				**Power park modules**			
Voltage parameters (pu)		**Time parameters (s)**		**Voltage parameters (pu)**		**Time parameters (s)**	
U_{ret}	0.05–0.3	t_{clear}	0.14–0.25	U_{ret}	0.05–0.15	t_{clear}	0.14–0.25
U_{clear}	0.7–0.9	t_{rec_1}	t_{clear}	U_{clear}	U_{ret}–0.15	t_{rec_1}	t_{clear}
U_{rec_1}	U_{clear}	t_{rec_2}	t_{rec_1}–0.7	U_{rec_1}	U_{clear}	t_{rec_2}	t_{rec_1}
U_{rec_2}	0.85–0.9 and U_{clear}	t_{rec_3}	t_{rec_2}–1.5	U_{rec_2}	0.85	t_{rec_3}	1.5–3.0

requirements and emphasising their differences. These differences are pointed out by some authors as major obstacles to the deployment of distributed energy renewable sources, such as PV [66,67]. Today, in an attempt to establish general rules to which new generators must adhere, the latest draft of the Network Code on Requirements for Generators (provided by ENTSO-E) gives guidance on the implementation processes for all generators wishing to connect below 110 kV and within the European Union; see Figure 2.10 and Table 2.1. Most developed countries are carefully observing the development of growing installed capacities and promoting initiatives to develop new disconnection rules for PV panels based on system requirements. These requirements need to be in line with the current draft version for the ENTSO-E network code. In general, contributions including both characterisation and classification techniques for power disturbances can be found in [68–70]; however, only a few contributions present power quality surveys of renewable power installations [71–74].

However, there is a lack of contributions focused on comparing real voltage dips to current requirements and estimating how severe and close to the rms-voltage limits the collected events are. According to the specific literature, representations usually reduce the analysis of events in terms of two main characteristics: minimum rms-voltage compared to time duration. These studies imply an important simplification of disturbances, neglecting the rms-voltage evolution during the event. In fact, apart from the residual rms-voltage, the severity of the faults is highly dependent on the voltage evolution along the disturbance as well as the point-on-wave initiation [75]. Therefore, a comparison between rms-voltage limits and real voltage dips based on the entire rms-voltage dip trajectory versus time initially appears to be more appropriate, and it is proposed by several authors as a relevant field of study for future work.

2.4 Renewable energy curtailment

Increasing wind and solar penetration levels may drive a power system to encounter transmission, distribution, or operational constraints, forcing the TSOs to take into account the performance of these spatially distributed renewable power plants [76]. Therefore, high levels of wind and solar power generation can be challenging to integrate into power systems because of their variability [77] and limits in predictability, leading to use curtailments, or less wind or solar power than is potentially available at that time [78–80]; however, curtailment of power stations has been a normal practice since the beginning of the electric power industry [81]. Taking into account the constraints imposed by flexibility in a specific power system, different solutions can be used to accommodate variable generation during these situations [82], such as dispatchable power plants, energy storage, interconnection capacity, or demand-side response [83]. The need for energy curtailment for variable generation is applied during a limited time, or transition phase, before some changes can be programmed to the grid – infrastructural, operational, or institutional. Even interannual oscillations in variable generation should be taken into account. Figure 2.11 shows the influence of the month on the global variable power generation in Spain, taking into account three typical technologies: wind, PV, and CSP. Data clearly show variability in the months and among the years.

There are many reasons for RES curtailment, including lack of transmission and distribution availability or system balancing challenges. Moreover, RES curtailment is a very complex issue because the coincidence in the time of the curtailment, network thermal congestion, or system-wide inertial stability problems carry out to the interaction among different sources of curtailment [84]. The main reasons in power systems with very high wind energy penetration targets are related to the lack of available transmission during a particular interval to incorporate some or all of the variable generation and system balancing issues [85]. In [86], a comparison of different levels of curtailment takes into account variable energy penetration ratios in selected countries/areas. For example, in the United States the Electric Reliability Council of Texas achieves upward primary frequency reserves

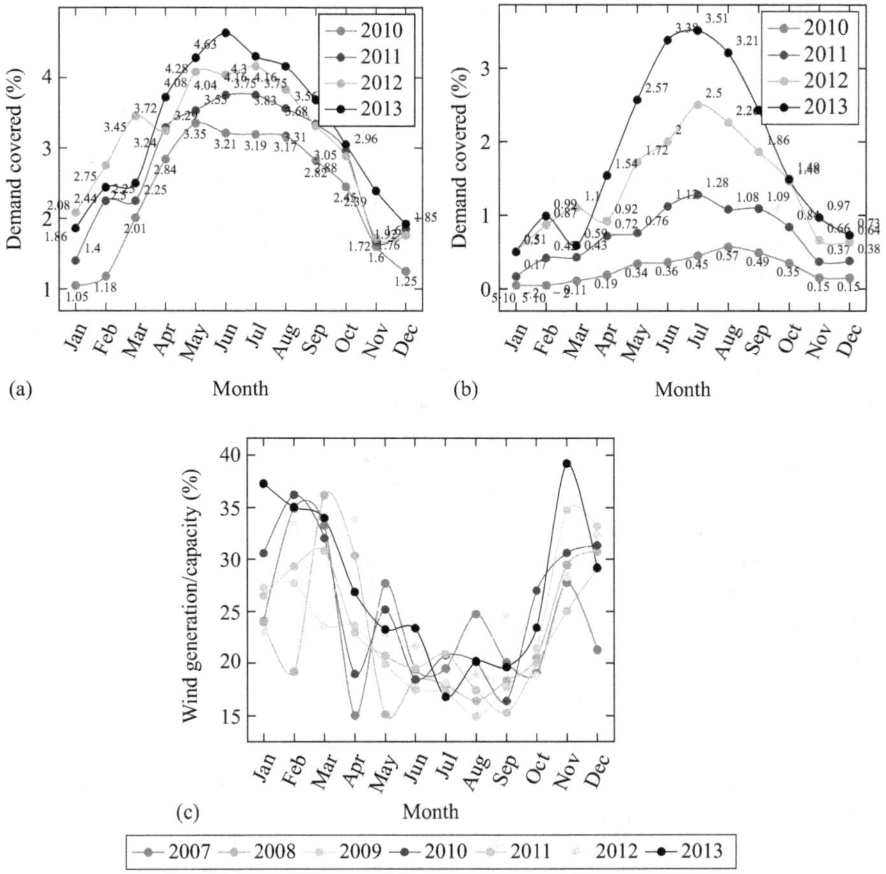

Figure 2.11 Variable generation in Spain

from curtailed wind generation. The value of these ancillary services implies that curtailment does not need to be seen as a waste of energy. China can be considered an extreme example, because in 2013 approximately 16.23 TWh of wind generation was curtailed (10.74% of total wind generation), and this value was even an improvement from the 20.82 TWh curtailed in 2012.

Transmission congestion is usually related to the development timescales of variable generation power plants. They can generally evolve much more quickly than the new required transmission projects, leading to a timing mismatch in construction between wind development and new transmission [87]. It may take 5–10 years to plan (or more, depending on the country), permit, and construct a transmission line, whereas a wind or solar project can be planned, permitted, and constructed in 1–3 years [87–89]. System balancing issues typically occur with high wind generation at night, when loads are low and thermal units are pushed down against their minimum operating constraints, and the generation excess cannot be exported to other balancing areas due to transmission constraints [90]. In some countries, such as Spain, variable generation, mainly wind, may be curtailed after other power plants are pushed down against their minimum operating constraints, reducing or even removing power imports reduced. The low capacity factor of variable generation and the uncertainty and variability associated with it, together with the system flexibility, are the main reasons for this kind of curtailment [83]. In countries with large amounts of wind power, this issue could be regulated somehow. For Spain's power system, this kind of curtailment is referred to in the *Non-Integrable Wind Power Excess* and is defined in the Operational Procedure 3.7 [91]. On the other hand, in the distribution system, curtailment can occur to avoid high penetrations or back-feeding, in which more energy is produced at the feeder level than consumed. In some cases, high penetrations of solar PV generation on feeders can lead to voltage control issues due to the variability of the resource [92].

Curtailments can be related to other reasons as well, such as grid stability and short-circuit power, mainly due to disturbances and voltage dips affecting the network. Such curtailment measures though are more likely to be observed in small, isolated systems, such as Ireland [93]. It has been identified that higher, instantaneous system, non-synchronous penetration levels may require wind power curtailment, unless no other measures to maintain the system stability are considered. In many countries this issue is partially solved due to the fact that most wind power plants and solar PV power converters can support LVRT [93]. In Spain, approximately 95% of wind generation is certified under LVRT requirements, and therefore curtailment due to voltage dips has not been an issue since 2009 [94].

In Spain, curtailments can be classified, according to market procedures, in two categories: real time and programmed. Real-time curtailments are obtained from the intraday markets. On the other hand, programmed curtailments are set before the day-ahead market is closed (Basic Operation Program, PDBF). The former is partially compensated as a function of daily market price, and the latter is not economically compensated [95]. In the case of wind generation, in Spain set points due to real-time curtailments are calculated and delivered – through the Control Centre of Renewable Energies (CECRE) – with maximum wind nodal production, and wind power plants affected must adapt their production to the given set point within 15 min. CECRE is an operating unit within the power control centre (CECOEL) that is connected to 98.6% of installed wind power capacity in Spain, whereas the remaining wind power (1.4%) is estimated. CECRE has been in operation since 2006, and it is considered a worldwide pioneering initiative to monitor and control renewable power plants, specifically wind power plants. CECRE does not tele-command generation equipment; this function is done by generation control centres. The telecommunication deployment of almost 800 wind power plants has been accomplished by aggregating power stations larger than 10 MW. Measurements such as active and reactive power, voltage, connectivity, temperature, and wind speed and wind direction are taken from wind power plants every 12 s and sent to a Renewable Energy Resource Control Centre (RESCC).

Curtailment data during 2013 for different countries are shown in Table 2.2 [78]. Wind curtailment typically ranged from approximately 1%–3% of wind generation

Table 2.2 Wind generation and curtailment data by country (2013) [78]

Country	China	Germany (2012)	Ireland	Italy	Spain	United States
Electricity generation (TWh)	5,372	577	26	290	284	4,066
Wind generation (GWh)	142,000	50,600	5,872	14,811	54,338	167,840
Wind/electricity generation	2.6%	9.8%	22.5%	5.1%	19.2%	4.1%
Wind curtailment (GWh)	16,230	358	196	152	1,166	–
Wind Curtailment/generation	11%	0.7%	3%	1%	2%	1%–3%

levels, although China exceeds this range, getting 11%. The implementation of curtailment in a power system is a complex issue because under certain conditions some events can produce severe deviations from the original set points of the TSO. In [93], a curtailment event with an overresponse in the level of the curtailment is described. It is shown that the Spanish TSO imposed different set points by taking into account the *Non-Integrable Wind Power Excess*. This example occurred in January 2010, and the final curtailment due to the overresponse in the wind power generation was more than four times the wind power curtailment originally required. Although this type of event could be classified as exceptional, the operation of the power system can be put at risk, even with from an economic point of view, because upward reserves are used for balancing, which increases the associated costs. Therefore, the evolution of the flexibility exhibited in the power system is related to the level of curtailment applied to variable generation. An example of this relationship is shown in [95] using the parameter, P_{flex_1}, in which values are referred to the registered values during a year. Two case studies representing curtailments in wind generation in two situations are examined: low power demand (off-peak periods) and high wind power contribution in the first case, and during high wind power forecast error stages in the second case.

The parameter $P_{\text{flex}_1} = \frac{P_{\text{max-year}} - P}{P_{\text{max-year}} - P_{\text{min-year}}}$ (pu), shows the availability of flexible generation during the curtailment compared to the ordinary power system operation. $P_{\text{max-year}}$ is the maximum yearly power value for that technology, $P_{\text{min-year}}$ is the minimum yearly power value for that technology, and P the power value for the considered time period.

Figure 2.12 shows the evolution of generation – wind and nuclear power – during Easter 2013. On March 31, due to the large amount of wind generation in conjunction with the significant amount of rainfall, saturation of the hydraulic reserves occurred. Moreover, due to the holiday season, this is one of the periods of the year with lower power consumption. So reserves for the pumped storage hydropower could not be used, and most of the hydraulic power plant became unmanaged. The result was a huge amount of curtailment in wind generation, reducing at the same time the nuclear generation, approximately 1,000 MW; see Figure 2.12(c). The evolution of the flexibility in the power system is shown in Figure 2.12(b), wherein eight steps are considered, taking into account the

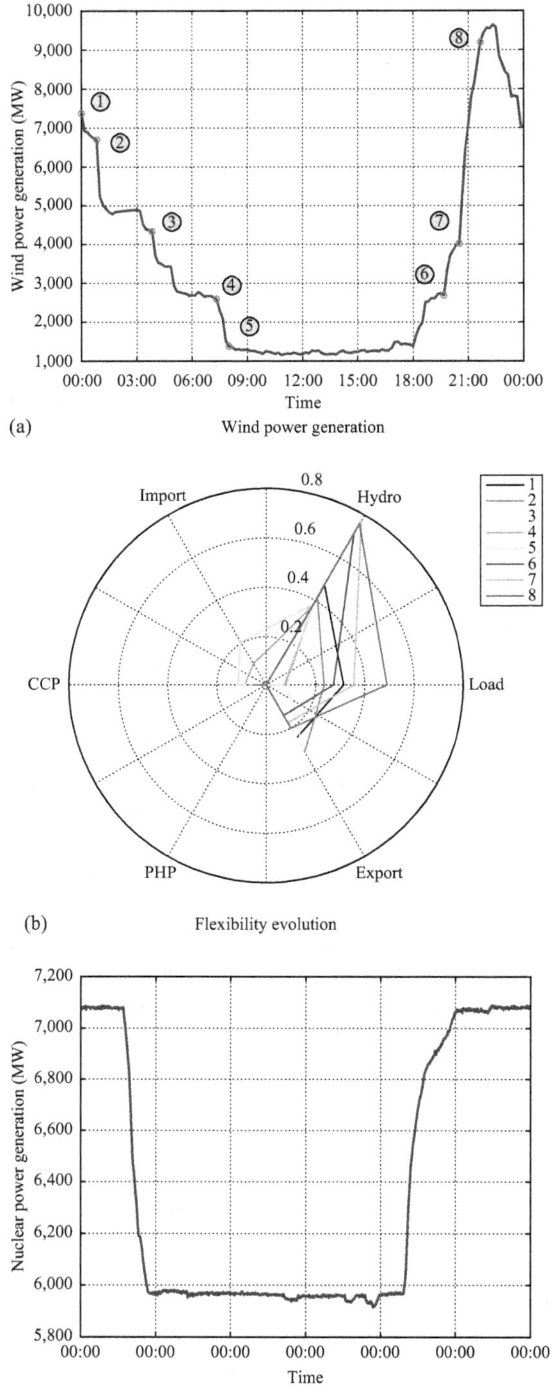

Figure 2.12 Evolution of wind and nuclear power generation during Easter 2013 in Spain

flexibility imposed by the hydropower generation, pumped hydropower generation, combined-cycle gas turbine, and the demand, together with the international imported/exported electricity. The figure clearly shows the evolution of P_{flex} with changes in the hydropower generation together with the international import/export. This phenomenon has occurred several times in Spain, although not as severe, due to the explosive cyclogenesis and heavy rainfall combined with high wind production; therefore, increasing the flexibility of the system is the best way to reduce curtailments. The methods of increasing system flexibility depend on the power system, with different possibilities taking into account the proportions of each method [96–98]. The main ones are grid capacity, physical additions to the system (such as storage or new transmission lines [89,99]), power system operational changes (such as economic dispatch or improved forecasting [100,101]), and institutional changes (such as access to new markets [101]).

Acknowledgements

The authors would like to thank the 'Ministerio de Economía, Industria y Competitividad' and the European Union FEDER, which supported this work under project ENE2016-78214-C2-1-R.

This work was supported by the US Department of Energy under Contract No. DE-AC36-08GO28308 with the National Renewable Energy Laboratory. Funding provided by [applicable DOE Office and Program, e.g., US DOE Office of Energy Efficiency and Renewable Energy Solar Energy Technologies Program]. The US Government retains and the publisher, by accepting the article for publication, acknowledges that the US Government retains a nonexclusive, paid-up, irrevocable, worldwide license to publish or reproduce the published form of this work, or allow others to do so, for US Government purposes. The authors would like to thank Katie Wensuc for the editorial work.

References

[1] European Union Committee, 27th Report of Session 2007–08, "The EU's Target for Renewable Energy: 20% by 2020", October 2008.

[2] P. V. L. W. M. Beurskens and M. Hekkenberg, "Renewable energy projections as published in the national renewable energy action plans of the European member states," European Environment Agency (EEA), Tech. Rep., November 2011.

[3] International Energy Agency, World Energy Outlook 2015, Energy Subsidies [online]. Available from: http://www.worldenergyoutlook.org/media/weowebsite/2015/Subsidies20122014.xlsx

[4] "World Energy Outlook 2012." International Energy Agency (IEA), Tech. Rep., 2012 [online]. Available from: www.iea.org

[5] J. Glassmire, P. Komor, and P. Lilienthal, "Electricity demand savings from distributed solar photovoltaics," *Energy Policy*, vol. 51, pp. 323–331, 2012.

[6] D. Carvalho, J. Wemans, J. Lima, and I. Malico, "Photovoltaic energy mini-generation: Future perspectives for Portugal," *Energy Policy*, vol. 39, no. 9, pp. 5465–5473, 2011.
[7] A. Battaglini, N. Komendantova, P. Brtnik, and A. Patt, "Perception of barriers for expansion of electricity grids in the European Union," *Energy Policy*, vol. 47, pp. 254–259, 2012.
[8] RENEWABLE ENERGY. Medium-Term. Market Report 2015. Market Analysis and Forecasts to 2020, IEA 2015 [online]. Available from: https://www.iea.org/Textbase/npsum/MTrenew2015sum.pdf
[9] The First Decade: 2004–2014, Renewable Energy Policy Network for the 21st Century [online]. Available from: http://www.ren21.net/Portals/0/documents/activities/Topical%20Reports/REN21_10yr.pdf
[10] C. E. Behrens, "Energy policy: 113th congress issues," in *Congressional Research Service*, 2013.
[11] "European Commission, 2010a. Energy 2020. A strategy for competitive, sustainable and secure energy." Brussels, Tech. Report, November 2011.
[12] "Monthly electricity statistics," International Energy Agency (IEA), Tech. Rep [online]. Available from: www.iea.org
[13] T. Sterner and M. Damon, "Green growth in the post-Copenhagen climate," *Energy Policy*, vol. 39, pp. 7165–7173, 2011.
[14] "World Energy Outlook," International Energy Agency (IEA), Tech. Rep., 2012 [online]. Available from: www.iea.org
[15] J. Lopesa, N. Hatziargyriou, J. Mutale, P. Djapic, and N. Jenkins, "Integrating distributed generation into electric power systems: A review of drivers, challenges and opportunities," *Electric Power Systems Research*, vol. 77, no. 9, pp. 1189–1203, 2007.
[16] The Danish Government's "Our future Energy," November 2011.
[17] Danish Energy Authority, Energy Statistics 2013 [online]. Available from: http://www.ens.dk/sites/ens.dk/files/info/tal-kort/statistik-noegletal/aarlig-energistatistik/energystatistics2013.pdf
[18] Store Project, "Energy needs in Denmark. Executive summary: "Overview of the Danish Power system and RES integration," [online]. Available from: www.store-project.eu
[19] PV-Tech, "Denmark to reach 2020 installation goal by Q4 2012," [online]. Available from: http://www.pv-tech.org/news/denmark_to_reach_2020_installation_ goal_by_q4_2012
[20] Energinet.dk, "Solar Power," [online]. Available from: http://www.energinet.dk/EN/KLIMA-OG-MILJOE/Miljoerapportering/VE-produktion/Sider/Sol.aspx
[21] 2015 European Statistics, Wind in Power (EWEA), February 2016 [online]. Available from: https://windeurope.org/wp-content/uploads/files/about-wind/statistics/EWEA-Annual-Statistics-2015.pdf
[22] T. Ackermann, N. Martensen, T. Brown, *et al.*, "Scaling up variable renewable power: The role of grid codes," *IRENA*, Tech. Rep., 2016 [online]. Available from: www.irena.org/publications

[23] "Global Market Outlook for Photovoltaics Until 2016," European Photovoltaic Industry Association, Tech. Rep., 2012 [online]. Available from: www.epia.org

[24] "The global wind report annual market update 2014," Global Wind Energy Council, Tech. Rep., 2014 [online]. Available from: www.gwec.net

[25] "Wind in power 2012 European statistics," The European Wind Energy Association EWEA, Tech. Rep., February 2012 [online]. Available from: www.ewea.org

[26] P. Sørensen, A. D. Hansen, and N. A. Cutululis, "Ancillary services from wind power plants," *10th Conference on Sustainable Development of Energy, Water and Environment Systems*, 27 September–2 October 2015, Dubrovnik, Croatia.

[27] F. V. Hulle, H. Holttinen, J. Kiviluoma, "Grid support services by wind and solar PV: A review of system needs, technology options, economic benefits and suitable market mechanisms," Synthesis report of the REserviceS project, REserviceS project, 2014 [online]. Available from: http://www.reservices-project.eu/wp-content/uploads/140724_REserviceS_D7.1_Synthesis-report1.pdf

[28] R. Passey, T. Spooner, I. MacGill, M. Watt, and K. Syngellakis, "The potential impacts of grid-connected distributed generation and how to address them: A review of technical and non-technical factors," *Energy Policy*, vol. 39, no. 10, pp. 6280–6290, 2011.

[29] A. J. Lamadrid and T. Mount, "Ancillary services in systems with high penetrations of renewable energy sources, the case of ramping," *Energy Economics*, vol. 34, no. 6, pp. 1959–1971, 2012.

[30] E.ON Netz GmbH, "Netzanschlussregeln-allgemein. Technische und organisatorische Regeln für den Netzanschluss innerhalb der Regelzone der E.ON Netz GmbH im Bereich der ehemaligen PreussenElektra Netz GmbH & Co. KG," Rev: December 1, 2001, Bayreuth, 2001.

[31] C. Clastres, T. H. Pham, F. Wurtz, and S. Bacha, "Ancillary services and optimal household energy management with photovoltaic production," *Energy*, vol. 35, no. 1, pp. 55–64, 2010 [online]. Available from: http://www.sciencedirect.com/science/article/pii/S0360544209003673

[32] D. Nock, V. Krishnan, and J. D. McCalley, "Dispatching intermittent wind resources for ancillary services via wind control and its impact on power system economics," *Renewable Energy*, vol. 71, pp. 396–400, 2014 [online]. Available from: http://dx.doi.org/10.1016/j.renene.2014.05.058

[33] N. Sangroniz, J. A. Mora, and M. D. Teixeira, "Review of international grid codes for wind generation," VIII CBQEE - Conferência Brasileira sobre Qualidade da Energia Elétrica, Blumenau, Aug. 2009.

[34] "Global Market Outlook for Photovoltaic 2013–2017. EPIA," European Photovoltaic Industry Association EPIA, Tech. Rep., 2013 [online]. Available from: www.epia.org

[35] A. Marinopoulos, F. Papandrea, M. Reza, S. Norrga, F. Spertino, and R. Napoli, "Grid integration aspects of large solar PV installations: LVRT capability and reactive power/voltage support requirements," in *PowerTech, 2011 IEEE Trondheim*. IEEE, 2011, pp. 1–8.

[36] European Network of Transmission System Operators for Electricity. Available from: https://www.entsoe.eu/publications/system-operations-reports/operation-handbook/Pages/default.aspx (Last accessed on: January, 31, 2017).

[37] ENTSO-E, NC RfG: Network Code Requirements for Grid Connection (adopted by Member States in Comitology in June 2015) [online]. Available from: https://www.entsoe.eu/Documents/Network%20codes%20documents/NC%20RfG/draft_ec_networkCodesJune.pdf

[38] Energinet.dk Ancillary Services Strategy [online]. Available from: https://www.energinet.dk/SiteCollectionDocuments/Engelske%20dokumenter/El/77566-11%20v1%20Energinet%20dk's%20ancillary%20services%20strategy.pdf

[39] V. Gevorgian and B. O'Neill, "Demonstration of Active Power Controls by Utility-Scale PV Power Plant in an Island Grid," in *International Workshop on Integration of Solar Power into Power Systems*. Vienna, Austria: Energynautics GmbH, October 2016, pp. 1–6.

[40] J. Merino, P. Mendoz-Araya, and C. Veganzones, "State of the art and future trends in grid codes applicable to isolated electrical systems," *Energies*, vol. 7, pp. 7936–7954, 2014.

[41] S. Martín-Martínez, E. Gómez-Lázaro, A. Vigueras-Rodríguez, J. A. Fuentes, and A. Molina-García, "Analysis of positive ramp limitation control strategies for reducing wind power fluctuations," *IET Renewable Power Generation*, vol. 7, no. 6, pp. 593–602, November 2013.

[42] F. Van Hulle, I. Pineda, P. Wilczek, Economic grid support services by wind and solar PV, REserviceS project - Recommendations, September 2014. Available from: http://www.reservices-project.eu/wp-content/uploads/REserviceS-project-recommendations-EN.pdf (accessed on 31 January 2017).

[43] E.ON Netz GmbH, "Grid Code High and Extra high voltage," Rev. April 1, 2006 [online]. Available from: http://www.eon-netz.com/, Bayreuth, 2006.

[44] R. Teodorescu, M. Liserre, and P. Rodríguez, *Grid Converters for Photovoltaic and Wind Power Systems*. UK: John Wiley & Sons, 2011.

[45] E. Camm, M. Behnke, O. Bolado, *et al.*, "Reactive power compensation for wind power plants," in *2009 IEEE Power & Energy Society General Meeting*. IEEE, 2009, pp. 1–7.

[46] J. Kiviluoma, H. Holttinen, Y. H. Wan, *et al.*, "Variability in large scale wind power generation," *Wind Energy*, vol. 19, no. 9, pp. 1649–1665, September 2016.

[47] M. P. Palsson, T. Toftevaag, K. Uhlen, and J. O. G. Tande, "Large-scale wind power integration and voltage stability limits in regional networks," in *Power Engineering Society Summer Meeting, 2002 IEEE*, vol. 2, pp. 762–769.

[48] A. Ellis, R. Nelson, E. Von Engeln, *et al.*, "Reactive power performance requirements for wind and solar plants," in *2012 IEEE Power and Energy Society General Meeting*. IEEE, 2012, pp. 1–8.

[49] M. Tsili and S. Papathanassiou, "A review of grid code technical requirements for wind farms," *IET Renewable Power Generation*, vol. 3, no. 3, pp. 308–332, September 2009.

[50] B. I. Craciun, T. Kerekes, D. Séra, and R. Teodorescu, "Overview of recent grid codes for PV power integration," *2012 13th International Conference on Optimization of Electrical and Electronic Equipment (OPTIM)*, pp. 959–965, May 2012. DOI: 10.1109/OPTIM.2012.6231767.

[51] J. Fortmann, R. Pfeiffer, E. Haesen, *et al.*, "Fault-ride-through requirements for wind power plants in the ENTSO-E network code on requirements for generators," *IET Renewable Power Generation*, vol. 9, no. 1, pp. 18–24, 2015.

[52] H. Geng, C. Liu, and G. Yang, "LVRT capability of dfig-based WECS under asymmetrical grid fault condition," *IEEE Transactions on Industrial Electronics*, vol. 60, no. 6, pp. 2495–2509, June 2013.

[53] "Network code requirements for grid connection applicable to all generators," [online]. Available from: https://www.entsoe.eu/fileadmin/user_-upload/_library/resources/RfG/131016_-_NC_RfG_implementation_guide-line.pdf, European Network of Transmission System Operators for Electricity, ENTSO-E, Tech. Rep., October 2013.

[54] T. Garcia-Sanchez, E. Gomez-Lazaro, E. Muljadi, M. Kessler, and A. Molina-Garcia, "Statistical and clustering analysis for disturbances: A case study of voltage dips in wind farms," *IEEE Transactions on Power Delivery*, vol. 31, no. 6, pp. 2530–2537, Dec. 2016.

[55] F. Jiménez, E. Gómez-Lázaro, J. A. Fuentes, A. Molina-García, and A. Vigueras-Rodríguez, "Validation of a double fed induction generator wind turbine model and wind farm verification following the Spanish grid code," *Wind Energy*, vol. 15, no. 4, pp. 645–659, 2012.

[56] W. Christiansen and D. T. Johnsen, "Analysis of requirements in selected grid codes," Section of Electric power Engineering – Technical University of Denmark (DTU), Tech. Rep., January 2006.

[57] H. V. Pico and D. Aliprantis, "Voltage ride-through capability verification of wind turbines with fully-rated converters using reachability analysis," *IEEE Transactions on Energy Conversion*, vol. 29, no. 2, pp. 392–405, June 2014.

[58] H. Villegas Pico and D. Aliprantis, "Voltage ride-through capability verification of wind turbines with fully-rated converters using reachability analysis," *IEEE Transactions on Energy Conversion*, vol. 29, no. 2, pp. 392–405, June 2014.

[59] IEC 61400-21 ed. 2, Wind turbine generator systems – Part 21: Measurement and assessment of power quality characteristics of grid connected wind turbines, 2008. Geneva, Switzerland: IEC.

[60] IEC 61400-27-1:2015 Wind turbine generator systems – Part 21: Electrical simulation models – Wind turbines, 2015. Geneva, Switzerland: IEC.

[61] S. Eftekharnejad, V. Vittal, G. Heydt, B. Keel, and J. Loehr, "Small signal stability assessment of power systems with increased penetration of

photovoltaic generation: A case study," *IEEE Transactions on Sustainable Energy*, vol. 4, no. 4, 960–967, October 2013.

[62] "Assessment of the system security with respect to disconnection rules of photovoltaic panels," [online]. Available from: https://www.entsoe.eu/publications/system-operations-reports/Documents/120530_Assessment_of_the_System_security_with_respect_to_disconnection_rules_of_PV_Panels.pdf, European Network of Transmission System Operators for Electricity, ENTSO-E, Tech. Rep., April 2012.

[63] REE, P.O. 12.3 Requisitos de respuesta frente a huecos de tensión de las instalaciones eólicas (in Spanish), Red Eléctrica de España, October 2006.

[64] "Royal Decree 1565/2010, by which regulates and modifies certain aspects of the activity of production of electric energy in special regime (n Spanish)," Tech. Rep., November 2010.

[65] BDEW, Generating Plants Connected to the Medium-Voltage Network [online]. Available from: http://www.bdew.de/internet.nsf/id/A2A0475F2FAE8F44C12578300047C92F/$file/BDEW_RL_EA-am-MS-Netz_engl.pdf

[66] D. Montoro, "Recommendations for unified technical regulations for grid-connected PV systems," [online]. Available from: http://www.pvsunrise.eu/fileadmin/PVS_docs/images/Recommendations_for_unified_technical_regulations_for_grid-connected_PV_systems__Public_pdf, SUNRISE project – European Photovoltaic Industry Association, the European Construction Industry Federation, the European Association of Electrical Contractors, International Union of Architects, Tech. Rep., 2009.

[67] B. Craciun, T. Kerekes, D. Sera, and R. Teodorescu, "Overview of recent grid codes for PV power integration," in *Optimization of Electrical and Electronic Equipment (OPTIM), 2012 13th International Conference on*, May 2012, pp. 959–965.

[68] L. Zhan and M. H. Bollen, "Characteristic of voltage dips (sags) in power systems," *IEEE Transactions on Power Delivery*, vol. 15, no. 2, pp. 827–832, 2000.

[69] M. H. Bollen and L. Zhang, "Different methods for classification of three-phase unbalanced voltage dips due to faults," *Electric Power Systems Research*, vol. 66, no. 1, pp. 59–69, 2003.

[70] E. Gómez-Lázaro, J. A. Fuentes, A. Molina-García, and M. Cañas-Carretón, "Characterization and visualization of voltage dips in wind power installations," *IEEE Transactions on Power Delivery*, vol. 24, no. 4, pp. 2071–2078, 2009.

[71] M. I. Muhamad, N. Mariun, and M. A. M. Radzi, "The effects of power quality to the industries," *Fifth Student Conference on Research and Development*, Selangor, Malaysia, Dec. 2007, pp. 1–4. DOI: 10.1109/SCORED.2007.4451410.

[72] E. Muljadi, Z. Mills, R. Foster, J. Conto, and A. Ellis, "Fault analysis at a wind power plant for one year of observation," in *Power and Energy Society General Meeting – Conversion and Delivery of Electrical Energy in the 21st Century, 2008 IEEE*, July 2008, pp. 1–7.

[73] M. Bollen, M. Stephens, S. Djokic, *et al.*, "Voltage dip immunity of equipment and installations," CIGRE, Tech. Rep., 2010.
[74] S. Elphick, V. Smith, V. Gosbell, and R. Barr, "The Australian long term power quality survey project update," *14th International Conference on Harmonics and Quality of Power*, pp. 1–7, 2010.
[75] A. Honrubia-Escribano, E. Gómez-Lázaro, A. Molina-García, and J. Fuentes, "Influence of voltage dips on industrial equipment: Analysis and assessment," *International Journal of Electrical Power & Energy Systems*, vol. 41, no. 1, pp. 87–95, 2012.
[76] D. Flynn, Z. Rather, A. Ardal, *et al.*, "Technical impacts of high penetration levels of wind power on power system stability," *Wiley Interdisciplinary Reviews: Energy and Environment*, 2017.
[77] E. Gomez-Lazaro, M. C. Bueso, M. Kessler, *et al.*, "Probability density function characterization for aggregated large-scale wind power based on Weibull mixtures," *Energies*, vol. 9, no. 2, pp. 1–15, February 2016.
[78] L. Bird, D. Lew, M. Milligan, *et al.*, "Wind and solar energy curtailment: A review of international experience," *Renewable & Sustainable Energy Reviews*, vol. 65, pp. 577–586, November 2016.
[79] D. Lew, L. Bird, E. Carlini, *et al.*, "Wind and solar curtailment: International experience and practices," in *International Workshop on Large-Scale Integration of Wind Power into Power Systems as well as on Transmission Networks for Offshore Wind Farms*. London, United Kingdom: Energynautics GmbH, October 2013, pp. 1–7.
[80] S. Martín-Martínez and E. Gómez-Lázaro, "Analysis of hydropower flexibility during wind energy curtailments: The Spanish case," in *IEEE PES General Meeting*. Denver, CO: IEEE Power & Energy Society, July 2015.
[81] L. Bird, J. Cochran, and X. Wang, "Wind and solar energy curtailment: Experience and practices in the united states," *NREL*, March 2014.
[82] H. Holttinen, P. Meibom, A. Orths, *et al.*, "Impacts of large amounts of wind power on design and operation of power systems, results of IEA collaboration," *Wind Energy*, vol. 14, no. 2, pp. 179–192, March 2011.
[83] S. Martín-Martínez, E. Gómez-Lázaro, A. Honrubia-Escribano, M. Cañas, and A. Molina-García, "Wind power curtailment analysis under generation flexibility requirements: The Spanish case study," in *IEEE PES General Meeting*. Denver, CO: IEEE Power & Energy Society, July 2015, pp. 1–5.
[84] D. J. Burke and M. J. O'Malley, "Factors influencing wind energy curtailment," *IEEE Transactions on Sustainable Energy*, vol. 2, no. 2, pp. 185–193, 2011.
[85] S. Fink, C. Mudd, K. Porter, and B. Morgenstern, "Wind energy curtailment case studies," *NREL Subcontract Report, NREL/SR-550*, vol. 46716, 2009, pp. 1–47.
[86] Y. Yasuda, L. Bird, E. M. Carlini, *et al.*, "International comparison of wind and solar curtailment ratio," in *International Workshop on Large-Scale Integration of Wind Power into Power Systems as well as on Transmission*

Networks for Offshore Wind Farms. Brussels, Belgium: Energynautics GmbH, October 2015, pp. 1–6.

[87] R. Piwko, D. Osborn, R. Gramlich, G. Jordan, D. Hawkins, and K. Porter, "Wind energy delivery issues [transmission planning and competitive electricity market operation]," *IEEE Power and Energy Magazine*, vol. 3, no. 6, pp. 47–56, 2005.

[88] J. C. Smith, D. Osborn, R. Zavadil, *et al.*, "Transmission planning for wind power, status and prospects," in *International Workshop on Large-Scale Integration of Wind Power into Power Systems as well as on Transmission Networks for Offshore Wind Farms*. Quebec, Canada: Energynautics GmbH, October 2010, pp. 371–382.

[89] J. C. Smith, D. Osborn, R. Zavadil, *et al.*, "Transmission planning for wind energy: Status and prospects," *Wiley Interdisciplinary Reviews: Energy and Environment*, vol. 2, no. 1, 1–13, January 2013.

[90] J. Zhang, B.-M. Hodge, J. Miettinen, *et al.*, "Analysis of variability and uncertainty in wind power forecasting: An international comparison," in *International Workshop on Large-Scale Integration of Wind Power into Power Systems as well as on Transmission Networks for Offshore Wind Farms*. London, United Kingdom: Energynautics GmbH, October 2013, pp. 1–6.

[91] P.O. 3.7 Programación de la generación de origen renovable no gestionable (in Spanish), Red Eléctrica de España, May 2009.

[92] L. Kane and G. Ault, "A review and analysis of renewable energy curtailment schemes and principles of access: Transitioning towards business as usual," *Energy Policy*, vol. 72, pp. 67–77, 2014.

[93] EirGrid and SONL. All island TSO facilitation of renewables studies, Dublin, Ireland, 2010.

[94] S. Martin-Martínez, A. Vigueras-Rodríguez, E. Gómez-Lázaro, A. Molina-García, E. Muljadi, and M. Milligan, *Advances in Wind Power*. Rijeka, Croatia: Intech, November 2012, ch. Wind Power Variability and Singular Events, p. 373.

[95] S. Martín-Martínez, E. Gómez-Lázaro, A. Molina-García, and A. Honrubia-Escribano, "Impact of wind power curtailments on the Spanish power system operation," in *IEEE PES General Meeting*. National Harbor, MD: IEEE Power & Energy Society, July 2014, pp. 1–5.

[96] Y. Yasuda, A. R. Årdal, E. M. Carlini, *et al.*, "Flexibility chart: Evaluation on diversity of flexibility in various areas," in *International Workshop on Large-Scale Integration of Wind Power into Power Systems as well as on Transmission Networks for Offshore Wind Farms*. London, United Kingdom: Energynautics GmbH, October 2013.

[97] D. Huertas, H. Farahmand, H. Holttinen, *et al.*, "Hydro power flexibility for power systems with variable renewable energy sources (RES): An IEA Task 25 collaboration," *Wiley Interdisciplinary Reviews: Energy and Environment*, vol. 6, no. 1 2016.

[98] A. Orths, H. Abildgaard, F. van Hulle, *et al.*, *Expert Group Report on Recommended Practices. 16 Wind Integration Studies*, H. Holttinen, Ed. Helsinki, Finland: International Energy Agency Implementing Agreement for Co-operation in the Research, Development, and Deployment of Wind Energy Systems, September 2013.

[99] J. C. Smith, A. Estanqueiro, E. Gómez-Lázaro, *et al.*, "Transmission planning for wind energy. Status and prospects," in *European Wind Energy Conference and Exhibition–IEAWIND Side Event*. Brussels, Belgium: European Wind Energy Association (EWEA), March 2011.

[100] H. Holttinen, M. Milligan, E. Ela, *et al.*, "Methodologies to determine operating reserves due to increased wind power," *IEEE Transactions on Sustainable Energy*, vol. 3, no. 4, pp. 713–723, October 2012.

[101] J. Cochran, M. Miller, M. Milligan, *et al.*, "Market evolution: Wholesale electricity market design for 21st century power systems," *National Renewable Energy Laboratory, IBM, VTT Technical Research Centre of Finland, Energinet.dk, Universidad de Castilla La Mancha, International Copper Association, Global Green Growth Institute, China National Renewable Energy Center*, pp. 1–57, October 2013.

Chapter 3

Nowcasting and short-term wind forecasting for wind energy management

Agustín Agüera-Pérez[1], José Carlos Palomares-Salas[1], Juan José González de la Rosa[1], José María Sierra-Fernández[1] and Álvaro Jiménez-Montero[1]

3.1 Wind forecasting in grid and market operations

During the last decades, wind energy has demonstrated to be an important alternative to the fossil fuels. According to the World Wind Energy Association [1], the total installed capacity around the world reached 435 GW at the end of 2015, achieving a growth rate of 17.2% during this year. Furthermore, in some electrical systems, wind power represents a significant percentage of the total energy production which must be taken into account for optimized market and grid operations. However, as opposed to the conventional electricity production, wind power is intermittent and difficult to predict, mainly due to the complex dynamics associated to the wind flow. This intermittency and the distributed production, derived from the geographical spread of wind farms, are two characteristics that make the integration of wind energy in the electrical system difficult. These characteristics convey a diversity of problems as wind power is more present in the electrical system: frequency deviations, balancing problems, transmission efficiency, increase of operating reserves, etc. Market operations associated to wind energy are also affected by the wind unpredictability. Thus, energy bids from wind energy producers involve important uncertainties that must be carefully managed, as deficits in production are strongly penalized. In these conditions, wind energy producers and market and system operators require accurate wind forecasts in order to optimize the grid management, the plans and schedules of the generation units, and their operations in the electrical market. All these elements have a repercussion in the final cost of the electricity.

The day-to-day integration of wind energy in the electrical system is fundamentally supported by wind forecasts with horizons from seconds or minutes to few days. This chapter focuses on the wind forecasts related to these horizons, describing how they are produced and their repercussion on the concerned

[1]Universidad de Cádiz, Algeciras, Spain

electrical system management processes. There is a diversity of interpretations about the expressions associated to the referred forecasting horizons in wind forecasting literature. We have considered recent wind forecasting reviews to illustrate this point. In [2–4], the expression 'short-term' is associated to horizons between 30 min and 6 h. However, in [5], Yan *et al.* apply the term to the range of 24–72 h. Other less precise definitions of 'short-term' are given by Foley *et al.* [6] ('hours up to two days') or by Zhang *et al.* [7] ('from hours to days'). The diversity of definitions in these reviews is also similar in the expressions 'medium-term' or 'long-term'. Hence, for clarity reasons, in this chapter we referred to the forecasting horizons in a quantitative way when necessary.

3.1.1 Uncertainty in wind energy production

In this subsection some information about wind production errors reported in different works is given. This information helps to illustrate the uncertainties that the electricity system have to deal with in the integration of wind energy. For instance, it seems that it is assumed that the hour-ahead production estimations for a single wind farm involve errors within the range of 10%–15% with respect to the actual wind production [8,9]. These errors could represent important deviations between the expected energy production and the actually delivered to the power network. As these deviations are penalized in most of the electricity systems, wind plant owners use conservative estimations in their plans which ultimately involve a non-optimized exploitation of the wind farms. Considering regional electricity systems, Weber states that the day-ahead error for the German market is more than 20% with respect to the averaged wind energy production [10]. This error could be significantly higher in specific cases. Thus, González-Aparicio and Zucker reported that 'on annual average, the maximum range of the error reached up to 70% of wind production' in the Spanish market during the period 2010–13 [11].

Uncertainty in wind power forecasts highly depends on the forecasting horizon. Weber quantified how prediction errors increase with the forecasting horizon in the German market. Concretely, the observed RMSE values, as percentage of the total installed power in Germany, were approximately 5.7%, 4.4% and 3.1% considering two days ahead, day ahead and intraday predictions, respectively [10]. Another remarkable effect is that wind power forecasting errors tend to be lower if larger areas are considered. In this sense, Borggrefe and Neuhoff explain that the day-ahead forecast error was reduced in Germany when the four transmission system operators were aggregated. The errors associated with the four separated areas were distributed between 6.6% and 8.9%. After the integration the error was reduced to 5.9% [12].

However, these rough numbers based on annual averages or installed capacity can hardly describe the structure of these errors and their repercussion on the day-to-day wind energy integration, because uncertainty in wind energy production depends on the load demand, current wind power production and percentage of wind power of the total production. In this sense, González-Aparicio and Zucker [11] made a thorough characterization of wind power forecasting errors in the Spanish electricity market from 2010 to 2013. They confirmed that wind power forecasting errors are spreader as higher loads are demanded by consumers. During the studied period, the installed wind power increased from 19.5 to 22.8 GW, and the hourly averaged production increased from 4.9 to 6.5 GWh. Despite they conclude that the averaged error remained similar, they corroborated that the uncertainty in predicting the absolute wind power output increases as wind penetration grows.

3.1.2 Effects of the wind forecasts uncertainty in the power system

The integration of wind power as an important part of the energy supply in a country or region imposes certain operating characteristics in the electrical system to assure security, as the intermittency affects the grid operation at different temporal and spatial scales [13]. Figure 3.1 summarizes the different elements and processes that can be affected by the insertion of wind energy in the power systems. The maximum level of wind energy integration would be associated to the capacity of the electrical system to absorb this intermittent production while keeping these elements and processes in security conditions. The magnitude of the integration problem is dependent on the proportion of wind power, the dispersion/concentration of wind farms and the correlation between wind production and demand [14].

The constraints imposed by grid lines and stations are decisive to determine the amount of wind capacity which could be accepted, having into account that energy must be transmitted from windy places which are generally far from the consumption areas. Thus, in first steps of regional wind energy development it is important to check the adequacy of the grid. For this purpose, it is essential to have very long-term estimations (in the range of years) of the wind resource available in the concerned region. These wind resource maps are used to demarcate zones where wind farms could be installed and to design the infrastructure to allow their

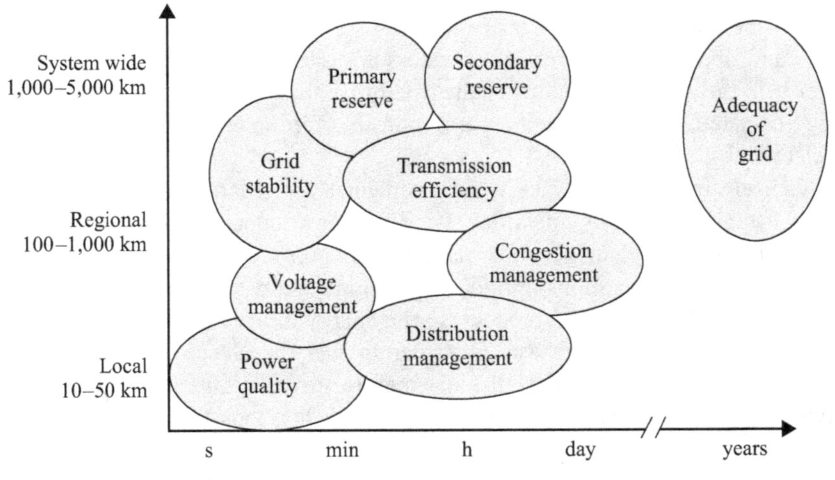

Figure 3.1 Processes and elements affected by the wind energy penetration, and their associated temporal and spatial scales

connection. Regarding the power networks design, there is a recent concern on interconnection of different transmission systems as a way to mitigate the problems derived from the unpredictability in power production [15].

Apart from long-term and network design considerations, wind penetration has an important impact in the daily management of the grid, because power quality, congestion management or transmission efficiency can be significantly affected by intermittency in wind production. This impact can be minimized (or at least smoothed) by the action of the operating reserves, which allows the system operator to make real-time adjustments to guarantee the grid security. In this sense, wind power variability and forecasting uncertainties force the system operator to increase the amount of reserve capacity in order to deal with unexpected variations in wind production. The operating reserves are divided according to the time required to their full activation. The primary reserve is generally associated to a response time from seconds to several minutes. The primary reserve is applied in frequency control and voltage management, thereby minimizing disturbances and preserving adequate power quality. The secondary reserve is operative in the range of minutes to approximately 1 h, and it is used to balance load and demand and grid congestion problems. Thus, system operators are particularly interested in these range of very short-term horizons (from seconds to 1 h) of wind forecasts, to correct deviations from the planned wind energy and to deal with unexpected fluctuations of wind power, for example, due to ramp events.

Medium-term wind forecasts (in the range of days) are especially interesting for market operations (see next subsection). However, a part of this market is devoted to contracting reserves and, in this sense, they also affect the system operation.

3.1.3 Wind uncertainty in market operations

Along with the wind production and the power network constraints, the electrical market is another factor that determines the level of integration of wind energy. Again, wind forecasts are necessary to optimize strategies adapted to the market design and operation. The market operators are interested in forecasting horizons longer than the system operators, as the first finish their operations hours before the energy dispatch, while the second are responsible for the real-time management of the electrical grid.

Each electrical market has its own timelines and characteristics, which ultimately establish the way in which wind power producers sell the electricity. Despite this diversity, in general, markets operations can be classified in three classes: day-ahead operations, intraday operations and reserve operations.

In the day-ahead markets, producers offer energy for the next day from 0:00 a.m. to 12:00 p.m. in an hourly base. Considering that the closure of the day-ahead market occurs at 12:00 a.m. of the day before dispatch[1], wind energy bids are unavoidably based on wind power forecasts with horizons from 12 to 36 h and,

[1]As said, each electrical market has its own timeline but, for explanation purposes, 12:00 a.m. represents a good generalization of the day-ahead market closure time of the different energy systems.

consequently, they involve important uncertainties. The market price is determined by marginal pricing of the supply and the demand. The market-clearing mechanism also has into account a set of constraints for security in grid transmission. As result of this process, a preliminary plan and distribution of electricity prices along the next day are obtained. Wind energy producers have very low short-run marginal costs, and therefore the amount of bidden wind energy affects the final price of electricity, especially in markets with high wind energy penetration [16]. However, the uncertainties in wind power forecasts force to adopt bidding strategies under certain confidence intervals, which ultimately are conservative as deficits in actual productions are strongly penalized.

If the day-ahead planning were perfect for the real-time operation, corrections in plans and additional market operations would not be necessary. However, unexpected failures, plant outages or changes in load and production conditions require short-term adjustments to guarantee the power system security. A part of these adjustments is derived from deviations in weather predictions. Obviously, these deviations directly affect renewable energy production plans and schedules based on wind and irradiance forecasts, but they also have a relevant effect in the expected load which depends on the temperature, luminosity or humidity conditions. In this context, market participants can use the intraday market to adapt their estimations or correct unfeasible schedules resulting for day-ahead markets. The intraday market starts after the closure of the day-ahead market and generally closes few hours before the real-time dispatch. Regarding wind energy producers, the intraday market allows re-bidding and correct positions according to forecasts with shorter horizons which, as commented before, are more accurate than those used in the day-ahead market.

The system operator is in charge of providing reserves able to face possible contingencies or unbalances in the real-time management of the grid. The provision of these reserves is done in the balancing market. In most of cases, only a part of the contracted reserve in the balancing market is finally used in grid operations, as reserve requirements are generally determined considering worse scenarios than actually occurs. The penetration of wind power increases the uncertainties in energy production and, consequently, forces the system operator to contract additional reserves to guarantee security. The reduction of wind power uncertainty (i.e., the availability of accurate wind power forecasts) would reduce the provision of reserves and the final cost of the electricity.

Figure 3.2 aims to illustrate all the above said by showing a simplification of a typical electrical market timeline (the specified hours are for guidance only). It can be seen how the composition of the real-time generation is the result of the day-ahead plan (established approximately one day before), the corrections introduced in the intraday market (established few hours before) and the final use of the reserves according to the real-time necessities (determined in the last minutes). In this illustration, the processes in which wind power forecasts are involved are delimited with dashed lines. Thus, the day-ahead bids and the reserve provision are based on wind power forecasts with horizons around a day, the intraday bids are based in forecasting horizons of few hours and the reserve management in forecasting

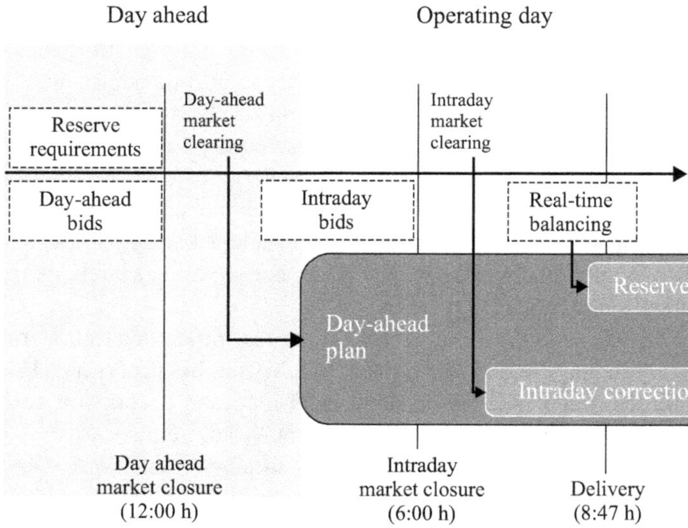

Figure 3.2 Generalization of an electrical market timeline

horizons of minutes to 1 h. For a specific case, Botterud *et al.* brings a detailed timeline of the Midwest Independent System Operator market (USA) [17].

It would be desirable that most of the short-term adjustments would be assumed by the intraday market. Thus, the use of reserve resources would be minimized and the final electricity cost would be benefited. A part of the current research deals with the adequate design of the electrical market. A common conclusion in these works is that flexibility and liquidity in the intraday markets allow a higher integration of wind energy [10–12].

3.2 Wind power forecasting systems

As discussed in the previous section, the accurate knowledge of wind production and its evolution is fundamental to achieve an effective integration of wind energy in the electrical system, as this information is necessary to match production and demand, to prevent grid failures and to operate in the electrical market. In this sense, the accuracy in wind power predictions helps to optimize plans of wind energy producers and to reduce the reserved capacity and, consequently, the electricity cost. The main tools thought to obtain the necessary wind energy information for the electrical system participants are wind power forecasting systems. However, wind control centres are becoming another critical source of information for an efficient wind energy integration. Both elements are complementary as wind power forecasting systems requires information about wind farms production to perform accurate predictions, being this information provided by wind control centres. In the same way, wind farms are operated and scheduled according to the wind predictions. Thus, wind control centres generally receive data from one more wind power forecasting systems.

3.2.1 Wind control centres

The information about the current wind energy production and the state of the wind farms is obtained via the wind control centres. Wind control centres not only monitor but also control the production and configuration of the wind farm. The monitor/control chain starts in each individual wind turbine and, in some cases, ends in regional control centres of the system operator. This is the case of the Spanish electrical system which will be the reference in this subsection.

Manufacturers generally include a supervisory control and data acquisition (SCADA) system as a part of the wind farm supply. The SCADA communicates with the sensors and actuators of the wind turbine, collecting the acquired data and allowing to send commands to configure the wind farm properly.

Most important wind energy companies manage wind farms distributed over large regions, even different countries. In order to make economically viable the surveillance and management of their wind farms, they have installed wind control centres to assemble the data from the different plants and/or their local SCADA systems. Thus, the personnel of the wind control centre is able to quickly detect incidents and operate the wind farms under control.

Each control centre is designed according to the amount of data which have to be managed. The basic elements in the control centre are the communication systems, the computer systems and the wind farms and wind data models [18]. The communication system is critical in the control centre performance. It must assure stability and low latency to obtain and send information. Typically, communications are performed via the internet by using virtual private networks, but in some cases fibre optics dedicated lines or satellite links are used. The computer system generally comprises one front end unit for every wind farm linked to the control centre. Each unit is associated to a console to show the received information. The information is generally summarized and represented in a video wall to have an overview of the state of the associated wind farms. The computer system also includes a cluster to store historical data and a web server to provide connections to the web clients. Managing different wind farms requires a standardization of the transmitted data, as each of them could implement different technologies, protocols or configurations. Wind farms models are thereby required to allow an easy and uniform management from the control centre. Wind forecasting models are also necessary to perform an efficient management of the wind farms, fundamentally focused on production control and market bids, but also for maintenance operations in which calm periods of several days could be necessary.

Spain is a reference in regulating the renewable energy production via control centres. In 2007, it was established that:

> all facilities of renewable energy whose rated power exceeds 10 MW must be assigned to a generation control centre, which acts as an interlocutor to the grid operator, transmitting information in real time and enforcing to execute the instructions received in order to ensure the reliability of the electric system [19].

According to this directive, the Spanish transmission system operator (Red Electrica de España) created the Control Centre of Renewable Energy (CECRE). The CECRE homologates and receives information from other control centres which, in turn, are connected to renewable energy plants. Currently, there are 47 homologated control centres which act as interlocutor with CECRE, and through which the system operator can send commands to renewable energy installations to assure a maximum integration of renewable energies while maintaining grid stability. During these years, CECRE has demonstrated its benefits for integrating renewable energies and, thus, from 2015, its operation range has been expanded to facilities of 5 MW [20].

Focusing on the wind farms, CECRE demands information about the output voltage and active and the reactive power, as well as other meteorological information of the emplacement such as wind speed, wind direction and temperature [21]. These data must be updated every 12 s. According to this information, and via the homologated control centres, the system operator sends curtailment commands to the wind farms which must be fulfilled in 15 min. If levels are continuously infringed, the system operator could even shut down the wind farm. If connection with CECRE is lost the wind farm is automatically curtailed at the 85% of the last power level [22].

3.2.2 Description of wind power forecasting systems

A wind power forecasting system is a tool which processes all the relevant data in order to generate estimations of the wind energy production. The consolidated wind power forecasting systems are run by companies which sell their results to system operators and wind energy producers. However, the most important wind energy companies and system operators with high penetration of wind energy generally run their own wind power forecasting systems.

Figure 3.3 shows a typical scheme of a wind power forecasting system. The energy production of a wind farm depends on the local wind conditions of the emplacement, but these conditions are driven by the weather at regional scale. Hence, accurate wind power forecasts should include information about the regional weather conditions. Numerical weather prediction (NWP) models represent the main tool to provide global and mesoscalar descriptions of the atmospheric conditions and their evolution. Typically, these models generate gridded predictions of atmospheric variables in steps of 1–3 h with a final forecasting horizon of some days (see Section 3.1). However the coarse resolution of the output grid and the absence of local considerations in the estimations advise against the direct application of NWP data for accurate wind energy estimations. Thus, a downscaling process is required in order to include local features in wind predictions. As a result of this process, the NWP data is refined and adapted to the topography surrounding the wind farm. In this refinement, it is desirable the inclusion of in situ weather stations to help in modelling local phenomena which could not be inferred from the mesoscalar data. Once local wind forecasts are obtained, they are processed along with wind farm data to obtain wind power forecasts, which is the final goal of a wind power forecasting system. In this point, wind farm models, information about the current production and

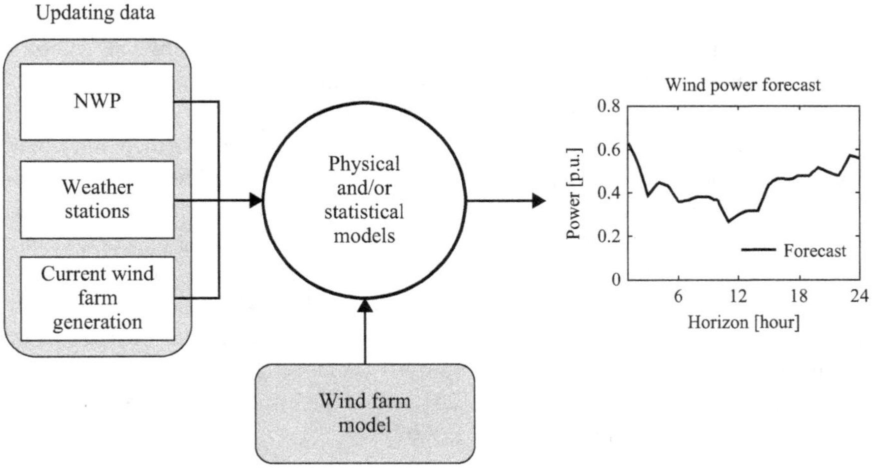

Figure 3.3 Typical scheme of a wind power forecasting system

historical data are critical elements for accurate estimations. Hence, wind control centres play a fundamental role in this stage.

The output of the wind power forecasting system is configured depending on the objective, since not all actors of the electricity system are interested in the same information. Thus, wind farm owners demand a local prediction with horizons from several hours to days. The information allows to establish schedules and to supply bids in the day-ahead electricity market and, in the same way, schedules adjustments and re-bidding in the intraday market. Transmission system operators require similar information for reserve requirements and wind power projections, but adapted to a regional framework. They also demand additional information related to shorter forecasting horizons (seconds to minutes) for real-time management of the grid. In this sense, NWP data become more relevant as forecasting horizons increases. On the contrary, wind farm and weather stations data prevail in very short-term forecasts. Errors in predictions of wind power forecasting systems will be lower for shorter forecasting horizons, but also for simple terrains where local conditions are weaker. In the same way and as commented along the text, wind power predictions at regional scales will be more accurate as the aggregated errors tend to cancel out, i.e., the local conditions are smoothed.

Following the description of the Spanish case in the previous subsection, the Spanish transmission system operator (REE) has developed its own wind power forecasting system: SIPREOLICO. SIPREOLICO processes NWP data from the European Centre for Medium-Range Weather Forecasts along with historical information about each wind farm to generate hourly wind power predictions for the following 48 h [23]. The combination of CECRE and SIPREOLICO brings an excellent framework for an efficient integration of wind energy in the Spanish power system.

Foley *et al.* compile a set of wind power forecasting systems, detailing the countries in which they are applied and the institutions and companies in charge of their management and development [6]. In this compilation it is shown how these wind power forecasting systems are mainly based on hybrid (physical and statistical) approaches. There are wind power forecasting systems which have been developed for a specific area or electrical system, and exploited by a unique organization (this is the case of SIPREOLICO). However, as result of the increasing demand of this information, some wind power forecasting systems have become more adaptable in order to offer their results as commercial products according to the customer requirements (see Previento [24] and Wind Power Prediction Tool [25] as two relevant examples). In this sense, several of these power forecasting have become commercial spin-off of research projects. For instance, ANEMOS Wind Power Predictions [26] is the result of collaboration among numerous European organizations since 2002.

3.2.3 Wind power forecasting system results: representation and validation

As said, wind power forecasts must be adapted to the necessity of the different actors of the electrical systems. In first steps of wind power forecasting, single estimations were the main option to provide this information. These single estimations can be assimilated to deterministic predictions as calculations lead to a unique solution. Improving single estimates is a direct way to enhance wind power forecasting system results. But, the single estimations of the wind energy production do not reflect the uncertainties derived from the different elements in wind power forecasting systems (NWP data, wind farm power curve, prediction algorithm). Thereby these results are not fully useful for decision-making strategies in the electrical system. In this sense, wind energy bids or reserve requirements can be benefited by a probabilistic or risk representation rather than a single scenario which, in a high percentage of cases, does not actually occurs. Thus, in recent years, wind power forecasts are expressed via wind power uncertainty analysis, i.e., evaluating the uncertainty inherent in the results of wind power forecasting systems [5].

There are three main ways to express the results in wind power forecasting uncertainty analysis: probabilistic forecasting, scenario forecasting and risk index [5,7] (Figure 3.4).

In probabilistic forecasting, wind power is treated as a random variable. In this representation, results are mainly expressed as probability density functions or quantile/interval forms. In the case of probability density functions, the problem lies on determining the type of distribution (Gaussian, Weibull, Beta, etc.) and adjusting its parameters. This representation is appropriated for very short-term prediction (from seconds to hours), and due to its simplicity requires a low computational cost. The quantile/interval form is focused on providing upper and lower limits for the energy production generally based on empirical distributions. In horizons from hours to days, these approaches perform better than the ones based on probability density functions, but large datasets and high computational costs are required.

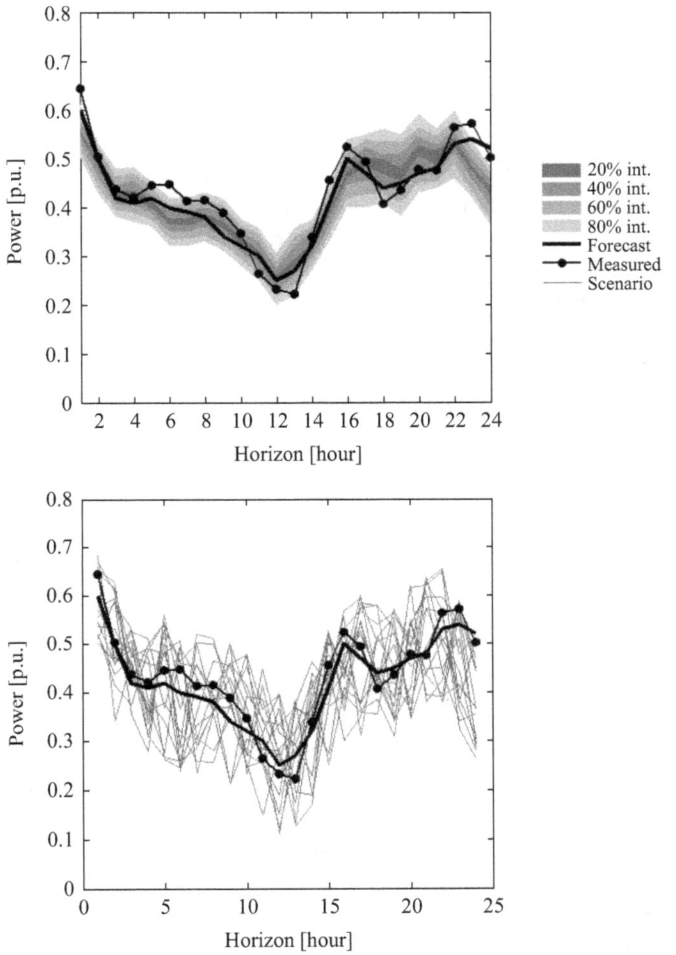

Figure 3.4 Wind power forecasts uncertainty analysis: interval representation (left) and scenarios representation (right)

Probabilistic forecasting generates predictions for each look-ahead time independently, without considerations of its evolution in time. However, decisions about market and grid operations would be benefited by a time-dependent description of the wind power uncertainty. For this reason, scenario descriptions are built as a set of single forecasts over a period (in other words, a set of possible scenarios), thereby providing a temporal evolution of the wind power uncertainty. The level of uncertainty is derived from the dispersion of the possible scenarios. In this sense, larger numbers of scenarios involve a better description of the uncertainty, but requiring more computation. Market and dispatching decisions are made based on the most-likely scenario, which is assumed to be the closest scenario to the actual situation. Thus, scenario descriptions allow a real-time updating as new wind power and meteorological data is received.

Risk index is the simplest way to express wind power uncertainty. It basically consists in a single-valued forecasting with additional information about its reliability, generally represented by a number or a colour code. Two examples of risk indexes are the Meteo-Risk Index [27] and the Normalized Prediction Risk Index [28].

Wind power forecasting systems are generally focused on providing estimations for individual wind farms. But the system operator is interested in regional wind power forecasts, i.e., the aggregated power output of the wind farms in certain area. Obtaining specific forecasts for each wind farm from NWP in order to compute the total power in a region is more difficult than a direct estimation of the regional production. Furthermore, the direct regional forecast is faster and more accurate; due to the commented smoothing effect of the aggregated wind production over an area. Regional forecasts are generally obtained upscaling on-line measurements from the concerned wind farms along with NWP data [2]. Uncertainties in regional forecasts are better described by the scenario representation, as it is suitable for stochastic optimization problems with temporal and spatial correlations [7,29].

The evaluation of the wind power forecasting system results depends on the manner in which results are expressed. Single wind power forecasts are generally assessed by comparison with a reference model. The persistence model is the most used reference model in these cases. The persistence assumes that the predicted wind speed (or wind power) in a look-ahead time will be similar to the last measured value. This naive approach establishes a lower limit from which the contribution of other models can be measured. An adaptation of the persistence method for regional forecasts is described in [30]. Common statistics are used to quantify the accuracy of predictions as mean absolute error, mean squared error or correlation coefficient. A compilation of the most frequent statistics used in wind forecasting and their formulas can be consulted in [2,3].

If results are expressed using probabilistic descriptions, the validation is more complex. Pinson *et al.* propose some properties to evaluate probabilistic forecasts: reliability, sharpness and skill score. Reliability informs about how the empirical proportion derived from actual values match the predictive quantiles. In a model with the highest reliability, the actual values would be distributed according to the predictive quantiles. Sharpness gives a measure of the concentration of the probability distribution. Good models should have a high sharpness, thereby being the probability concentrated in a short-range of possible values. Skill score has different implementations, as it is defined depending on the uncertainty representation and the scoring rules selected to test the results [31].

3.3 Physical approaches for wind forecasting

Physical approaches for wind forecasting aim to reproduce the atmospheric processes related to the wind by solving the equations of conservation of mass, momentum and energy in the air flow. One of the benefits of the physical forecasts is that, in opposition to statistical approaches, results can be obtained from initial and boundary conditions without considering historical series. Furthermore, the calculations of physical models are performed in a given domain, generating gridded results, and thereby providing spatial wind descriptions which are difficult to obtain from statistical approaches. However, these models are computationally expensive, and they involve high delays in the generation of results, particularly when high resolution is required. Wind energy forecasts for wind farms would require a high resolution in order to consider local features, but the usability of predictions also require short computational time. Thus, wind energy forecasts for wind farms are rarely based exclusively in physical approaches, especially for forecasting horizons shorter than few hours. However, forecasts obtained by physical models (as those provided by NWP models) are a cornerstone in wind energy management.

3.3.1 Numerical weather prediction

The term 'NWP model' is associated to dynamical models which perform meteorological predictions, not only related to wind power. All NWP approaches are essentially similar as they solve the hydrodynamical and thermodynamical equations of the system atmosphere-surface. They only differ on scales, parameterizations and resolution of the output grids. The NWP models can be divided in global models and regional models, also called limited area models.

3.3.1.1 Global models

Global NWP models calculate the current state of the atmosphere and its evolution at planetary scale. The level of data resources and computational power necessary to carry out these predictions is only accessible for international and state institutions. The two most frequent global NWP data sources used in wind power predictions are the Global Forecasting System (GFS) managed by the National Centre of Environmental Prediction (NCEP) and the forecast products of the European Centre for Medium-Range Weather Forecasts (ECMWF). These institutions typically disseminate the results via web server, allowing an easy and computer-friendly access to their forecasts. As a result of the involved resources and the accessibility, global NWP models are used as input in most of the physical and statistical wind power forecasting models.

The NWP models are not only focused on providing wind information, but also on producing estimations of a wide range of meteorological variables. Thus they implement different modules in order to deal with different geophysical elements: ocean, surface, ice, radiation, etc. The evaluation of these elements to perform global meteorological forecasts also requires a geophysical description of the scenario, including topography and land surface properties. In the case of global models, the scenario is the Earth. Hence, in order to maintain appropriated computational times, the geophysical description of the Earth must be done with a coarse resolution. Concretely, the domain is divided into cells with a resolution of dozens of kilometres with a basic topographical/geophysical description: mean height elevation above sea level, percentage of land/sea and an approximate description of the surface coverage. As a result of the resolution in the geophysical description, global NWP models cannot reproduce local atmospheric features.

An important element in the global NWP models is the data assimilation system which allows an updating feedback between real and calculated meteorological data. The data assimilation system introduces the real information that is composed of remotely sensed data acquired by space-borne instruments, and in situ measurements from surface weather stations, ships, buoys, radiosonde stations and aircraft. The data assimilation system, after a quality control, processes all the valid measurements to build a sequence of past situations which support the initialization of the model.

In the cases of GFS and ECMWF, both deliver their results four times a day at 00, 06, 12 and 18 UTC. Each delivery is composed of a sequence of forecasts which in the case of ECMWF cover 6 days in steps of 3 h, and 4 additional days in steps of 6 h. GFS delivers 5 days of hourly forecasts, 5 additional days of 3-hourly forecasts and 6 additional days of 12-hourly forecasts. The spatial resolution depends on the chosen product, but the highest resolution is comparable in both models ranging around 10 km. There are other global NWP models as the Unified Model of the UK Meteorological Office, the Analysis and Prediction of the Australian Bureau of Meteorology or the GSM managed by the Japan Meteorology Agency, but their use is in wind power forecasting is exceptional respect GFS and ECMWF.

The results of global NWP data have two important drawbacks for a direct application to wind power forecasting. Firstly, results are delivered with a delay of few hours as result of the computing time. Thus, forecasts of the shorter horizons are passed in the time of delivery. Secondly, the coarse resolution and topographic simplification requires additional calculations for local applications, which is the case of wind farms. For instance, in [32], ECMWF advise about the weakness of their near-surface wind forecasts in mountainous areas resulting from the 'highly varying subgrid orography'.

3.3.1.2 Limited area models

The limited area models (LAM) are basically similar to the global models commented in the previous subsection, but they are designed to be applied in a concrete area. The main purpose of these models is twofold: refining of the global models output and including local features. The applied equations are similar although they can include specific modules to deal with local phenomena, as turbulence, clouds or eddies. LAMs are able to provide useful forecasts for wind power production but, as the resolution and local details increase, more computation and time is required to produce the results and, consequently, these approaches are not efficient for shortest horizons.

The most extended LAM is the Weather Research and Forecasting (WRF) model, developed by NCEP and able to work with GFS data. The dissemination of WRF model has been benefited by the previous diffusion the fifth generation mesoscale model (MM5) which can be considered its origin. Furthermore the code is freely downloadable in WRF website, and enhanced code can be also proposed by the WRF community, being the model in permanent evolution. As a result of this dissemination multitude of WRF configurations have appeared for the different modules and computational schemes, as the land surface model, the planetary boundary layer or the microphysics scheme. In [33], the WRF model, fed with GFS data, is run to test how these different configurations affect the WRF wind forecasts. WRF is able to achieve resolutions of tens of meters but, as said, consuming important amounts of time and computation. Thus, this level of resolution is only applied in long-term wind resource assessments in which computational times are not as critical as in wind forecasting. There are numerous organizations that run WRF to produce forecasts for concrete areas. Some of these organizations, as well as the downloadable code and additional information of the model, can be found in [34].

The High-Resolution Limited Area Model (HIRLAM) is other LAM with extended use, although its diffusion is not comparable to WRF. HIRLAM has been developed by a European consortium and it is used by different meteorological agencies of Europe. The model is adapted to use ECMWF data as initial and boundary conditions, producing grids up to 5 km. As a third step, the HIRLAM output can be also processed by HARMONIE to obtain resolutions of 2.5 km [35].

As in the case of the global models, there are other LAMs with lower diffusion, as COnsortium of Small-scale MOdelling (COSMO) led by Federal Office of Meteorology and Climatology MeteoSwiss, the Lokal-Modell managed by the GErman Weather Service or the North-American Mesoscale Forecast System (NAM) run by NCEP.

3.3.1.3 Competitive ensemble forecasting

As NWP results are solutions of deterministic equations, they produce single-value forecasts. As commented in Section 2.3, this expression of wind predictions is not fully adapted to the wind energy integration problems which are benefited by an uncertainty evaluation of these results. Competitive ensemble forecasting is the way to obtain uncertainty information from NWP. Ensemble forecasts are produced by running NWP models under different initial conditions or different parameterizations, what leads to a set of different results. This set of predictions is then processed under the framework of wind power uncertainty analysis to adapt the results to the desired way of expression [3].

Regarding global NWP forecasts, both ECMWF and GFS implement ensemble products. Some works have analysed and compared their performance regarding wind speed and directions [36–38]. LAMs also can be run under different configurations to provide ensemble forecasts. In [39], the WRF model is run to perform ensemble wind forecasts.

3.3.2 Physical approaches focused on wind forecasting

NWP data are the base of the wind power predictions, especially at regional scale and horizons longer than few hours. However, in most of cases NWP forecasts are post-processed to a better adaptation to the wind energy context by using statistical approaches, in order to not add additional delays. Foley *et al.* [6] compile 15 wind forecasting systems which process NWP data, showing than only two of them are purely based on physical considerations: Prediktor (developed by Riso, Denmark) and SOWIE (developed by GmbH, Germany). The rest of them are based on statistical or hybrid approaches.

Other physical approaches for wind power forecasting consider computational fluid dynamics to translate the NWP data to a concrete local scenario. In [40], the complex numerical simulations needed to perform these simulations. In order to reduce the computation time, Li and Liu propose a pre-calculation of flow fields at local scale to conform a database. Then, each NWP situation is associated to a local pre-calculated flow field, thereby reducing the computational time required to generate the local prediction [41].

3.4 Statistical approaches for wind forecasting

In the statistical approach a vast amount of data is analysed and meteorological processes are not explicitly represented. The link between historical power production and weather is determined and then used to forecast the future power output. Unlike physical methods, statistical methods involve only one-step to convert the input variables into power output. Hence, the methods used are described as 'black box'. Generally a statistical relationship is developed between the weather forecast or prediction and the potential power output from the wind farm. Statistical approaches perform better in shorter horizons in which current conditions strongly determine the future state of the wind. However, as prediction horizons are larger, the mesoscalar atmospheric conditions must be evaluated and, in this case, NWP data become more important.

There are myriads of articles focused on statistical approaches for wind forecasting. In this section, a review of models for short-term forecasting that have been published in the international literature is presented. It has been limited only to published works where the models used are based on statistical methods, machine-learning approaches and/or hybrid methods. The objective of this review is not to make a thorough compilation of all the published works in this area, but providing a general overview of the different approaches that are being applied in statistical wind power prediction. We have tried to select works that used the persistence as reference model, allowing a comparison among approaches.

The first models used in wind speed or wind power forecasting were based on Box–Jenkins methodology (also known as ARIMA models). These models are still being used nowadays as prediction models or reference models in many cases, though limited to a horizon prediction a few hours. Thus, Torres *et al.* [42] have adjusted several ARMA models to predict the hourly average wind speed up to 10 h in advance in five locations with different topographic characteristics of Navarra (Spain), and obtaining a prediction error decrease of up 20% compared to persistence model.

The classical time-series analysis is not the only way to model the statistical relation among data. The main machine learning approaches used, so-called soft computing models, are artificial neural networks (ANN) and fuzzy systems. These methods are often referred to as artificial intelligent methods and are known as learning approaches because the relationship between inputs and expected output is learned by using historical time series. These methods, unlike physical methods, involve only a step to convert the input variables and output, so are often described as black boxes.

Cadenas and Rivera [43] conducted a study on the prediction of the hourly average wind speed in the region of La Venta (Oaxaca, Mexico) using diverse configurations of ANN and comparing the results based on two forecast error measures: the mean absolute error (MAE) and the mean square error (MSE). Gong Li and Jing Shi [44] also made predicting the hourly average wind speed at two locations of North Dakota (USA) using three architectures of ANN, namely, backpropagation, radial basis function and adaptive linear, and comparing them according to three performance measures: MAE, root mean square error (RMSE) and mean absolute percentage error (MAPE). The same procedure as described is applied by Gong Li *et al.* [45] to predict the hourly average wind speed up to 48 h, but this time including a technique called Bayesian Model Averaging (BMA) based on expectation-maximum algorithm, used to select the weights of the model, and the Box–Cox transformation to convert time series in normal distributions.

Many wind speed measurements in different locations allows to study of the correlation between the mean values of wind speed in different weather stations. To this end, Alexiadis *et al.* [46] propose a system based on a neural network for predicting the wind speed in Thessaloniki Bay (Greece) with a forecast horizon of up 2 h with which obtained an improvement of up to 27% compared to the persistence model.

The first major work where a broad comparison is made among different models was performed by Sfetsos [47]. It provides prediction of the hourly average wind speed on the island of Crete (Greece) using the following models: persistence, ARMA, five architectures of ANN (adaptive linear, backpropagation, radial basis function, Elman recurrent and neuronal logic) and models based on adaptive network-based fuzzy inference system (ANFIS). The objective was to obtain the best model with respect to persistence model. The RMSE performance function is used in this study. The best result was obtained by one of the ANN-based models, achieving an improvement of 4.89% over the persistence is obtained for 1-h ahead wind speed prediction. Haque *et al.* [48] also carry out a performance analysis of short-term wind speed prediction techniques based on ANN (backpropagation and radial basis function) and ANFIS, and comparing them according to three error rates: MAE, MAPE and RMSE. Furthermore, the forecasting performance of the models used in this study is improved by similar days method, which considers similar historical weather information corresponding to the forecasting day in order to determine similar wind speed days for processing. Research conducted by Sfetsos in [49] provided a novel method for the forecasting of mean hourly wind speed data using two time-series approaches (iterative and simultaneous procedures). Specifically, the proposed methodology is based on the multistep forecasting of 10 min averaged data and the subsequent averaging to generate mean hourly predictions. Datasets from two different locations (zone 1: Monastery of Odigitria, Crete (Greece), zone 2: Rutherford Appleton Laboratory, Oxfordshire (United Kingdom)) have been used, thus obtaining consistent improvements by means of iterative method (zone 1 – 19.17%) and simultaneous method (zone 2 – 10.09%) with respect to persistence model.

Campbell and Adamson [50] performed the average wind speed forecasting in different prediction horizons (10 min, 1 h, 2 h, 5 h and 12 h) on a wind farm located in County Donegal (Ireland), by applying the following models: persistence, ARIMA, Moving Average (MA) and ANN, and comparing them based on the RMSE index. Bivona *et al.* [51] conducted a study for hourly average wind speed forecasting in two areas of Sicily (Italy), using the Seasonal ARIMA and ANN models and comparing them in function of the following indicators of accuracy: MAE, RMSE, MAPE and mean absolute scaled error (MASE). On the other hand, Qing Cao *et al.* [52] provide a comparative analysis of the wind speed forecasting accuracy of univariate and multivariate ARIMA models with their recurrent neural network counterparts for a predicting horizon of 15 min. The data employed in the study were obtained on the same weather station in five different heights and have been used as multivariate inputs to building of the assessed models, where the latter were analysed and compared in function of the following error metrics: mean square percentage error (MSPE), MAE, MAPE and correlation coefficient. The findings indicate that multivariate models perform better than univariate models and that the recurrent ANN models outperform the ARIMA models.

Hybrid forecasting of time-series data is considered to be a potentially viable alternative compared with the conventional single forecasting modelling approaches such as ARIMA and ANN. Cadenas and Rivera [53] conducted a study on the prediction of the hourly average wind speed in three different regions of Mexico using ARIMA, ANN and Hybrid models (ARIMA-ANN), comparing the results based on mean error (ME), MAE and MSE, and concluding that the Hybrid models predict the wind speed with a higher accuracy than the ARIMA and ANN models in the three examined sites. In addition, Cadenas *et al.* [54] also made mean wind speed forecasting for the shorter prediction horizon (10 min) using the single exponential smoothing method (SES) and ANN models, comparing the results based on MSE, MAE, mean percentage error (MPE) and MAPE, and concluding that SES method showed the former better results.

Hui Liu *et al.* [55] propose two ARIMA-ANN and ARIMA-Kalman hybrid models to predict the hourly average wind speed up to 3 h in advance, and they are compared with each other and with the ARIMA model through performance indices MSE, MAE and MAPE, showing that the ARIMA-Kalman hybrid model has the better results. Another study by Hui Liu *et al.* [56] also made mean wind speed forecasting up to 3 h in advance using another hybrid model named EMD-ANN that is based on the empirical mode decomposition (EMD) and ANN. This model is compared to persistence, ARIMA and ANN models through of the measurements of performance MSE, MAE and MAPE, obtaining that the EMD-ANN hybrid model provides the better results.

Another powerful technique called support vector machine (SVM) and belonging to soft computing that is widely applied in a variety of classification and function estimation problems, carries great potential for the application of short-term wind speed forecasting. Zhou *et al.* [57] presented a systematic study on fine tuning of SVM model parameters for one-step ahead wind speed forecasting and these different configurations were compared with the persistence approach according to the RMSE parameter. The SVM parameters considered include the training sample, SVM order, regularization parameter, and Kernel parameters (three SVM Kernels, namely linear, Gaussian and polynomial), and it is found that the SVM models can outperform the persistence model in the majority of cases. Shi *et al.* [58] investigated two ARIMA-ANN and ARIMA-SVM hybrid models to predict the wind speed and wind power generation for the following prediction horizons: 1, 3, 5, 7 and 9 h. These hybrid models are compared with ARIMA, ANN and SVM models based on two metrics of forecasting accuracy (MAE and RMSE), concluding the study that the hybrid approaches are viable options for forecasting both wind speed and wind power generation time series, but they do not always produce superior forecasting performance for all the forecasting time horizons examined.

Forecasting models currently employed can be divided in two groups: a first group that only employ time series data and predict future values taking into account past history; and a second group that uses forecasted values from a NWP model. Nowadays, prediction models use the values forecasted by a NWP model and current or historic measures of wind speed and/or wind power generation for its construction. For example, Salcedo-Sanz *et al.* [59] make the hybridization of the fifth-generation mesoscale model (MM5) with ANN for mean hourly wind speed forecast at specific points of a wind park, obtaining the best model configuration based on MAE and MSE. Ramirez *et al.* [60] propose two new wind-power forecasting systems up to 72 h in advance where the input variables used in both systems were the same: forecasted meteorological variable values obtained from a NWP (MM5) and electric power-generation registers of the wind farm. The first forecasting system is called FORECAS and combines inputs to predict the wind power production through a power curve model represented by a multilayer perceptron ANN. The other model used, namely SGP system, uses a combination of the Kalman filter, ARMA model and several ANN models, selecting the best combination of the models in a fuzzy inference system. Both the FORECAS and the SGP systems obtained good results and improved very significantly the persistence model forecasts. Khalid and Savkin [61] present a method to improve the short-term wind power prediction (10 min) at a given turbine using information from NWP and from multiple observation points, comparing the results with the persistence and Grey predictor model in terms of MAE and RMSE.

Table 3.1 briefly summarizes the most important characteristics of predictive models belonging to the family of models based on time series, proposed in the international literature. In this table the variable to predict (wind speed or wind

Table 3.1

Reference	Forecast variable	Model	Prediction horizon	Assessment criteria	Improvement respect to persistence
[42]	Wind speed	ARIMA	1–10 h	RMSE	12%–20%
[43]	Wind speed	ANN	1 h	MSE, MAE	–
[44]	Wind speed	ANN	1 h	MAE, RMSE, MAPE	–
[45]	Wind speed	ANN, BMA	1–48 h	MAE, RMSE, MAPE	–
[46]	Wind speed and wind power	ANN	15 min–2 h	ME	4%–27%
[47]	Wind speed	ASIMA, ANN, ANFIS	1 h	RMSE	4.89%
[48]	Wind speed	SD, ANN, ANFIS, Hybrid	1 – 24 h	MAE, RMSE, MAPE	63%
			1 – 72 h		59%
[49]	Wind speed	ARIMA, it ANN, sim ANN	1 h	RMSE, ME	0.94%–19.17%
[50]	Wind speed	ARIMA, ANN	10 min–12 h	RMSE	63.48%
[51]	Wind speed	SARIMA, ANN	1–24 h	MAE, RMSE, MAPE, MASE	–
[52]	Wind speed	ARIMA, ANN	15 min	R, MSPE, MAE, MAPE	–
[53]	Wind speed	ARIMA, ANN, ARIMA-ANN	1 h	ME, MSE, MAE	–
[54]	Wind speed	SSE, ANN	10 min	MSE, MAE, MPE, MAPE	–
[55]	Wind speed	ARIMA, ARIMA-ANN, ARIMA-Kalman	1–3 h	MSE, MAE, MAPE	–
[56]	Wind speed	ARIMA, ANN, EMD-ANN	1–3 h	MSE, MAE, MAPE	8%–65%
[57]	Wind speed	SVM	1 h	RMSE	13%–40%
[58]	Wind speed	ARIMA, ANN, SVM, ARIMA-ANN, ARIMA-SVM	1–9 h	MAE, RMSE	8.65%
[59]	Wind speed	SNN	1 h	MSE, MAE	–
[60]	Wind power	FORECAS, SGP	12–24 h	RMSE	55%
			24–48 h		54%
			48–72 h		50%
[61]	Wind power	GM	10 min	MAE, RMSE	29%–72%

power generation), type of model used, prediction horizon, evaluation criteria and the improvement obtained regarding to the persistence model (if indicated by the reference) have been indicated.

3.5 Enhancing predictions with nowcasting

A generalized opinion in wind energy literature assumes that, with the increasing penetration of wind power, new forecasting approaches must be developed. One of the obvious enhancement lines is the improvement of NWP, reducing the computation times and increasing the spatial and temporal resolution. The consolidation and development of new wind power uncertainty forecasts is other research line that is being explored. But there is also an extended trend that focus on capturing spatio-temporal dynamics from real measurements as an interesting research line to obtain relevant information for wind energy management that cannot be provided by NWP models [2,5,7,62]. This research line would take advantage of the increasing amount of data from wind farms, meteorological stations from institutions or amateur observers, and the introduction of sensors in the context of the smart grid development. This amount of data, and their on-line data availability, could allow to obtain updated representations of the wind conditions over a region.

The achievement of accurate real-time wind descriptions is fundamental for the scenario uncertainty analysis, as they would help to determine the actual state and development of real-time dynamic strategies [5]. Supported by a high density sensors network, these real-time wind descriptions could inform about the local features and sudden changes in wind conditions that cannot be inferred form NWP data. For example, the current characteristics of local wind gusts, the early detection of extreme events or the irruption of atmospheric fronts in a concerned area could be captured by these sensor networks, providing high-frequency updates and allowing predictions based on the observed atmospheric motion and inertial effects [5,62]. In the same context, wind ramp forecasting would also be directly benefited by the availability of these real-time and high-resoluted wind descriptions. Ramp events can produce drastic variations on the power production of wind farms (even producing plant outages) and they represent a topic with increasing interest for the wind power community [2,63]. The development of the smart grids, micro grids and nano-grids, and the subsequent increase of distributed production and self-consumption, will also require new tools with regional focus and higher temporal resolution for their optimized management [7].

The benefits of the spatio-temporal analysis for wind power forecasting based on observational data in a region have been demonstrated in many cases. Tastu *et al.* studied the spatio-temporal wind dependencies observed in western Denmark, verifying significant correlations between different zones for time delays between 1 and 5 h [62]. An application of spatio-temporal correlations for robust unit commitment is proposed by Li *et al.* [29]. The procedure is based on obtaining ellipsoidal uncertainty sets from observed wind generation data in 15 wind farms distributed over 250,000 km^2.

Nowcasting is a methodology appropriated to generate wind information adapted to the above commented issues. According to the World Meteorological organization (WMO), 'nowcasting comprises the detailed description of the current weather along with forecasts obtained by extrapolation for a period of 0 to 6 h ahead. In this time range it is possible to forecast small features such as individual storms with reasonable accuracy. A forecaster using the latest radar, satellite and observational data is able to make analysis of the small-scale features present in a small area such as a city and make an accurate forecast for the following few hours' [64]. Thus, once enough density of measurements is available, the application of nowcasting techniques would produce information with high interest for wind energy integration. For example, Torres *et al.* applied nowcasting to improve the performance of ECMWF forecasts in the context of wind energy [65]. In [66], the data assimilation system VDRAS developed by the National Center for Atmospheric research is tested for wind energy nowcasting purposes. The updating cycle of 6–15 min allowed the detection of wind power ramps and improved the mesoscale models in horizons shorter than 1 h.

But the usefulness of nowcasting can be expanded by the recent proliferation of low-cost and low-power sensors and the consequent availability of large amount of meteorological data. Furthermore, as these sensors are mostly connected via internet, they present two valuable advantages: information is adapted to a computational framework and data can be transmitted in real-time. The coordination and exploitation of all this information is in line with emerging topics as Big Data and crowdsourcing. Some works have studied the potential of these techniques in producing relevant information from atmospheric measurements [67]. The fundamental problems of dealing with large amount of these meteorological data sources are the diversity of the managed measurements (units, updating frequency, codification, etc.) and the lack of quality assurance (errors in instruments, bad emplacements, etc.) [68]. In this sense, according to the WMO guidelines, wind measurement present very strict requirements for a proper data acquisition: measurement height of 10 meters at least, data acquisition frequency of 4 Hz, no obstacles in the surroundings, even emplacements, etc. It would be expected that wind measurements from stations below the quality level required by WMO were useless in wind estimations. However, it has been demonstrated that a group of 'low-quality' stations distributed over a region can provide acceptable wind information if their measurements are processed together [69]. It has been also demonstrated that the use of these records as exogenous variables can improve wind forecasts based on ANN models [70]. In [71], it is tested how these basic stations, in cooperation with other ones that match the WMO requirements, are able to produce regional wind descriptions with a basic processing, even improving the performance of NWP models. The last work also tested the effect of the density of stations in the accuracy of the calculated wind speeds and direction. Results confirm that higher densities lead to more accurate estimations, being the improvement more marked if the density exceeds 1 station per 2,000 km^2.

The blooming an integration of these data sources would impulse and consolidate the use of energy nowcasting. As said, the development of nowcasting techniques based on big data or crowdsourcing, could benefit wind energy integration. But they also could represent an important tool for emerging configurations of power systems, as micro-grids and nano-grids. These grids range in a scale between one building to a small community of neighbours. The integration of wind energy in this context demands low-cost and easy-to-implement wind power forecasting products, and the developments in nowcasting could provide satisfactory solutions.

References

[1] World Wind Energy Association. *Quarterly Bulletin*. 2016; (1):6–7.

[2] Jung J., Broadwater R. P. 'Current status and future advances for wind speed and power forecasting'. *Renewable and Sustainable Energy Review*. 2014; 31:762–777.

[3] Ren Y., Suganthan P. N., Srikanth N. 'Ensemble methods for wind and solar power forecasting—A state-of-the-art review'. *Renewable and Sustainable Energy Reviews*. 2015; 50:82–91.

[4] Okumus I., Dinler A. 'Current status of wind energy forecasting and a hybrid method for hourly predictions'. *Energy Conversion and Management*. 2016; 123:362–371.

[5] Yan J., Liu Y., Han S., Wang Y., Feng S. 'Reviews on uncertainty analysis of wind power forecasting'. *Renewable and Sustainable Energy Reviews*. 2015; 52:1322–1330.

[6] Foley A. M., Leahy P. G., Marvuglia A., McKeogh E. J. 'Current methods and advances in forecasting of wind power generation'. *Renewable Energy*. 2012; 37:1–8.

[7] Zhang Y., Wang J., Wang X. 'Review on probabilistic forecasting of wind power generation'. *Renewable and Sustainable Energy Reviews*. 2014; 32:255–270.

[8] Ahlstrom M. L., Zavadil R. M. 'The role of wind forecasting in grid operations and reliability'. *2005 IEEE/PES Transmission & Distribution Conference & Exposition: Asia and Pacific*, Dalian, 2005; pp. 1–5. DOI: 10.1109/TDC.2005.1547203

[9] Venayagamoorthy G. K., Rohrig K., Erlich I. 'One step ahead: Short-term wind power forecasting and intelligent predictive control based on data analytics'. *IEEE Power and Energy Magazine*. 2012; 10(5):70–78.

[10] Weber C. 'Adequate intraday market design to enable the integration of wind energy into the European power systems'. *Energy Policy*. 2010; 38(7): 3155–3163.

[11] González-Aparicio I., Zucker A. 'Impact of wind power uncertainty forecasting on the market integration of wind energy in Spain'. *Applied Energy*. 2015; 159:334–349.

[12] Borggrefe F., Neuhoff K. *Balancing and intraday market design: Options for wind integration* [online]. 2011. Available from http://www.diw.de/sixcms/detail.php?id=diw_01.c.387259.de [Accessed 19 Feb 2017].

[13] Holttinen H., Meibom P., Orths A., *et al.* 'Impacts of large amounts of wind power on design and operation of power systems, results of IEA collaboration'. *Wind Energy.* 2011; 14:179–192.

[14] Ibrahim H., Ghandour M., Dimitrova M., Ilinca A., Perron J. 'Integration of wind energy into electricity systems: Technical challenges and actual solutions'. *Energy Procedia.* 2011; 6:815–824.

[15] Spiecker S., Vogel P., Weber C. 'Evaluating interconnector investments in the north European electricity system considering fluctuating wind power penetration'. *Energy Economics.* 2013; 37:114–127.

[16] Zeineldin H. H., El-Fouly T., El-Saadany E., Salama M. M. A. 'Impact of wind farm integration on electricity market prices'. *IET Renewable Power Generation.* 2009; 3(1):84–95.

[17] Botterud A., Wang J., Miranda V., Bessa R. J. 'Wind power forecasting in U.S. electricity markets'. *The Electricity Journal.* 2010; 23(3):71–82.

[18] Gallardo-Calles J. M., Colmenar-Santos A., Ontañon-Ruiz J., Castro-Gil M. 'Wind control centres: State of the art'. *Renewable Energy.* 2013; 51:93–100.

[19] 'Real Decreto 661/2007, de 25 de mayo, por el que se regula la actividad de producción de energía eléctrica en régimen especial'. BOE num. 126, 26 May 2007, pp. 22846–22886.

[20] 'Real Decreto 413/2014, de 6 de junio, por el que se regula la actividad de producción de energía eléctrica a partir de fuentes renovables, cogeneración y residuos'. BOE num. 140, 10 June 2014, pp. 43876–43978.

[21] 'Resolución de 24 de julio de 2012, de la Secretaría de Estado de Energía, por la que se aprueba la modificación de los procedimientos de operación del Sistema Eléctrico Peninsular (SEP) P.O.-3.1; P.O.-3.2; P.O.-9 y P.O.-14.4 y los procedimientos de operación de los Sistemas eléctricos Insulares y Extrapeninsulares (SEIE) P.O. SEIE-1 P.O. SEIE-2.2; P.O. SEIE-3.1; P.O. SEIE-7.1; P.O. SEIE-7.2; P.O. SEIE-8.2; P.O. SEIE-9 y P.O. SEIE-2.3 para su adaptación a la nueva normativa eléctrica.' BOE num. 191, 10 August 2012, p. 57494.

[22] Ceña Lázaro A., Jiménez Sarciada J. 'The Spanish experience in the integration of electricity from wind power plants into the electrical system'. *Proceedings of the Seventh International Workshop on Large Scale Integration of Wind Power on Transmission Networks for Offshore Wind farms*; Madrid, Spain, May 2008. pp. 1–8.

[23] Sánchez I. 'Short-term prediction of wind energy production'. *International Journal of Forecasting.* 2006; 22(1):43–56.

[24] Energy and Meteo Systems. Previento – Wind Power Prediction [online]. Available from www.energymeteo.com/en/services/previento.php [Accessed Aug 2016].

[25] ENFOR. Wind Power Prediction Tool [online]. Available from www.enfor.dk/products/wppt.aspx [Accessed Aug 2016].

[26] ANEMOS. ANEMOS Wind Power Prediction [online]. Available from www.windpowerpredictions.com/index.html [Accessed Aug 2016].

[27] Pinson P., Kariniotakis G. 'On-line assessment of prediction risk for wind power production forecasts'. *Wind Energy.* 2004; 7:119–132.

[28] Pinson P., Nielsen H. A., Madsen H., Kariniotakis G. 'Skill forecasting from ensemble predictions of wind power'. *Applied Energy.* 2009; 86(7–8): 1326–1334.

[29] Li P., Guan X., Wu J., Zhou X. 'Modeling dynamic spatial correlations of geographically distributed wind farms and constructing ellipsoidal uncertainty sets for optimization-based generation scheduling'. *IEEE Transactions on Sustainable Energy.* 2015; 6(4):1594–1605.

[30] Agüera-Pérez A., Palomares-Salas J. C., González de la Rosa J. J., Moreno-Muñoz A. 'Spatial persistence in wind analysis'. *Journal of Wind Engineering and Industrial Aerodynamics.* 2013; 119:48–52.

[31] Pinson P., Nielsen H. A., Møller J. K., Madsen H., Kariniotakis G. N. 'Non-parametric probabilistic forecasts of wind power: Required properties and evaluation'. *Wind Energy.* 2007; 10:497–516.

[32] Persson A., Andersson E., Tsonevsky I. *User Guide to ECMEWF Forecast Products.* ECMWF, 2015.

[33] Cheng W. Y. Y., Liu Y., Liu Y., Zhang Y., Mahoney W. P., Warner T. T. 'The impact of model physics on numerical wind forecasts'. *Renewable Energy.* 2013; 55:347–356.

[34] WRF. The Weather Research & Forecasting model [online]. Available from www.wrf-model.org/index.php [Accessed Aug 2016].

[35] HIRLAM. Available from http://hirlam.org/ [Accessed Aug 2016].

[36] Nielsen H., Nielsen T., Madsen H., *et al.* '*Wind power ensemble forecasting.*' *Proceedings of the 2004 Global Wind Power Conference*, Chicago, Illinois; 2004.

[37] Nielsen H., Madsen H., Nielsen T., *et al.* Wind Power Ensemble Forecasting Using Wind Speed and Direction Ensembles from ECMWF or NCEP. Technical Report, Denmark Project PSO; 2005.

[38] Pinson P., Hagedorn R. 'Verification of the ECMWF ensemble forecasts of wind speed against analyses and observations'. *Meteorological Applications.* 2012; 19(4):484–500.

[39] Deppe A. J., Gallus W. A., Takle E. S. 'A WRF ensemble for improved wind speed forecasts at turbine height'. *Weather Forecasting.* 2013; 28: 212–228.

[40] Zajaczkowski F. J., Haupt S. E., Schmehl K. J. 'A preliminary study of assimilating numerical weather prediction data into computational fluid dynamics models for wind prediction'. *Journal of Wind Engineering and Industrial Aerodynamics.* 2011; 99(4):320–329.

[41] Li L., Yongqian L., Yongping Y. 'Short-term wind speed forecasting based on CFD pre-calculated flow fields'. *Proceeding of the CSEE.* 2013; 33(7):27–32.

[42] Torres J. L., García A., De Blas M., De Francisco A. 'Forecast of hourly average wind speed with ARMA models in Navarre (Spain)'. *Solar Energy.* 2005; 79:65–77.

[43] Cadenas E., Rivera W. 'Short term wind speed forecasting in La Venta, Oaxaca, México, using artificial neural networks'. *Renewable Energy*. 2009; 34:274–278.

[44] Li G., Shi J. 'On comparing three artificial neural networks for wind speed forecasting'. *Applied Energy*. 2010; 87:2313–2320.

[45] Li G., Shi J., Zhou J. 'Bayesian adaptive combination of short-term wind speed forecasts from neural network models'. *Renewable Energy*. 2011; 36:352–359.

[46] Alexiadis M. C., Dokopoulos P. S., Sahsamanoglou H. S. 'Wind speed and power forecasting based on spatial correlation models'. *IEEE Transactions on Energy Conversion*. 1999; 14:836–842.

[47] Sfetsos A. 'A comparison of various forecasting techniques applied to mean hourly wind speed time series'. *Renewable Energy*. 2000; 21:23–35.

[48] Haque A. U., Mandal P., Kaye M. E., Meng J., Chang L., Senjyu T. 'A new strategy for predicting short-term wind speed using soft computing models'. *Renewable and Sustainable Energy Reviews*. 2012; 16:4563–4573.

[49] Sfetsos A. 'A novel approach for the forecasting of mean hourly wind speed time series'. *Renewable Energy*. 2002; 27:163–174.

[50] Campbell P. R. J., Adamson K. 'A novel approach to wind forecasting in the United Kingdom and Ireland'. *International Journal of Simulation*. 2005; 6:1–10.

[51] Bivona S., Bonanno G., Burlon R., Gurrera D., Leone C. 'Stochastic models for wind speed forecasting'. *Energy Conversion and Management*. 2011; 52:1157–1165.

[52] Cao Q., Ewing B. T., Thompson M. A. 'Forecasting wind speed with recurrent neural networks'. *European Journal of Operational Research*. 2012; 221:148–154.

[53] Cadenas E., Rivera W. 'Wind speed forecasting in three regions of Mexico, using hybrid ARIMA-ANN model'. *Renewable Energy*. 2010; 35:2732–2738.

[54] Cadenas E., Jaramillo O. A., Rivera W. 'Analysis and forecasting of wind velocity in chetumal, quintanaroo, using the single exponential smoothing method'. *Renewable Energy*. 2010; 35:925–930.

[55] Liu H., Tian H., Li Y. 'Comparison of two new ARIMA-ANN and ARIMA-Kalman hybrid methods for wind speed prediction'. *Applied Energy*. 2012; 98:415–424.

[56] Liu H., Chen D., Tian H., Li Y. 'A hybrid model for wind speed prediction using empirical mode decomposition and artificial neural networks'. *Renewable Energy*. 2012; 48:545–556.

[57] Zhou J., Shi J., Li G. 'Fine tuning support vector machines for short-term wind speed forecasting'. *Energy Conversion and Management*. 2011; 52:1990–1998.

[58] Shi J., Guo J., Zheng S. 'Evaluation of hybrid forecasting approaches for wind speed and power generation time series'. *Renewable and Sustainable Energy Reviews*. 2012; 16:3471–3480.

[59] Salcedo-Sanz S., Pérez-Bellido A. M., Ortiz-García E. G., Portilla-Figueras A., Prieto L., Paredes D. 'Hybridizing the fifth generation mesoscale model

with artificial neural networks for short-term wind speed prediction'. *Renewable Energy*. 2009; 34:1451–1457.

[60] Ramirez-Rosado I. J., Fernández-Jimenez L. A., Monteiro C., Sousa J., Bessa R. 'Comparison of two new short-term wind-power forecasting systems'. *Renewable Energy*. 2009; 34:1848–1854.

[61] Khalid M., Savkin A. V. 'A method for short-term wind power prediction with multiple observation points'. *IEEE Transactions on Power Systems*. 2012; 27:579–586.

[62] Tastu J., Pinson P., Kotwa E., Madsen H. and Nielsen H. Aa. 'Spatio-temporal analysis and modeling of short-term wind power forecast errors'. *Wind Energy*. 2011; 14:43–60. doi:10.1002/we.401

[63] Gallego-Castillo C., Cuerva-Tejero A., Lopez-Garcia O. 'A review on the recent history of wind power ramp forecasting'. *Renewable and Sustainable Energy Reviews*. 2015; 52:1148–1157.

[64] World Meteorological Organization. Nowcasting. Available from http://www.wmo.int/pages/prog/amp/pwsp/Nowcasting.htm [Accessed Aug 2016].

[65] Torres A., Prada J., Dorronsoro J. R. 'Nowcasting meteorological readings for wind energy prediction'. *EWEA 2014 Europe's Premier Wind Energy Event*. 2014, Barcelona, Spain PO.ID 156. http://www.ewea.org/annual2014/conference/programme/info2.php?id2=909&id=52%20&ordre=1#top

[66] Sun J., Boulder C. O., Zhang Y., Wiener G., Haupt S. 'A rapid update wind analysis and nowcasting system based on mesoscale model, radar, and surface data'. *94th American Meteorological Society Annual Meeting*. Atlanta, USA, February 2014.

[67] Muller C. L., Chapman L., Johnston S., *et al*. 'Crowdsourcing for climate and atmospheric sciences: Current status and future potential'. *International Journal of Climatology*. 2015; 35:3185–3203.

[68] Goodchild M. F., Li L. 'Assuring the quality of volunteered geographic information'. *Spatial Statistics*. 2012; 1:110–120.

[69] Agüera-Pérez A., Palomares-Salas J. C., González de la Rosa J. J., Ramiro-Leo J. G., Moreno-Muñoz A. 'Basic meteorological stations as wind data source: A mesoscalar test'. *Journal of Wind Engineering and Industrial Aerodynamics*. 2012; 107–108:48–56.

[70] Palomares-Salas J. C., Agüera-Pérez A., González de la Rosa J. J., Moreno-Muñoz A. 'A novel neural network method for wind speed forecasting using exogenous measurements from agriculture stations'. *Measurement*. 2014; 55:295–304.

[71] Agüera-Pérez A., Palomares-Salas J. C., González de la Rosa J. J., Sierra-Fernández J. M. 'Regional wind monitoring system based on multiple sensor networks: A crowdsourcing preliminary test'. *Journal of Wind Engineering and Industrial Aerodynamics*. 2014; 127:51–58.

Chapter 4

Solutions and active measures for wind power integration

Argo Rosin[1], Imre Drovtar[2] and Jako Kilter[3]

4.1 Introduction

Energy systems in the world are experiencing a revolutionary change. Among other trends, the integration of energy markets to a single common market (Common European Energy Market) is of considerable importance. It increases the share and integration of weather-influenced and dispersed power generation into the electricity system, expanding the need for energy storages, flexible demand, energy efficiency and climate policy related targets, new types of market participants (e.g., energy service companies, aggregators, virtual power plants [VPPs]) and raises the awareness of consumers and demand for new types of services. As the geographical and physical borders between different energy (gas, electricity and heat) markets are disappearing, more unpredictable energy flows together with exponentially growing data flows are appearing within the energy system. The management of energy systems has to adapt to the new revolutionary situation and move towards smart grids. Smart grids are able to provide both the consumers and producers a variety of new intelligent services with a much lower ecological footprint. As the consumers do not need electricity or gas, but rather heat and light, smart grids are seen as a bridge between consumers and energy producers to deliver the most efficient end service. At the same time, smart grids enable the producers not only to deliver energy for consumption on demand but also to utilise the flexibility of consumers to receive valuable storage and system services when needed. This chapter aims to describe problems arising from large-scale wind integration into the power system and to introduce possible solutions and active measures to help solve those problems from the smart grid point of view.

[1]Department of Electrical Power Engineering and Mechatronics, Tallinn University of Technology, Tallinn, Estonia
[2]Graanul Invest Ltd and Department of Electrical Power Engineering and Mechatronics, Tallinn University of Technology, Tallinn, Estonia
[3]Elering Ltd and Department of Electrical Power Engineering and Mechatronics, Tallinn University of Technology, Tallinn, Estonia

4.2 Energy policy

Energy consumption and production account for over 60% of the world's greenhouse gas emissions [1]. At the 2015 United Nations Climate Change Conference, a wide-scale agreement was reached on deep cuts of greenhouse gas emissions, while sustaining economic development, maintaining energy security and increasing the share of modern solutions in the energy technology in use. According to the International Energy Agency (IEA), the global economy grew approximately 3% in 2014, while for the first time in 40 years, the carbon dioxide (CO_2) emissions related to the energy sector remained unchanged. What is even more remarkable is that this happened outside the economic crisis, indicating clearly that the energy sector is moving towards decoupling economic growth from emissions. At the same time, in renewable technologies, investments for new power generation capacities installed in 2014 remained around 245 billion euros. Nevertheless, the greatest obstacle moving towards low-carbon society is the fossil fuel subsidies that according to IEA amounted worldwide to 497 billion euros in 2013.

The EU's energy policy complies with the Paris Climate Agreement. The Paris Climate Agreement incorporates a global action plan to avoid dangerous climate change that will require a global capping of greenhouse gas emissions as soon as possible and achieve climate neutrality in the second half of this century [2].

Under the climate and energy framework for the period 2020–2030, the EU countries have agreed to reduce greenhouse gas emissions at least by 40% compared to the 1990 level, to increase the share of renewable energy (RE) consumption at least by 27% and to raise the energy efficiency compared to the business as usual scenario by at least 27%. The targets set in the framework aim to achieve a more competitive, secure and sustainable energy system within the EU and meet its long-term greenhouse gas reductions target. The targets are based on an economic analysis about the decarbonisation of the energy system. Instead of investing in replacing the old infrastructure with a carbon-intensive technology, the framework aims to replace it with a low-carbon technology. Investing in the low-carbon technology should also boost the economy through research and development of new technologies and the resulting business opportunities. These investments should not be considered as additional but rather as redistribution across fuels, sources and sectors and across supply and demand.

EU's main plan to fulfil the obligations taken on with the Paris Agreement is combining the implementation plan with its internal priorities for development, by creating new jobs, boosting economic transformation and growth through prioritised investments, competitiveness, research, innovation and energy transition to a low-carbon economy. In order to achieve the goals, EU's function as a leader is not only to promote and implement regulatory policies that reduce emissions but also to enable factors that promote investment and innovation in new technologies in all sectors. The transition to a low-carbon economy needs to be properly managed, taking into account the differences in the energy mix and economic structure across the EU [2]. That means it is also necessary to anticipate and mitigate the possible impacts of the transition in specific regions (e.g., Eastern Europe) and socio-economic sectors at different levels of the society.

According to the IEA analysis [3], Europe will require cumulatively 2 trillion euros of investments to replace ageing infrastructure and to meet the decarbonisation aims in the power sector. Renewables will account for 75% of investments into new power plants in 2035. Despite the overcapacity of the thermal units today, an additional 100 GW of new thermal power plants will be required by 2025 to maintain the security of supply and reliability of the power systems.

For example, the drop in wholesale electricity prices can be linked directly to the EU policies: increased competition due to the liberalised electricity markets, unbundling electricity generation from the system operations, but also the growth of electricity generation capacity with low marginal costs (i.e., wind and solar in addition to the existing nuclear and hydro capacities) [4]. According the World Wind Energy Association, two-thirds of the top 15 countries by total wind installation are in Europe, constituting 29% of the total wind installations globally in 2015 [5]. Additional 540 billion euros will be invested by 2035 in wind power (onshore and off-shore), increasing furthermore the share of wind power in Europe; at the same time, it raises the question how the decreasing market price will incentivise further investments into wind power.

In addition to the decreasing energy prices, the levels of taxes and levies that should be an additional source of investment for financing promotion of energy efficiency and RE production vary in wide ranges between different member states, reaching a difference of up to 160%. In order to meet the commitment taken on under the Paris Agreement, it can be expected that today's EU legislation that fails to harmonise the energy taxation over all Member States will be changed soon. Member States tend to use these taxes and levies for a variety of purposes, i.e., for general revenue raising (e.g., health and education), or for financing external costs of energy production and consumption, but also for financing energy-specific policies. The data about the tax exemptions and subsidies offered by different Member States particularly to energy intensive industries are inconsistent, and therefore it is prepared to be analysed thoroughly to review the technologies being subsidised and to harmonise them in accordance with the long-term aims of the EU.

4.3 Technology overview and prospective changes in the power grid

The following section focuses on the challenges that arise with wind technology integration and on the prospective changes taking place in the power grid. On the one hand, the prospective changes complicate the wind integration process by adding another dimension of uncertainty into the system control; on the other hand, those changes offer numerous benefits that could be used for increasing the wind (but also other weather-influenced) generation integration. This section reviews the technologies of the wind power plant and its grid connection possibilities and grid connection topologies with the necessary relay protection. The prospective changes taking place in the power grid are based on the potential of using electric transportation as a source of temporary energy storage and the possibilities of the electricity consumers to support the power grid through demand side management or by being a prosumer.

4.3.1 Overview of wind power plant technologies

Wind power plants usually consist of multiple wind turbines with necessary unit and plant control systems, collector systems and other ancillary equipment. Due to the requirements set by the grid codes, network operators, and connection agreements, most of the transmission network connected wind power plants should be able to contribute to voltage control and active power control, to ride through voltage dips, comply with power quality requirements, and not to have adverse interactions to other nearby plants. Somewhat relaxed requirements may exist to wind plants connected to medium- or low-voltage networks.

In the literature, wind turbine technologies are classified into five principal types [6,7]:

- Type 1: Fixed speed wind turbine
- Type 2: Limited variable speed wind turbine
- Type 3: Variable speed wind turbine with partial-scale power electronics converter
- Type 4: Variable speed wind turbine with full-scale power electronic converters
- Type 5: Variable speed wind turbine with mechanical transmission

The first two are uncommon for modern wind power plant solutions and the last one is considered as a possible future solution but described in the literature as a separate type. More common technologies used in wind power plants are types 3 (Figure 4.1(a)) and 4 (Figure 4.1(b)) and are therefore covered here in detail. For additional information, the readers are referred to use the extensive and relevant literature available on this topic.

In the following, the characteristics and performance of the two main types of wind turbines are briefly described. Type 3 technology is more commonly known as doubly fed induction generator or doubly fed asynchronous generator. Here, the rotor of the generator is connected to a variable frequency AC source, which is supplied from a voltage-source converter connected to back-to-back with a grid side converter and exchanges power directly with a grid as required. This solution enables enhancement of the control ability of the unit and separate control of active and reactive power, much like a traditional synchronous generator. As compared to the older types, a soft starter or a reactive power compensator is not required. Type 4 wind turbines refer to a solution where the generator is connected to the grid through a full-scale frequency converter. This solution provides a lot of flexibility in the design and operation as the speed of the generator is not coupled with the grid frequency, allowing the unit to rotate at its optimal speed. One additional advantage

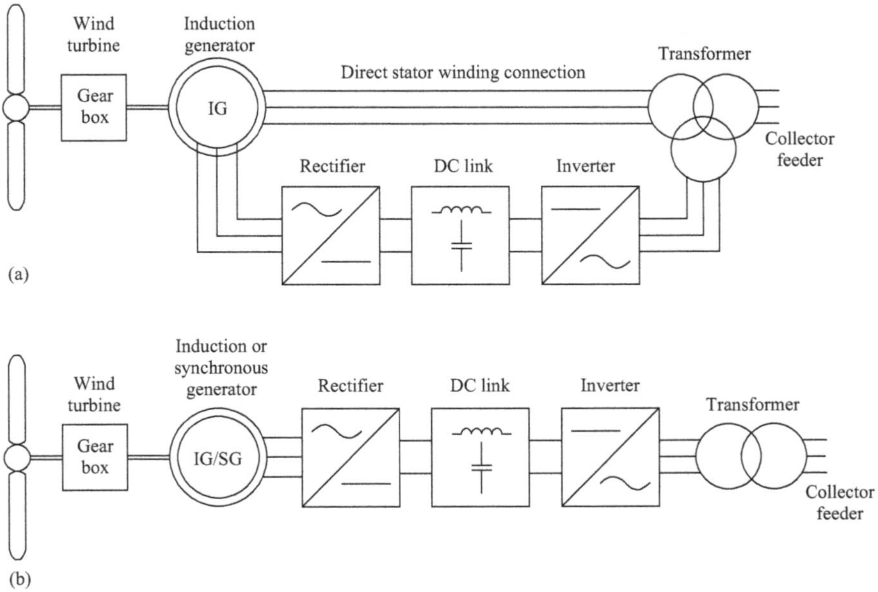

Figure 4.1 Principal schemes of (a) Type 3 and (b) Type 4 wind turbines

is that a gearbox may not be needed, enabling the use of frequencies below that in the gird. In this technology, the rotating machines used in the industry are wound rotor synchronous machines, permanent magnet synchronous machines and squirrel cage induction machines. The converters used are both responsive and efficient, and may offer a possibility to compensate reactive power with or without the generator in operation.

In addition to wind turbines, the wind power plant includes additional components and should be considered as a system not only as a collection of wind turbines. These additional components include a collector system, transformers, compensation devices, control system, auxiliary systems, etc. In general, the aim would be to compose a wind power plant that is economically efficient and technically compliant to grid requirements.

The purpose of the wind power plant collector system is to transmit the energy produced from the wind turbines to the utility network at minimum costs, provided all technical requirements are fulfilled. Typically, a collector system is operated at medium- or high-voltage level depending on the transmitted power; its configuration is similar to that in the distribution network where mostly radial connections are in use. As compared to the utility networks, the difference is that the energy flows to the opposite direction and in the case of a wind power plant, availability is more important than reliability. When planning the utility network configuration, the main criterion is the reliability of the network and the outage to the utility may be more costly than energy not delivered, as regulators may have set sanctions, etc.

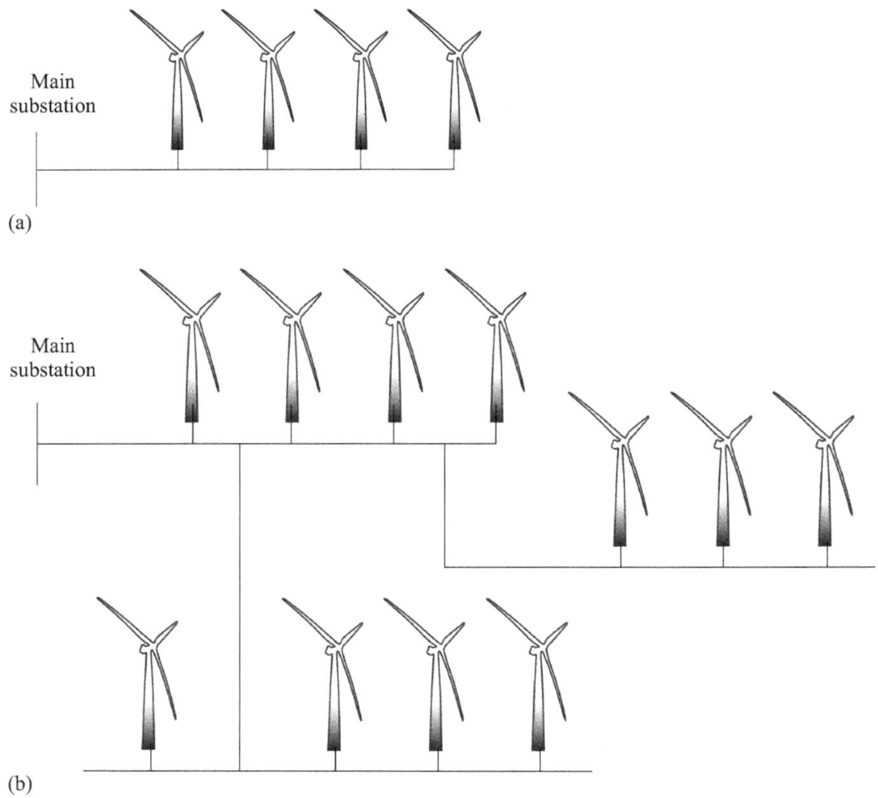

Figure 4.2 Wind power plant collector system: (a) radial feeder and (b) extended radial feeder configuration

For a wind power plant owner, the outage in the plant means only loss of revenue. Therefore, it is required to consider the objective for minimising lifecycle costs in the design of a plant and the costs for raising availability should be compared to the value of unavailability.

The collector system feeders are almost always configured radially. This configuration is implemented using the 'daisy chain' principle where the feeder is brought to each turbine and then continued from there to another (Figure 4.2(a)). An additional option is to implement a configuration where multiple radial extended feeders exist as origin from the main feeder (Figure 4.2(b)). Selection of the configuration at the end depends on the geographical location and layout of turbines in a specific plant. In most cases, the simplest option is selected. There may be parallel or closed loop configurations available but the system is always operated radially since in loop operation there is a need to increase the cross section of the lines, which means higher costs. In current practice, additional chains are not implemented in wind power plants and loop operation as the cost for production losses does not justify building additional connections.

One of the main components in the wind power plant is the connection substation. In this substation, the wind power plant is connected to the utility network through step up (MV to HV or EHV levels) transformer(s). Selection of a busbar configuration and the number of transformers is again a question of cost optimisation. This substation may also enclose additional equipment necessary to fulfil the plant performance requirements regarding reactive power compensation, for example, shunt capacitors and reactor banks, a static VAR compensator or a static synchronous compensator. The most simple and most applicable solution is to use one MV busbar and one transformer. The utility side part of the substation depends on the utility practice and it may include two busbars and redundant connections to ensure reliability. The number of transformers is usually a question when the rating of the wind power plant is higher than several hundred megawatts. Here, the reliability is of importance because an outage of the only transformer will cause an outage of the whole plant. At the same time, having two transformers increases the cost of the substation significantly. An option common with two transformers is that the rating of one of the transformers is not the full power of the plant but only half of it. Thus, it is possible to balance the costs and reliability requirements and allow the plant to operate even if there is an outage of one of the transformers.

In addition, power quality and especially harmonics are of interest in the wind power plant collector system. The level of the harmonic currents is usually limited by the grid codes or connection agreements, which are applicable at the connection point. To determine the level of harmonics in the connection point, it is required to conduct harmonic analysis. Here, the individual harmonics from the wind turbines combine and the level of them depends on their phase relationships, the level of wind turbine output power, existing background distortion, etc. Performance of the plant also depends on the characteristics of network harmonic impedance. Depending on the level of harmonics, filtering may be required.

An important aspect of wind power plant operation and connection to utility grid is its control and protection part. In this respect, it is essential to equip the plant with adequate relay protection functions to detect and isolate faults. Wind plant feeders are normally equipped with phase and ground overcurrent relays. Protection functions used in the main substation include typical protection functions, for example, bus differential protection, transformer differential protection, etc. To coordinate utility network protection, it is required to take into account that the wind power plant is not a normal short-circuit current source. Depending on the philosophy of utility system protection, additional protection or control functions, for example, over- and under-frequency and voltage protection functions may need to be included to provide capabilities for support of the system at wider contingencies.

The functions and capabilities of wind power plant control depend on the requirements set to the plant. The main requirements include reactive power, active power, fault-ride through and other functionalities. All these requirements are applicable at the connection point, usually at the high-voltage side of the connection transformer, and therefore the plant should be equipped with a plant controller the purpose of which is to enable these control functionalities on the plant level. Commonly, a separate controller is used, which has inputs from the connection point. Based on these inputs, it sends control actions to individual wind turbines, considering also the characteristics of the collector system.

In the definition of reactive power capability, usually the required reactive power output is stated by such functions as *Q* control, *U* control or *PF* control. For example, the wind plant may be required to be able to guarantee 0.95 lagging and 0.95 leading at the connection point. Another criterion here may be its availability at all plant power levels, which needs somewhat sophisticated control and additional equipment.

Active power control may be specified by such characteristics as ramp rate speed, curtailment options, frequency response including inertial controls, start-up and shut-down. The latest requirements in the grid codes and connection agreements are related to the inertial controls that have to do with the frequency control philosophy in the systems if the level of synchronous mass is decreasing. This is especially relevant in smaller power systems, for example, the UK and Ireland, which are not synchronously connected to larger synchronous areas and where the synchronous mass may become low at certain times of a year. Other control functionalities have been requested by grid codes and network operators for quite some time and they have become standard solutions.

Fault-ride through option has been under discussion for years and all grid codes today require that wind power plants connected to the system should be able to ride through voltage drops, both symmetrical and asymmetrical, in the network. To determine the levels, special fault-ride through curves are available, which determine the range and duration of voltage change the unit has to stay connected to the grid and where a trip of the unit is not allowed. Lately, voltage swell ride-through requirements have been included to the grid codes in some countries. Some grid codes also specify requirements for fault current contribution during grid faults.

Wind power plants are mostly located onshore and have a direct connection to local utility networks. However, a recent development is to install wind power plants offshore with an AC connection to the grid if the plant is near the shore and with a DC connection [8] from the plant to the AC grid connection point if the distance is long, i.e., in general, over 100 km to the nearest utility connection point. But it depends also on the cost benefit analysis on a case-by-case basis. When the HVDC connection is used, the VSC-based HVDC has become the preferred solution because of its ability to continuously transfer any power level, ease of integration with wind turbines in islanded grids, black-start capability, future option for multi-terminal connections, etc. Abundant discussions can be found in the recent literature on the future development of the DC technology and multi-terminal DC grid in the context of wind power plants and their expandability.

The offshore wind power plant technology and operation to standard onshore solutions differ in the HVDC technology and its control and protection system coordination, wind turbine design in harsher climate conditions than the onshore units, and collector system protection. A collector system can be designed using the same principles as in onshore plants. It is important to consider the controls between the HVDC and the wind power plant where, for example, it is necessary to consider both aspects in active power control, and HVDC should be able to start up the offshore grid, etc. These considerations are discussed in detail in [8,9].

4.3.2 Impact of electric transportation and electric vehicles

According to the European Commission [10], the European Union's (EU) road transportation contributes about 20% of the total emissions of CO_2. These emissions derived from road transport have increased almost by 23% over two decades since 1990, and without the financial crisis that started in 2010, even a higher growth could have been expected. Transportation is the only major sector in the EU where greenhouse gas emissions are still rising. In addition to the climate concerns, the high traffic concentrations inside cities are sources for health problems caused by poor air quality. In order to improve the air quality in cities and to decrease the CO_2 emissions in the transportation sector, a possible solution could be the large-scale integration of electric transportation in the urban areas. This means a wide development of quick charging (<30 min) infrastructure that would cover most of European countries. The coinciding impacts of load increase related to large-scale usage of EVs and/or plug-in hybrid electric vehicles (PHEV) are an important factor to be taken into account in power system planning. At the same time, it could serve as a potential distributed energy storage for balancing services if managed in a smart manner.

An additional question that arises is the impact of increasing shares of EV on the existing electrical infrastructure. Current networks are designed to provide an average household additional reserve capacity around 2 kW, while the simultaneous charging of vehicles may increase the demand up to 10 kW, causing heavy stress and overloading on the supply. Solutions will be necessary to increase the 'hosting capacity' of existing networks for EVs. The term 'hosting capacity' is not only used to express the ability of the network to integrate renewable generation units but can be also used to evaluate any other quantity that needs to be integrated into existing networks before a technical limit is exceeded [11].

The large share of EVs and PHEVs is of special interest for northern countries where the power system loads within one year can differ significantly. Average loads of spring–autumn and summer seasons can be 25%–40% lower than the loads during winter. High level of EV or PHEV integration might cause a problem during winter peak loads and possible additional loads during that period on localised levels. Usually in the power system planning, power generation adequacy during winter peaks is taken into account; however, an additional load has an impact both on the technical infrastructure during peak loads and on possible additional system services during off-peak periods. Integrating these loads into the power system thus does not only mean additional infrastructure improvements but also an opportunity to use them for the benefit of the system, for example, balancing weather-influenced generation with smart charging or providing valuable system services on a widely distributed geographical region. Studies have shown that in Germany over 80% of time the EV fleet would be standstill and most probably also connected to the power system.

A study [12] (© 2013 IEEE. Reprinted, with permission, from [12]) has shown that with the EV integration remaining between 10% and 30% from the total passenger car fleet, an additional load would have minor impact on the power system operation and peak load from a central point of view. If in general, vehicle charging is driven by the market price, the largest impact on the power system load will be dispersed over the night period as a load-levelling effect that could be addressed as a potential source of energy storage for renewable energy generation (REG), illustrated in Figure 4.3 with an EV fleet of 100,000 cars.

As the market price is a direct indicator of the actual state of the power system, it should have a certain coincidence also with the renewable generation. The proposed methodologies to evaluate the impact of EV charging on peak load could be easily adopted to create a charging pattern that follows the generation pattern. However, taking into account the uncertainty factor, especially for weather-influenced generation, this charging pattern would be more of a theoretical concept than an implementable solution as compared to the price based load control. Nevertheless, even the price based load control provides a good overview about the potential source of load sink or storage for the increasing wind integration. Interestingly, the traffic density information and studies, for example, [13–15], indicate that up to 89% and more of the vehicles can simultaneously be in a standstill mode or connected to the power grid.

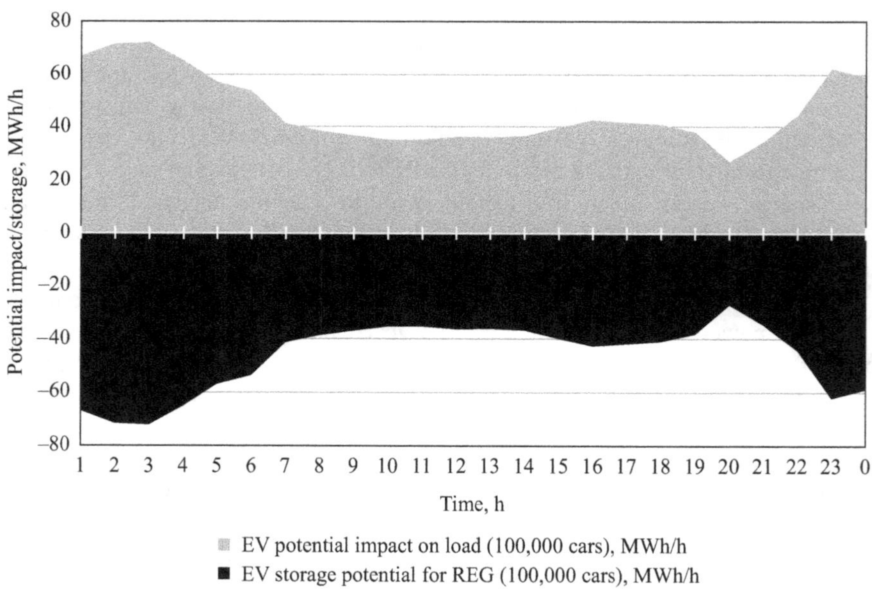

Figure 4.3 EV fleet potential as additional load or as load levelling storage for REG

Another aspect of EVs is the possible vehicle-to-grid (V2G) solution that provides a theoretical source of balancing for the weather-influenced generation by either returning electricity to the grid or by throttling their charging rate [16,17]. Possible V2G solutions have been thoroughly studied, but as neither commercial solutions nor the trials to implement are available or because of too high consumer costs, they are not discussed here. In addition, the potential revenue of providing the EV's battery for grid support does not outweigh the shortened life time of the battery and the coinciding high cost of replacing the battery pack, making it currently unattractive for an average EV user. Nevertheless, it can be assumed that with the increase of battery capacities and reduction of battery prices, the interest in V2G technologies in the market may increase and become a viable source of an additional support mechanism for higher wind integration in the future.

4.3.3 Impact of consumers

The increasing share of stochastic and unpredictable generation in the system inevitably increases the need for additional balancing services to cover the possible mismatches between demand and supply. Today, electrical loads are more predictable and controllable, giving a new opportunity to manage the power system all together through DSM. At the same time, the increasing interest in installing renewable energy sources (RES) at the end user is leading in parallel to a revolutionary change: electricity consumers are also able to produce electricity, turning them into 'prosumers' with a whole different role in the power system [18]. The current system already today includes hundreds of thousands of potential end consumers who would be able to provide different system and ancillary services through demand side response (DSR) or as prosumers for the system operators or market participants.

4.3.3.1 Demand side response

As described in [19], DSR offers flexibility for the TSO to maintain security of supply, optimise the utilisation of infrastructure, investments and system adequacy. The vast potential is lying around the grid unused, although it can in many cases be a more competitive alternative than generation. In many cases, loads are used as a last resort in emergency situations [19,20]. At the same time, the need for reserves to maintain a secure system is growing by the increase of weather-influenced generation. So far these reserves are supplied with synchronous machines but the increasing need for balancing would require large investments into new generation that are needed for its power rather than for its produced energy [21]. DSR could be the necessary missing resource to cover the need for additional balancing at a reasonable competitive cost. At the same time, it is predicted that it would also allow more secure renewable integration [19].

Palensky and Dietrich in [22] point out that DSM's main advantage is that it is financially and economically more promising to intelligently control consumers for providing balancing services than to build new power plants or install energy storage units. DSR and demand shifting could be used more effectively for system-wide services through a platform, which would give the consumers the necessary information, such as actual need for regulating power in addition to simple price incentives for scheduling its processes. However, as grid tariffs prevail in the cost of electricity, the cost and possible economic gain from providing energy and system services does not reach the traditional end user. At the same time, when having aggregated effect, these local consumers could provide the necessary system support at critical time periods if not for the TSO, then for the DSO, for example, to prevent outages in the power system or participate in the system control (balancing), as well as for optimal development of the network described in [23–25].

Based on the interviews with consumer representatives and the TSO, there are three main factors to be addressed to qualify as a DSR service provider. The first criterion (4.1) describes the minimum regulating power acceptable by the TSO:

$$P_{\mathrm{ACL}} \geq P_{\mathrm{TS}} \tag{4.1}$$

where P_{ACL} describes the consumer aggregated changeable load in MW and P_{TS} is the TSO's minimum required regulation step in MW.

The second criterion (4.2) takes into account the response time of the consumer:

$$\Delta t_{\mathrm{r}} \leq \Delta t_{\mathrm{TR}} \tag{4.2}$$

where Δt_{r} is the consumer's time for the requested load change in minutes and Δt_{TR} describes the TSO's maximum timeframe for the requested up-regulation in minutes.

The final criterion (4.3) takes into account the duration of the regulation:

$$\Delta t_{\mathrm{SAL}} \geq \Delta t_{\mathrm{TS}} \tag{4.3}$$

where Δt_{SAL} is the consumer's willingness to change the load in hours and Δt_{TS} is the TSO's minimum regulation duration in hours.

The consumer's DSR pricing principle is rationally based on the lost profit (or on the risk of losing production). Lost profit is used to determine the price for load shedding and the risk of losing production is used to determine the price for load shifting. Therefore, whenever the regulation price exceeds the industry's price cap for the service provision, the DSR service could be provided. Large industrial consumers are expected to act in a rational way, i.e., it can be expected that they will provide the DSR service if they find it cost effective [24].

The main principle for a possible shift of certain processes or sub-processes at a consumer is based on the existence and capacity of intermediate storages after the controllable load. The aim of this limitation is to remove the DSR impact on the final production or normal operation of the establishment. Intermediate storages were selected, for example, in the industry because they can be used as on-site energy storages to shift the electricity consumption of certain parts of the production process. The 'electrical' storage capacity of such units is determined by the physical size of the intermediate storage and its coherence with the following sub-process.

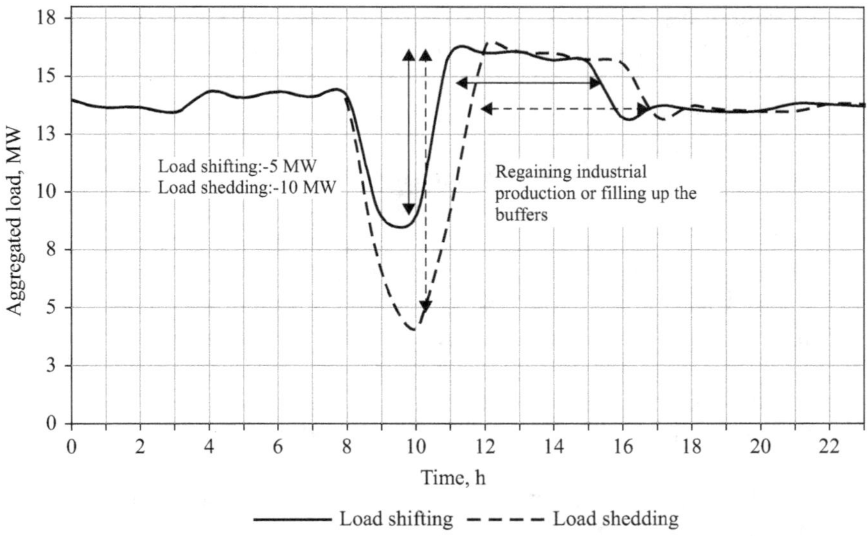

Figure 4.4 Aggregated demand profile of the consumer providing regulating service for the TSO (before, during and after service provision). © 2013 IEEE. Reprinted, with permission, from [26]

It should be noted that while with load shifting it is possible to catch up the production quota by increasing the demand for a couple of hours, then with load shedding, the increased consumption after regulation can only fill up the buffers in the production line but the production quota will not be caught up. The aggregated demand profile of the consumers providing the regulation service (with load shifting and load shedding) before, during and after service provision to catch up on the production quota is illustrated in Figure 4.4.

4.3.3.2 Aggregated prosumers

In a smart grid where consumers have become prosumers, the energy and information flows in the power system will be bidirectional. Prosumers will be given a more active role than mere energy consumers and due to their number and flexibility are likely to become major players in the new business model for energy generation.

To enable prosumers exposure to electricity markets in an efficient way, the business functions can be taken care of by an independent organisation or an existing market participant (electricity supplier/retailer), called an aggregator. Load aggregation is not the only possibility for ancillary system service provision that could be offered by prosumers. Another important possibility is the aggregation of renewables (i.e., mCHP) and energy storages (including storages of EVs) in VPPs [27]. A VPP combines elements of demand response and distributed generation in a way that allows them to be dispatched as if they act as a single power plant. The intelligent control system plays an important role in this process as the place where all information runs together, acting as a control, data processing and data communication centre. Often VPPs are called also as distributed energy management systems (DEMSs).

The coordinated operational mode delivers extra benefits such as the ability to deliver peak load electricity or load-aware power generation at short notice. Such a VPP (or DEMS) can replace a conventional power plant while providing higher efficiency and more flexibility [28]. VPPs can deliver needed energy at peak usage times, and can store any surplus power, giving the energy aggregator more options than in a single power plant. Other advantages include improved power network efficiency and security, cost and risk savings in transmission systems, increased value from existing infrastructure assets and reduced emissions from peaking power plants. Importantly, VPPs can also enable more efficient integration of RES into the grid by balancing their variability. For example, if one wind power source generates a bit more energy than predicted and another generates a bit less, they will compensate for each other, resulting in a more accurate forecast and making it easier to sell the capacity in the market or to use it in power systems operation [29]. VPPs and DEMSs can match load fluctuations through forecasting, advance metering and computerised control, as well as performing real-time optimisation of energy sources [30].

4.4 Technical and economic impacts of large-scale wind integration

Several studies (e.g., [31,32]) report that large-scale wind integration may significantly reduce the electricity price on the electricity market. In a short term, it may be positive for the society and even increase the reliability of supply due to the dispersion of wind power plants in the power system. However, in a long-term perspective, it jeopardises the economic viability of conventional generators that might give incentives for restructuring the generation portfolio to meet the requirements of a new power system configuration with less base capacities. As shown by the author in [33], by introducing intermittent generation, for example, solar and/or wind, and by avoiding fuel and variable costs of generation units that can be dispatched on demand, the overall marginal costs of the system tend to decrease. As new generation capacities in deregulated markets are built based on their profitability, as a result, the system security could be decreased. Instead, additional fast switching units or alternative services (i.e., storages and DSR) will be required to cover the possible mismatches between demand and generation caused by stochastic outputs of weather-influenced stochastic power plants.

4.4.1 Technical challenges

Changes in generation portfolio around the world from largely synchronous based generation to converter-based generation bring changes to power system characteristics. Depending on the level of the change, the converter-based technology integration may lead to major technical challenges in the normal operation of a power system as well as during disturbed or even emergency operations. These kind of changes are becoming more and more relevant not only to few system operators but also to whole synchronous areas in the future. Therefore, it is essential to recognise and consider them not only in the current but also in the future perspective, for example, in 2030–2050 system planning and operation. According to the European Network of Transmission System Operators for Electricity (ENTSO-E), RES are expected to have a dominant role in new generation capacity over the next 10–20 years. Installed wind and solar capacities are forecast to increase by 80% and 60%, respectively [34].

In the current power systems, the effect of wind power is mainly determined by the observable system size and wind power penetration level. In most cases, voltage and frequency control with available compensating reserves is of interest. This means that it is relevant to guarantee availability of sufficient reactive power and active power reserves whether in the same area or wider, which depends on the availability of interconnections and reserve power. The focus in [6] is on the short- and long-term effects of wind power integration on the power system in terms of the reserves and network and power capacity, and its impacts on power system stability.

Large-scale integration of wind power has a significant role in changing the operation principles of the current system and has a direct effect on system strength. Here, the terms of local fault level and total system inertia for a synchronous area are the main system characteristics under discussion. Based on these, the following aspects are of concern when large-scale converter-based generation is connected to the system [35]:

- Level of system inertia and impact on frequency management;
- Stability challenge with conventional and converter based technologies;
- Stable operation of converters;
- Level of fault current and influence on system protection;
- System strength and influence on the voltage quality of the supply.

Frequency management in the synchronous area is dependent on the system size and available level of system inertia determined by the availability of synchronous generation. As this is changing in the future, the level of system inertia in the context of the rate of change of the frequency (df/dt) arising from the loss of the largest unit is one of the parameters to consider. For example, there are already two synchronous areas in Europe (the UK and Ireland) where from time to time, constraining actions in relation to having too low total system inertia had to be taken. There are various possibilities to enable coordinated frequency management and each of these possibilities, for example, synthetic inertia provision, inertia contribution from HVDC, synchronous compensators and (very fast) DSR, and their influence on the system context have to be carefully assessed.

With an increasing level of converter-based technologies, it is essential to perform dynamic studies to assess if the selected control strategies and parameters are positively contributing to the overall system stability or not. For example, in order to improve transient stability, the reactive power controls selected for voltage control mode for converters have to be in the order of the same speed of response as synchronous unit automatic voltage regulator controls. In addition, recent research in [36,37] has shown that there are instabilities possible in high frequency range at high level (e.g., 70%) on non-synchronous units in the system.

Most of the wind power will be connected to the grid through converters. As the level of converters is increasing, it is essential that all these new converters and existing ones operate together in an accepted manner. The main requirement at the moment in many countries set for the generating units and therefore also for the wind power parks is that they should stay connected to the grid at voltage drop in the connection point. Moreover, coordinated control and avoidance of adverse interactions between multiple wind parks and nearby HVDC or other equipment may be needed. As a next step, in addition to voltage ride through capability, the converters should be capable of positively contributing to the power system strength during faults in the system and improving the local voltage quality.

Protection systems of the power system are based on the fault currents usually delivered by the synchronous units. With the increase of the level of non-synchronous generation, the level of current required may not be available in the future. As a consequence, detection of faults in current technical solutions could become cumbersome and more sophisticated mitigation measures are needed. In the short-term basis, new converter technologies providing fast fault current contribution are an effective mitigation measure. In the long-term basis, protection schemes used in the power system are expected to see an evolution of the relay algorithms.

Voltage quality of supply depends on the level of system strength, which in turn depends on the number, size and location of synchronous units in the system. To date, a major focus of the wind and solar as well as the HVDC industry has been on minimising and eliminating harmonic and imbalance current contributions from the converters to limit their potential negative influence on the voltage quality. A critical consideration is the level of voltage harmonics in the system. Lowering the system fault level increases the level of voltage harmonics proportionally. As a rough estimate, the level of voltage distortion will double if the fault level is halved. This has to be considered during system design and when operating the system. Synchronous units inherently behave as a sink for low-order harmonics. Without them, the levels of low-order harmonics will increase. Traditionally, a converter connected to the power system is at least to some extent responsible for emitting harmonics and therefore measures like filters and control to limit them have been used. To mitigate voltage quality issues, the future converters connected to the system should have a higher share and active participation in the delivery of better voltage quality. To achieve this, changes are needed in the grid codes and in converter designs to encourage both harmonic currents and imbalance current contributions from inverters to counter existing system voltage imperfections. This is challenging, but some lessons already learnt by small independent power systems may prove applicable [38]. In the future, these advanced approaches may be considered as possible types of auxiliary services although the complexities would be considerable.

Experience of system operation as described is limited for large power systems. Offshore wind power plants and offshore AC collection networks can be considered as good examples of a future converter-based grid. Harmonic resonance and converter control are discussed and illustrated in [39]. In conclusion, it should be emphasised that coordinated converter control is relevant in the future and absolutely needed for stable system operation.

Energy supply for electricity is undergoing fundamental changes. This triggers the need for fundamental developments in the field of system control, stability, protection, voltage quality, frequency control, etc. A major new challenge lies in coping with extremely weak power systems and enabling capability of stable operation of RE dominated systems with or without synchronous generation. For this, it is essential for all parties to enhance their understanding of the different phenomena at stake and exploit optimally the current and future technological possibilities in manufacturing. More specifically, mitigating capabilities of the generating units and HVDC systems will be needed to solve low total system inertia, inadequate fault levels for converters, inadequate fault currents for protection, and inadequate contributions to voltage quality.

4.4.2 Impacts on existing power plant economics and electricity market

To evaluate the impacts of large-scale wind integration on the Baltic region's electricity market, a simplistic 'copper plate' electricity market model was created for the year 2025 in [40]. The Baltic region was chosen as an example because the electricity market was liberalised quite recently; earlier, a majority of the power generation capacities were based on fossil fuels. As the Baltic countries were accepted to the European Union, they took on several obligations, among others the EU2020 and other environmental and energy efficiency related targets as well as the deregulation of energy markets. With these steps and supporting mechanisms targeted to renewable energies, there was a period in the early 2000s where the installation and plans for installing wind energy technology boomed. In 2010, there were plans to install wind parks exceeding 4 GW, i.e., more than twice of the peak load (1.5 GW), only in Estonia. As a result, several studies and analyses were conducted to evaluate the potential technical and economic impacts of such large integration of wind power.

In the example, the wind integration levels were chosen taking into account the total possible capacity that can be connected to the current power system with no need to build a new infrastructure or to curtail the output of the wind power plants [41]. The following wind integration levels for each Baltic country were used to evaluate the economic impact of large-scale wind integration:

(a) Low wind integration level of 250 MW offshore capacities and 250 MW onshore capacities – total installed wind capacity in the Baltic States power system 1,500 MW;
(b) High wind integration level of 450 MW offshore capacities and 450 MW onshore capacities – total installed wind capacity in the Baltic States power system 2,700 MW, which represented the maximum possible integration of wind power with the current grid infrastructure and without wind curtailment [41].

Equal wind integration levels for both onshore and offshore technologies in all three Baltic countries were selected in order to analyse and compare the impacts on the selected wind portfolio of each country in terms of conventional units and electricity market. The selected integration levels also illustrate the impacts that doubling the wind power evokes on the electricity market and its participants.

The analysis was based on the electricity pool principle described in many power system economics handbooks, for example, [42,43]. All the producers bid on the market according to their short-run marginal costs (SRMC). After all, the bids submitted are arranged from the smallest to the highest. This list of arranged bids gives the merit order curve of dispatching units at the trading hour. The system's marginal price (SMP) or electricity price at a specific trading hour is determined by the last unit that covers the demand forecast [18].

Based on the theory provided in [42,43], the SRMC is expressed as follows (4.4):

$$\mathrm{SRMC} = \frac{c_{\mathrm{fuel}} + c_{\mathrm{ext}}}{\eta_{\mathrm{PP}}} + c_{\mathrm{var}} \tag{4.4}$$

where SRMC is the short-run marginal cost of the unit in €/MWh; c_{fuel} is the forecasted price of the fuel used in €/MWh$_{\mathrm{fuel}}$; c_{ext} is the external cost of production in €/MWh$_{\mathrm{fuel}}$ calculated by multiplying the CO_2 tax in €/t_{CO_2} and the fuel's carbon content in t_{CO_2}/MWh$_{\mathrm{fuel}}$; η_{PP} is the power plant's efficiency in per cent; c_{var} is the variable cost of production in €/MWh.

Results obtained from the model are used for the economic evaluation, which is based on comparing the long-run marginal cost (LRMC) of the power plant with the revenues collected from the market. For the power plants to be economically viable, the revenues from the market should be greater than their long-run costs (4.5):

$$\frac{\sum_{i=1}^{8760} MR_i}{\sum_{l=1}^{8760} E_l} > \mathrm{LRMC} \Rightarrow \begin{cases} \text{FALSE, economically not viable} \\ \text{TRUE, economically viable} \end{cases} \tag{4.5}$$

where LRMC is the long-run marginal cost of the unit in €/MWh; MR_i is the hourly revenue collected from the market in €; E_l is the hourly supplied energy to the market in MWh.

In addition to the SRMC, the LRMC takes into account the capital (operation and maintenance) and investment costs. The following formulation of the LRMC was used in the model (4.6):

$$\mathrm{LRMC} = \mathrm{SRMC} + \frac{c_{\mathrm{inv}}}{E_a} + \frac{c_{\mathrm{fix}}}{E_a} \tag{4.6}$$

where LRMC is the long-run marginal cost in €/MWh; SRMC is the short-run marginal cost in €/MWh; c_{inv} is the discounted annual investment cost for the unit in €; c_{fix} is the annual fixed cost occurring irrespective of the unit's operation in €; E_a is the total energy produced in one year in MWh.

Table 4.1 Utilisation hour changes in conventional technologies due to an increase of wind power

Technology	Oil shale CFB*	Coal CFB	CCGT	Biomass
Average decrease in utilisation hours	8%–9%	6%–14%	23%	0%–19%

*CFB – circulating fluidised bed combustion technology.

The analysis indicated that the wind technologies influence the economics of conventional power plants over three indicators: changes in the market price, changes in the utilisation hour and the average price received from the market.

The increasing capacity of wind power in the system inevitably decreases the electricity prices on the market because it alters the merit order curve of dispatching. In this case, the distortion of the merit order curve is caused by the increased capacity from the wind power, usually with negligible marginal cost. The last unit that determines the SMP might be shifted in the merit order curve and thus also the price lowered at some trading hours, which decreases the overall average price on the market. In this specific case, the average annual market price was lowered from 74 €/MWh (with 1,500 MW of wind power) to 65 €/MWh if the wind integration level increased to 2,700 MW, decreasing approximately by 11%.

As the increase of wind capacities in the power system decreases the average market price by altering the merit order curve, it also decreases the utilisation hours of conventional technologies. However, as the analysis indicated, the change in the utilisation hour strongly depends on the type of fuel used in the power plant. To be more specific, it depends on the price of the fuel used rather than on the CO_2 content. This gives significant leverage for power plants running on cheap or local fuels, such as coal and oil shale in this case. Although the model includes also externalities, such as carbon tax (25 € per ton of CO_2 emitted), the level of the taxes is not high enough to promote more expensive but less polluting fuels such as natural gas. If the tax was 90 €/t_{CO_2}, fuel switching would seem a lot more attractive [41]. The ranges of technology-specific changes occurring in utilisation hour are given in Table 4.1. Only newly installed units and main technologies used in the Baltic region are included because the economic data available for older units are inconclusive in order to provide a reliable analysis on whether their economic feasibility is affected.

It should be noted that the impact on biomass technology depends on whether it is receiving subsidies or not. A more detailed analysis on the impacts resulting from subsidies and higher environmental taxes is presented in [44] ().

Changes in utilisation hours together with average market price changes thus affect the overall economic feasibility of different technologies. The economic viability of the technology is evaluated by comparing the gap between the LRMC and the average price received from the market. The economic feasibility of different power plants in the Baltic region is represented in Figure 4.5.

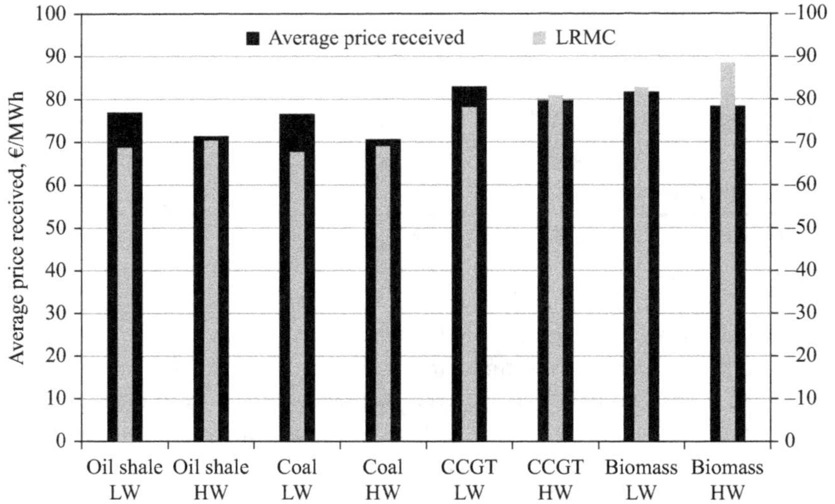

Figure 4.5 Economics of different power plants on the Baltic region electricity market; LW – low wind integration, HW – high wind integration

The results show that all the technologies analysed in the modelled market situation are running on their economic feasibility limits. The increase of the wind integration level in the system might make them economically infeasible. Most probably this will have no impact on the investment decisions of existing or planned power plants but it will certainly have an impact on their revenues and will jeopardise the investments in new conventional capacities.

As of the beginning of 2016, the wind integration levels in the Baltic States have increased to approximately 800 MW. At the same time, due to the influence of the Nord Pool Spot market, which was not accounted for in the model and where the average annual system price in 2015 was ~21 €/MWh, the electricity prices in the Baltic region have remained relatively modest: 31 €/MWh in Estonia and 42 €/MWh in Latvia and Lithuania. Such low prices inevitably pave a way for technologies with low marginal costs such as wind technology and decrease the possible investments in conventional thermal power plants.

4.4.3 System frequency regulation and increasing wind capacity impacts on regulating reserves

This section focuses on the impacts that large-scale wind integration has on the operating reserves required to ensure a smooth operation of the power system. Operating reserves in the scope of this section should be considered as the actually available power that could be used for frequency regulation and other ancillary services. It is the transmission system operator's (TSO) duty to keep the balance of its control area between generation and consumption. All the operations performed either by the TSO or system protection devices to maintain the balance between generation and consumption can be called frequency regulation. Frequency regulation can be performed either increasing or decreasing the outputs of power plants or at severe emergencies it is also possible to switch loads.

Turbine speed governors respond first and foremost to the frequency deviations occurring in the system. The governors manipulate with the vapour flow and therefore also the output power of the turbine and generator. As there are a large number of turbines connected to a power system, the speed governors are adjusted to operate with a specific droop. The primary regulation is done continuously by all turbine-generator units connected to the system. Secondary regulation, on the other hand, is handled by large fast switching units. In order to ensure the availability of primary and secondary reserves for the next possible contingencies, a tertiary reserve power is used, which can be called upon, for example, in 15 min, depending on the synchronous control principles.

Increasing wind capacities in the power system cannot affect the primary and secondary reserve requirements, because all generators connected to the system provide primary reserves irrespective of the amounts of wind capacity in the system and secondary regulation is done from one location, leaving only the $N-1$ contingency reserve to be managed. However, increasing levels of wind power may increase the requirement for tertiary reserves, as the forecast errors associated with net load change are likely to increase [41].

Milligan *et al.* in [46] describe and compare the regulating reserve mechanisms in different synchronous areas and countries. They report that in Spain the deployment of tertiary reserves has increased with increased wind capacities in the system. Paulus and Borggrefe also confirm in [25] that integrating wind capacities will lead to an increased demand for balancing power. The demand for up and down regulating power in Germany could accumulate into an increase of 33% and 41%, respectively, by 2030. According to the study of Wind Power in Estonia [41], a similar situation can be expected also in Estonia and in the rest of the Baltic States. For Estonia, the subsequent requirement for reserves to deal with the wind forecast errors is summarised in Table 4.2. Two integration levels, 900 and 1,800 MW, correspond to the reference values used in [41].

The data in Table 4.2 above consider that the maximum up-regulation reserve cannot exceed the forecast wind production, and similarly the maximum necessary

Table 4.2 Requirement of reserves (MW) to deal with the wind forecast errors [41]

Forecast intervals of wind production (p.u.)	**0.0–0.2**	**0.2–0.4**	**0.4–0.6**	**0.6–0.8**	**0.8–1.0**
Installed wind capacity 900 MW	162	270	243	189	54
Installed wind capacity 1,800 MW	216	324	270	324	162

down-regulation reserve cannot exceed the difference between the installed and the forecast wind power. Taking into account that increasing wind capacities influence the decisions of investing into new power plants, a gap may widen between the reserves required and actual reserves that the system will be able to provide. This gap, however, could be utilised by aggregated energy consumers for providing the necessary reserves through DSR. DSR, including energy storages, provides opportunities for the RE sector to balance their unpredictable production.

4.5 Measures to support large-scale wind integration

Continuously increasing percentage of wind power integrated into a power system presents additional challenges to system operators, especially in small interconnected power systems. This section of the book introduces several methods that could be used to reduce the forecast errors that cause main problems for large-scale wind integration. In addition to balancing forecast errors, the potential of smart DSR is discussed more thoroughly from the EV and DSM perspective.

4.5.1 Aggregated thermal storages for balancing of power generation forecast errors

In an open energy market, the RE producer has to plan energy production one day ahead. Forecasts have to be made for short periods, for example, hourly forecasts. There will be a decline of profit from the difference between the forecast and the real production referred to as forecast imbalance (FI). The calculation of the energy of FI (E_{FI}) in (4.7) shows the difference between the energy production forecast (E_{F}) and the real energy production (E_{R}) [45] (© 2016 IEEE. Reprinted, with permission, from [45]):

$$E_{\mathrm{FI}} = E_{\mathrm{F}} - E_{\mathrm{R}} \tag{4.7}$$

This section introduces a possibility to reduce the profit loss from FI with DSR. Loss of profit will appear at a particular hour when E_{R} is different from E_{F}, which indicates that there is a need for fast and direct co-operation between the RE producer and the DSR provider. Under direct co-operation, DSR must have fast reaction time for balancing RE production. DSR provider will benefit from the service and can utilise cheaper energy, which is the result of FI. A good example for the RE producer is a wind park that will always have FI because the wind is not well predictable. DSR with fast reaction time can be provided, for example, by electric boilers with heat storage or industrial processes that can be shifted in time. This section provides a solution for wind parks to have DSR in direct cooperation to reduce costs related to production forecasting errors. Wind park production is forecast in two steps. First, the availability of each turbine forecast on hourly basis is taken into account. Second, the availability plan of the wind park is given for production estimation that takes into account turbine stop times, wind meteorological forecast in the wind park location and turbine production curves.

Turbine production estimation also considers historical data for error correction. Nevertheless, the energy production forecasts are not 100% accurate. Table 4.3 illustrates the share of E_{FI} in three real wind parks named WP-A, WP-B and WP-C. The three wind parks have a significant share of energy sold or bought from a balancing market [45] (© 2016 IEEE. Reprinted, with permission, from [45]).

To describe the accuracy of the forecast, a normalised mean absolute percentage error (n_{MAPE}) given in [47] is presented in (4.8) [45]:

$$n_{MAPE} = \frac{1}{n}\sum_{t=1}^{n}\left|\frac{P_a - P_f}{P_{inst}}\right| \times 100\% \tag{4.8}$$

where P_a is the actual wind park output power; P_f is the wind park power forecast; P_{inst} is the installed capacity of the observed wind park; and n is the number of hours in the period.

DSR provider will forecast production as well and can reduce or increase its production according to the RE producer FI in equal opposite amounts. DSR providers are also responsible for planning their load accurately, i.e., FI of the DSR provider equals 0 in case DSR services are not needed. Taking into account (4.8), the DSR can balance the FI of the RE producer with an opposite load imbalance, as shown in (4.9) [45] (© 2016 IEEE. Reprinted, with permission, from [45]):

$$E_{DSR} = E_{RE} \times (-1) \tag{4.9}$$

where E_{DSR} is FI of the DSR provider and E_{RE} is FI of the RE producer. In case the RE producer and the DSR service provider have composed a common balancing region, then from the balance management point of view, total FI is equal to the sum of the FI of the RE producer and the DSR service provider FI, as shown in (4.10) [45] (© 2016 IEEE. Reprinted, with permission, from [45]):

$$E_{Tot} = E_{RE} + E_{DSR} \tag{4.10}$$

where E_{Tot} is the energy of the total FI.

Table 4.3 shows the share of the produced energy that could be balanced by DSR. To have 100% DSR for that energy, DSR capacity must be equal to the

Table 4.3 Share of forecast imbalance of wind parks from real production. © 2016 IEEE. Reprinted, with permission, from [45]

Wind park	**E_{FI} from total E_R, %**	
	$E_F < E_R$	**$E_F > E_R$**
WP-A	16	17
WP-B	13	14
WP-C	10	23

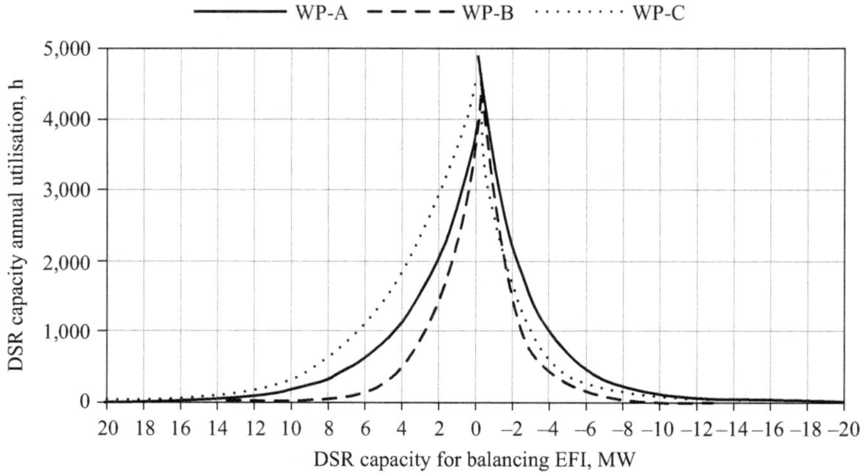

Figure 4.6 Annual utilisation of DSR capacity for three wind parks. © 2016 IEEE. Reprinted, with permission, from [45]

Table 4.4 Average forecast imbalance from wind parks. © 2016 IEEE. Reprinted, with permission, from [45]

Wind park	Average E_{FI}, MWh	
	$E_F < E_R$	$E_F > E_R$
WP-A	−1.36	1.47
WP-B	−0.86	0.91
WP-C	−0.94	2.20

nominal power of the wind park. From a practical point of view, the maximum needed capacity of FI from wind parks is not reasonable, as the maximum is seldom used. Financial calculations in each individual case must be made to find an optimal DSR capacity. To find the optimal capacity, the annual utilisation of the DSR in an amount agreed has to be found. It is reasonable to examine DSR capacity utilisation on the graph, which shows duration in hours when E_{FI} is to be balanced in the nominal DSR capacity shown in Figure 4.6 for three different wind parks sized from 10 to 40 MW [45] (© 2016 IEEE. Reprinted, with permission, from [45]).

To acquire good utilisation, in that analysis, DSR capacity equal to an hourly average FI from the wind parks was used. Table 4.4 shows an average FI from the three wind parks.

To provide DSR for the total FI, the capacity required is more than 10 times higher than with DSR provided for the average FI. Based on Figure 4.6, it can also

Table 4.5 Duration of average forecast imbalance. © 2016 IEEE. Reprinted, with permission, from [45]

Wind park	**Duration t_{FI} of average E_{FI}, h**	
	$E_F < E_R$	$E_F > E_R$
WP-A	2,732	2,531
WP-B	2,686	2,564
WP-C	2,431	2,906

be stated that at almost every hour, a need of DSR for the wind parks is present. Table 4.5 shows the time period (t_{FI}) in a year when FI can be balanced with the DSR nominal capacity equal to an average FI of the wind parks. Utilisation time is an important parameter in the calculation of the economic effect of DSR [45] (© 2016 IEEE. Reprinted, with permission, from [45]).

To estimate the effect of the proposed DSR capacity, the annual E_{RE} that can be balanced by DSR (E_{CFI}) is found using (4.15) and a share from the total FI (E_{share}) is found using (4.16) [45] (© 2016 IEEE. Reprinted, with permission, from [45]):

$$\text{if } E_F > E_R \wedge E_{FI,i} < E_{PFI}, \text{ then } E_{FI} = \sum_{i=1}^{n} E_{FI,i} \tag{4.11}$$

$$\text{if } E_F > E_R \wedge E_{FI,i} \geq E_{PFI}, \text{ then } E_{FIn} = \sum_{i=1}^{n} E_{PFI,i} \tag{4.12}$$

$$\text{if } E_F < E_R \wedge E_{FI,i} > E_{NFI}, \text{ then } E_{FI} = \sum_{i=1}^{n} E_{FI,i} \tag{4.13}$$

$$\text{if } E_F < E_R \wedge E_{FI,i} \leq E_{NFI}, \text{ then } E_{FIn} = \sum_{i=1}^{n} E_{NFI,i} \tag{4.14}$$

$$E_{CFI} = E_{FI} + E_{FIn} \tag{4.15}$$

$$E_{share} = \frac{E_{CFI}}{\sum_{i=1}^{n} E_{FI,i}} \tag{4.16}$$

where i is an hour in the overall period defined by n; E_{PFI} is the maximum positive balanced energy by DSR; E_{NFI} is the minimum negative balanced energy by DSR and E_{FIn} is the sum of applied E_{PFI} or E_{NFI}, as given in (4.12) and (4.14), respectively. E_{FI} is calculated as a sum of the energy of FI in the particular hour i, calculated with (4.11) and (4.13), respectively, in the analysis, E_{NFI} equals an average negative FI and E_{PFI} equals an average positive FI by a wind park. The results are given in Table 4.6 [45] (© 2016 IEEE. Reprinted, with permission, from [45]).

Table 4.6 Balancing capacity and per cent from total FI at DSR size equal to an average wind park FI. © 2016 IEEE. Reprinted, with permission, from [45]

Wind park	**Balancing capacity E_{CFI}, MWh**		**Share from total FI E_{share}, %**	
	$E_F < E_R$	$E_F > E_R$	$E_F < E_R$	$E_F > E_R$
WP-A	−4,976	4,632	41.8	36.0
WP-B	−3,148	2,872	41.8	35.9
WP-C	−2,952	8,186	35.8	42.4

As can be seen from Table 4.6, DSR with nominal capacity equal to average FI can reduce approximately 36%–42% of the total FI. At relatively small DSR capacity, the share of FI reduced is quite high. In the proposed DSR with direct cooperation between the RE producer and the DSR provider, some special aspects must be considered and applied. These include a need to provide DSR within the same hour when FI is appearing, so that DSR will react to FI errors almost in real time. DSR is required to have an accurate load forecast, participation in day-ahead energy market and efficiency in energy conversion. For those particular restrictions, a suitable DSR provider can be an electric boiler with heat storage. In the described approach of DSR implementation, energy storage in different forms is required. Further analysis focuses on the energy storages required to correct the FI of wind parks. Energy storage capacity required depends on the time necessary to cover the FI. For example, the participants in the Nord Pool Spot have to provide a day-ahead market production plan at 12.00 p.m. for the next day [48]. A DSR provider can adjust its energy load plan according to energy storage fulfilment once every 24 h. This, however, is not sufficient, as it fails to take into account the period of time when the next production plan will come into force. If the plan is made at 12.00 p.m., the period to the end of the day must also be taken into account. This means that the size of energy storage must be large enough to balance FI at least for 36 h. DSR provider should have energy storage available for the end of the day when planning is made for 12 and 24 h in the new planned period. With a forecast change, it is possible to plan to empty or to fill heat storage to keep an optimal storage amount. If the longest period to store the energy by DSR into energy storage is 36 h, the energy required to store (E_s) in storage can be found by (4.17) and (4.18) [45] (© 2016 IEEE. Reprinted, with permission, from [45]).

$$E_S = \frac{P_{DSR} \times t}{\eta} \tag{4.17}$$

and

$$P_{DSR} = FI_a \times 2 \tag{4.18}$$

Table 4.7 Heat storage capacity required for FI of the wind parks. © 2016 IEEE. Reprinted, with permission, from [45]

Wind park	Nominal power for DSR P_{DSR}, MW	Heat storage capacity E_s, MWh	Volume of the heat storage V_S, m^3
WP-A	2.94	125	2,014
WP-B	1.82	77	1,247
WP-C	4.40	186	3,015

where P_{DSR} is nominal power for DSR; FI_a is the forecast imbalance agreed to be balanced by DSR; t is the time period required by FI; and η is the efficiency of the energy storage. FI_a must be multiplied with 2 to have storage on both sides for surplus energy and for lack of energy. In case P_{DSR} equals FI_a, the DSR provider can only balance the energy when the RE producer E_F is higher than E_R; in that case, DSR will stop using energy equal to FI_a. In case E_F is lower than E_R, DSR is required to increase its energy usage. To cover this case as well, it is required to increase the load of DSR equal to FI_a (in maximum). As FI requires an increasing DSR load, it is optimal to keep storage half full so that an increased load can be stored. In the described example, the used energy storage can be an electric boiler in aggregated form. The required heat storage capacity is given in Table 4.7; boiler efficiency is considered to be 85%, 5% loss will come from the heating unit (e.g., heating electrode) and 10% of loss is considered as loss from heat storage during 36 h. For FI_a, absolute values of maximum wind parks E_{FI} are taken from Table 4.4. Based on the energy storage requirement shown in Table 4.7, we can calculate the storage volume by (4.19) [45] (© 2016 IEEE. Reprinted, with permission, from [45]):

$$V_S = \frac{Q_S}{\rho \cdot c_p \cdot (T_{a_2} - T_{a_1})} \tag{4.19}$$

where V_S is the volume of the heat storage in m^3; ρ is water density; c_p is specific heat at constant pressure; T_{a_2} is required maximum temperature; and T_{a_1} is allowed minimum temperature.

Heat storage size required for the FI of the wind parks is presented in Table 4.7 as well. In the calculation, the following parameters are used: $\rho = 1{,}000$ kg/m^3, $c_p = 4{,}180$ J/(kg·K), $T_{a_2} = 60$ °C and $T_{a_1} = 7$ °C. Temperatures are chosen as typical parameters for electric boilers for tap water heating. Under direct cooperation with a RE producer, DSR can have a positive effect only if it reduces costs for the RE producer. Positive effect results from the situation where the RE producer suffers under the decline of profit from the energy inappropriately planned and forecast to the market. To find out the economic effect, it is required to determine price differences between unplanned mean balancing market price (BMP) and planned mean day-ahead prices (DAP). It must be pointed out that BMP is not always more expensive than DAP and a difference for the producer depends on whether one buys or sells energy with BMP. In case BMP is cheaper than DAP and the RE producer has to buy from a balancing market, due to FI, the RE producer will have an increase of profit instead of a decline. Further, if BMP is higher than DAP and the RE producer has to sell some of the energy with BMP due to underestimation of the energy production, the RE producer can gain an increase of profit. The RE producer without manageable electricity production must use service from a balancing market, unaware of whether it is more beneficial to forecast more or less [45] (© 2016 IEEE. Reprinted, with permission, from [45]).

RE producer profit decline c_{re} was calculated as shown in (4.20):

$$c_{\text{RE}} = \left(\sum_{i=1}^{n} (\text{BMP}_{S,i} - \text{DAP}_i) \cdot E_{\text{FI},i}; E_{\text{F},i} > E_{\text{R},i} \right) + \left(\sum_{i=1}^{n} (\text{BMP}_{P,i} - \text{DAP}_i) \cdot E_{\text{FI},i} \cdot (-1); \quad E_{\text{F},i} < E_{\text{R},i} \right) \tag{4.20}$$

where $\text{BMP}_{P,i}$ is the energy price under a need to purchase from a balancing market; $\text{BMP}_{S,i}$ is the energy price under a need to sell to a balancing market; i is a particular hour when a transaction is made; and n is the number of hours in the analysed period. DSR can reduce the decline of profit; however, its effect must be found hour by hour and after BMP is known. To show the economic effect that DSR can have, for the FI of the wind parks, an average loss of profit per one MWh of FI is calculated, summarised in Table 4.8 [45] (© 2016 IEEE. Reprinted, with permission, from [45]).

The calculated DSR capacities for wind parks are as follows: WP-A 1.47 MW, WP-B 0.91 MW and WP-C 2.2 MW. Economic effect is reachable and the DSR service provider can decrease the lost profit from the FI of the wind parks from 30% to 44%. By increasing the revenue, some of the economic effect must be transferred to the DSR provider as well. The proposed solution offers a profitable effect for the RE producers and the DSR providers. About 30%–44% of the profit decline from the total FI profit decline can be reduced with the DSR capacity equal to an average wind park FI. The DSR provider must control its consumption in two ways, i.e., to increase or decrease energy demand, as the forecast of the wind park can be lower or higher than the real energy production. Two-way consumption control requires double DSR capacity from the DSR

Table 4.8 Profit loss and average profit decline c_{re} from wind parks FI per MWh. © 2016 IEEE. Reprinted, with permission, from [45]

Wind park	Profit loss decline, %	Average profit decline C_{RE}, EUR/MWh	
		$E_{\text{F}} < E_{\text{R}}$	$E_{\text{F}} > E_{\text{R}}$
WP-A	30	4.39	4.12
WP-B	33	6.12	6.19
WP-C	44	12.58	3.05

provider as compared with the wind park FI, which is agreed to be balanced. As a result, the solution proposed shows a positive economic effect of DSR for the wind parks and enables DSR use on a small scale, as the wind parks exist in many different sizes [45] (© 2016 IEEE. Reprinted, with permission, from [45]).

4.5.2 *Pumped hydro energy storage for balancing of power generation forecast errors*

According to Ter-Gazarian [49], the most suitable utility scale and conventionally feasible energy storage technology for balancing large-scale wind energy is the pumped hydro energy storage (PHES) system. PHES facility holds the ability to accumulate or generate several hundreds of megawatt hours of energy. It can handle all rapid wind power fluctuations, reaching tens of megawatts within a minute, has very high cycle efficiency and no CO_2 emissions during operation. These reasons make PHES outrun other conventional technologies used in the grid, such as compressed air energy-storage, batteries or flywheel, especially when the task is to balance the unpredictable, rapid and vast wind energy production forecast errors.

A test system comprising 250 and 900 MW of wind power plants serves as an example. Wind production baseline data and forecast errors during a calendar year were obtained from the Estonian TSO's homepage for the period of 2012 summer to 2013 summer. The assumed n_{MAPE} for the studied test system is to be considered as 6.07%. The PHES-based balancing system was analysed from two perspectives: manoeuvrability and 24 h balance commitment. The values used for benchmarking the PHES capability were taken from the wind production and forecast errors and are as follows:

- Maximum positive error in wind power generation (surplus power change in the grid) occurs at 250 MW of capacities with the amount 165.62 MW/h (2.76 MW/min) and at 900 MW of capacities with the amount of 577.76 MW/h (9.63 MW/min);
- Maximum negative error in wind power generation (deficiency power change in the grid) occurs at 250 MW of capacities with the amount of (–)164.1 MW/h ((–)2.74 MW/min) and at 900 MW of capacities with the amount of (–)572.44 MW/h ((–)9.54 MW/min).

The PHES system benchmarked consists of three pump-turbines (100 MW and 2 × 175 MW) and one additional turbine (50 MW). First, it is important to recall the main principle of a power system: maintaining the balance between generated power and demand at every point of operation. In that light, all PHES system operational mode scenarios must be examined. For example, if there is a sudden deficiency of generated power in the grid due to vast negative error of wind generation forecast (unforeseen), a PHES station can balance it by multiple means, depending on its operational point. Such opportunities are starting power generation (at standstill), decreasing power consumption (in pumping mode) or switching from pumping to generation (if negative error is still growing). In the case of

Table 4.9 Finding manoeuvrability for corresponding PHES configurations

Mode change	Changeover time, min	Correspond. capacity, MW	Maximum output change speed, MW/min
Standstill to full-load generation	2	500	250
Standstill to full-load pumping	10	450	(−)45
Full-load generation to full-load pumping	15	500/450	(−)145
Full-load pumping to full-load generation	8	450/500	325
Full-load generation to standstill	5	500	(−)100
Full-load pumping to standstill	6	450	75

uncontrollable surplus power entering the grid (positive forecast error of wind production), a PHES station can switch to pumping from full-load generation to full-load pumping (depending on the amount of surplus energy entering the grid). Changeover times must be found and the corresponding balancing capabilities must be assessed for all of those scenarios. In order to do so, technical parameters for Francis pump-turbines must be assessed according to the proposed underground PHES project layout. The calculations and results are presented in Table 4.9, together with technical parameters for Francis pump-turbines (according to the operational mode) [49].

In addition, it is assumed that the point of operation for the PHES station is random at a certain time. For example, deficiency may occur at standstill, full-load pumping or generating. As can be seen, the balancing capabilities are most limited in the pumping mode, therefore manoeuvrability of balancing capabilities must be assessed in the pumping mode. In contrast, corresponding manoeuvrability of thermal power plants (TPP) remains between 2.5 MW/min and 7 MW/min for negative and positive errors, respectively. It is important to recognise that in the context of power system balance, the positive error of wind power generation must correspond to the decrease of the PHES and the TPP power generation and vice versa.

With reference to technical manoeuvrability, it can be seen in Table 4.9 that at 250 MW of wind power installed, the TPP unit is almost capable of balancing the wind power output variation due to forecast inadequacy: at least in the case of surplus energy in the grid (the TPP can ramp down fast – 7 MW/min). Still, in the case of deficiency in the grid, problems occur (maximum deficiency in the grid 2.74 MW/min compared to TPP's ramp up capability of 2.5 MW/min). Nevertheless, with higher wind capacities (900 MW), the situation is very different – the balancing capabilities will become exhausted. There is also a theoretical possibility to regulate with multiple thermal units together, but in such situation, the inefficiencies and decrease in lifetime occur. At the same time, different economic expenditures are increased. However, with the PHES station, the situation is totally different, its manoeuvrability is many times higher and it can be balanced by only some of its four pump-turbine units. Therefore, as compared to large thermal units, the technical balancing capabilities of PHES are superior for balancing forecast errors of wind power output.

From the balance commitment point of view for 24 h, the maximum daily energy production error information is essential. Both the maximum daily surplus and maximum deficiency wind energy are needed as straightforward information of the quantity of energy for a PHES system to be stored for balancing deficiency and of the quantity it must be able to store for accumulating surplus energy in the pumping mode. In any type of a PHES project, the energy capacity required for both up and down balancing depends on the storage capacity of the reservoir. Therefore, it is required to have information about the minimal volume amount of required water to store maximum possible surplus energy in the grid (24 h) and minimal required empty space volume to compensate (generate) maximum possible deficiency in the grid.

The maximum surplus and maximum deficiency energy amounts found for the corresponding 24 h must be calculated into the corresponding amounts of water volume by the PHES characteristics to guarantee the 24 h commitment. It is the main goal of this task and for a PHES station as a whole in order to substitute TPPs for balancing of wind energy forecast errors in the grid. To be able to absorb the maximum daily (24 h) surplus energy in the grid, caused by maximum positive wind energy generation forecast error, the PHES must work in the pumping mode and have in stock the corresponding amount of water. Equation (4.21) that describes PHES in the pumping mode and the amount of energy that can be saved – E_{pumping} given in Joules [J] (equal to ~2.78×10^{-10} MWh) consists of the following characteristics: ρ – density of water [kg/m^3] (1,000 kg/m^3); g – gravitation constant (9.81) [m/s^2]; h – the height difference between the turbine and the higher reservoir where the water is being pumped [m]; V – the amount of water pumped (going through the turbine) in order to save energy [m^3], and η_{pumping} – pumping efficiency of the PHES system (assumed to be 85%).

$$E_{\text{pumping}} = \frac{\rho \cdot g \cdot h \cdot V}{\eta_{\text{pumping}}}. \tag{4.21}$$

The minimal needed amount of 'free space' to be held in stock in order to absorb the amount of calculated maximum daily surplus energy in the grid is calculated as follows (4.22):

$$V_{\min} = \frac{E_{\max\,24h\text{surp}} \cdot \eta_{\text{pumping}}}{\rho \cdot g \cdot h}. \tag{4.22}$$

To be able to respond to the maximum daily (24 h) deficiency energy in the grid caused by maximum negative wind energy generation forecast error, PHES

Table 4.10 Finding manoeuvrability for corresponding PHES configurations

	Maximum wind energy forecast error at 250 MW of wind capacity, MWh	**Minimum water or free space (in storage reservoir) at 250 MW of wind capacity, m^3**
Maximum surplus	2,569	1,564,855
Maximum deficiency	(−)1,481.20	1,202,016
	Maximum wind energy forecast error at 900 MW of wind capacity, MWh	**Minimum water or free space (in storage reservoir) at 900 MW of wind capacity, m^3**
Maximum surplus	8,963	5,455,307
Maximum deficiency	5,167	4,193,081
Maximum volume needed	9,648,400 m^3	

must work in the generation mode and have stock of free water in the reservoir. Equation (4.23) that describes PHES in the generation mode and the amount of energy that can be produced $E_{generation}$ consists of the same characteristics as (4.21), with the difference that V is the amount of water being released (through the turbine) to produce energy [m^3] and $\eta_{generation}$ is the generation efficiency of the PHES system (assumed to be 88%).

$$E_{generation} = \rho \cdot g \cdot h \cdot V \cdot \eta_{generation}. \tag{4.23}$$

To calculate the minimal needed amount of 'free water' in stock in the reservoir to produce the amount of calculated maximum daily deficiency energy in the grid, the equation has to be adjusted as follows (4.24):

$$V_{min} = \frac{E_{max\ 24hdef}}{\rho \cdot g \cdot h \cdot \eta_{generation}}. \tag{4.24}$$

From the available wind production and forecast error information, the maximum surplus and maximum deficiency of energy for both wind integration scenarios for all hours must be summed after 24 h intervals and compared. The results for the calculated maximum daily (24 h) errors in wind energy production are given in Table 4.10.

In order to guarantee balancing power for both surplus and deficiency generation of wind power, balancing capabilities in the form of minimal needed water and empty volume overlap – the minimal volume of needed water is 5,455,307 m^3 and the minimal volume of needed empty space is 4,193,081 m^3 in the reservoir.

Since the PHES requires very specific geographic location with significant height differences, at first glance it seems as a solution for areas where naturally occurring water reservoirs can be found near coastal areas and at great heights. However, in Estonia (a relatively flat area), a solution was proposed to build a PHES underground. The idea is to operate sea as an upper reservoir together with a lower reservoir situated underground.

Storage medium would be seawater, which in the Baltic Sea is harmless to metal constructions due to marginal salt concentration (0.5%–1.5%). The main idea behind such concept is to remove all requirements for special topographical expectations that conventional PHES requires, leaving only the main requirement for altitude difference between the two reservoirs. The proposed solution would not just create a big hole in the seafloor but rather exploit the leftovers of a mining operation, i.e., mining of granite. The solution would suit for flat countries rather than for those with high wind generation like Denmark, north Germany, coastal area of Poland, etc. The technical parameters of the proposed solution are as follows [47] (© 2014 IEEE. Reprinted, with permission, from [47]):

- PHES installed capacity of turbines:
  - maximum storage/input power: 450 MW;
  - maximum generation/output power: 500 MW;
- three reversible vertical-axis Francis pump-turbines: 1 × 100 MW; 2 × 175 MW;
- one additional vertical-axis Francis turbine: 50 MW;
- maximum flow rate as pumping and generating: 120 m^3/s;
- water head (between upper reservoir level and turbines): 500 m;
- working volume of the lower reservoir: 5,200,000 m^3;
- water density can be taken as normal: 1,000 kg/m^3 (due to marginal salt concentration in the Gulf of Finland).

Total (round trip) efficiency: 75% (pumping 85%; generating 88%). Still, the solution itself is much more costly in comparison to traditional PHES systems, the cost can be significantly reduced if abandoned or to be abandoned mines turn out to be suitable for an underground reservoir. The calculations provided beforehand with formulas (4.21)–(4.24) can be easily modified to suit also for an underground PHES system with slight modifications that take into account the fact that the storage is located underground and not on a mountain.

4.5.3 Demand side management for providing balancing power

As described in [25], DSR could offer the TSO additional ancillary reserve resource to securely integrate weather-influenced generation, maintain security of supply, and optimise the utilisation of infrastructure and investments. However, in many cases, loads are used as a last resort in emergency situations [21,25]. On the other hand, additional ancillary services require large investments into new generation needed for power, not for the energy produced [50]. The following section of the book discusses the possibility to utilise large-scale EV integration and smart DSR to integrate more wind capacities into the power system by using the service's potential to coincide or relieve the impact of overproduction.

4.5.3.1 EV charging as alternative storage for RE

An average EV battery capacity could sustain one regular household's daily electricity demand. However, as the majority of today's commercially available EVs are unable to generate electricity towards the grid without any significant

Table 4.11 Main data describing EVs in 2016

	Mitsubishi iMiEV	Nissan Leaf	Tesla model S
Power consumption, Wh/km	100	120	161–180
Range, km	160	200–250	372–500
Top speed, km/h	130	150	210–250
Acceleration 0–100 km/h, s	13	11.5	3.3–5.4
Power bank, kWh	16	24–30	60–90
Charging capacity (standard AC), kW	3.6	3.3–6.6	11

modification, their controlled charging would be able to provide only valuable storage capacity or balancing load when there is significant overproduction in the system.

Table 4.11 summarises the main manufacturer data describing commercially available EVs from three different classes: eco, standard and luxury.

The authors assume that by 2030 the average EV would have a power bank with the capacity of 100 kWh. According to IEA [51], in Finland, approximately 60% of the cars have an engine block heater; with a slight modification, these block heaters could also be used for EV charging as well as for integration into the power system as energy storages. This way most of the EV owners have an access to a standard charger, which would serve for charging plug-in vehicles, as indicated by several studies [13,52]. While the EVs are connected into the domestic electricity network, with controlled charging, their on-demand energy 'storage' capacity would remain between 3 and 11 kWh/h.

From the factory datasheet it can be assumed that the average power consumption of an EV would remain around 170 Wh/km; however, Vee in [53] has proven that due to the geographical and climatic conditions in Estonia, the actual consumption of an EV could be up to 40% higher than indicated by the manufacturer. This can be also extrapolated for other northern countries. Because of the extra energy required by the heating during winter periods, the Mitsubishi iMiEV's power consumption per km peaked at 190 Wh (compared to the 100 Wh indicated by the producer) during a test period. For further calculations, the authors have assumed that the power consumption of an EV during winter time would be 235 Wh/km and the average daily mileage is expected to be around 40 km in the UK [14]. In other seasonal periods, it can be expected that the EV's average energy consumption per km would not drop below 200 Wh, as the cars require a certain level of heating or cooling during driving.

To calculate the possible charging impact on the load, it has been divided into three levels: standard charging, comfort charging or preheat and quick charging. Standard charging is assumed to be influenced mainly by the electricity market price. All of these three charging principles offer a potential energy storage possibility for the weather-influenced generators.

Standard charging takes into account the market price fluctuations and the traffic intensity. Market price impacts are based on the assumption that the prime time for charging is between 20:00 and 6:00. However, this can be modified according to the user's specifications, for example, prime time for charging according to the market price could be replaced with prime time for storage of excess energy in the power system according to the load curve of the weather-influenced generation. Equation (4.25) is used to calculate the hourly load impact or also the storage capacity (MWh/h) of standard charging outside the prime time charging period:

$$\underset{h=7\ldots19}{E_{C_h}} = \frac{\sum_{l=20}^{6} d \cdot \rho_{T_h} \cdot E_d}{\sum_{i=7}^{19} 1_i} \cdot \Delta V_{P_h} \tag{4.25}$$

where h is the hour of the day; l defines the prime time hours; i is the time period outside the prime charging time; E_{C_h} is the charging energy of the hour; d is the total daily distance covered by EVs (km); ρ_{T_h} is hourly traffic density; E_d is the energy consumption per distance covered (2.35×10^{-5} MWh/km); and ΔV_{P_h} is the relative change in the number of parking EVs to the average parking EVs over the observed hours of the day (%/h). The values of h, l and i used in the equation should match the prime time selection.

Equation (4.26) is used to calculate the hourly impact of standard charging during the prime time.

$$\underset{h=20\ldots6}{E_{C_h}} = a \cdot d \cdot E_d \cdot D(MP)_h \tag{4.26}$$

where h is the hour of the day; E_{C_h} is the charging energy of the hour; a is the share of EV owners (%) who have access to a standard charging station (block heater); d is the total daily distance covered by *EV*s; E_d is the energy consumption per distance covered; and $D(MP)_h$ is the charging distribution coefficient during the prime charging time (%/h). The values of h used in the equation should match the prime time selection.

Comfort charging is the energy used for preheating the car and the battery 2 h prior to usage in order to maximise the battery usage for driving and not to heat the cabin in cold battery condition. The preheating is assumed to have an energy intensity of 3×10^{-3} MWh per car in 1 h and could serve as an extra storage possibility during autumn and winter periods (e.g., the Nissan Leaf's heater power remains between 1.5 and 3 kW). Equation (4.27) describes how to calculate the hourly energy required for preheating:

$$E_{P_h} = (\Delta V_{(h+2)} \cdot E_H)\big|_{h<(h_n-1)} + (\Delta V_{(h+1)} \cdot E_H)\big|_{h>(h_m-2)} \tag{4.27}$$

where h is the hour of the day; E_{P_h} is the preheating energy of the hour; ΔV_h is the change of vehicles in traffic during an observed hour (1/h); E_H is the hourly consumption of the preheaters; h_m is the hour when the increase of traffic intensity starts; and h_n is the last hour of the cycle when the traffic intensity has increased.

Quick charging takes into account the car users who do not have a full-time access to standard chargers and thus need to utilise the quick charging infrastructure. This load is assumed to be proportional to the traffic intensity and is calculated with (4.28):

$$\underset{h=0...23}{E_{Q_h}} = (d \cdot E_d - E_C) \cdot \rho_{T_h} \tag{4.28}$$

where h is the hour of the day; E_{Q_h} is the charging energy of the hour; d is the total daily distance covered by EVs; E_C is the total energy charged via the standard chargers; and ρ_{T_h} is the hourly traffic density (%/h).

According to the proposed equations (4.25)–(4.28) that take into account traffic density, prime time for standard charging (according to the market price), charging from quick chargers and preheating of the cars during winter time, the additional load required by the EV fleet consisting of 100,000 cars during a winter day would remain around 1,113 MWh. The hourly distribution of the load over a 24 h period is illustrated in Figure 4.7.

If the charging is driven by the market price, the average additional load that large-scale EV integration adds to the power system remains between 30 and 70 MWh/h, depending on the number of EVs. Higher load increase can be observed during off peak times but during peak hours 1.3%–3.8% of additional load should be expected on a power system similar in size and capacity of Estonia [12] (© 2013 IEEE. Reprinted, with permission, from [12]). If the charging is assumed to be

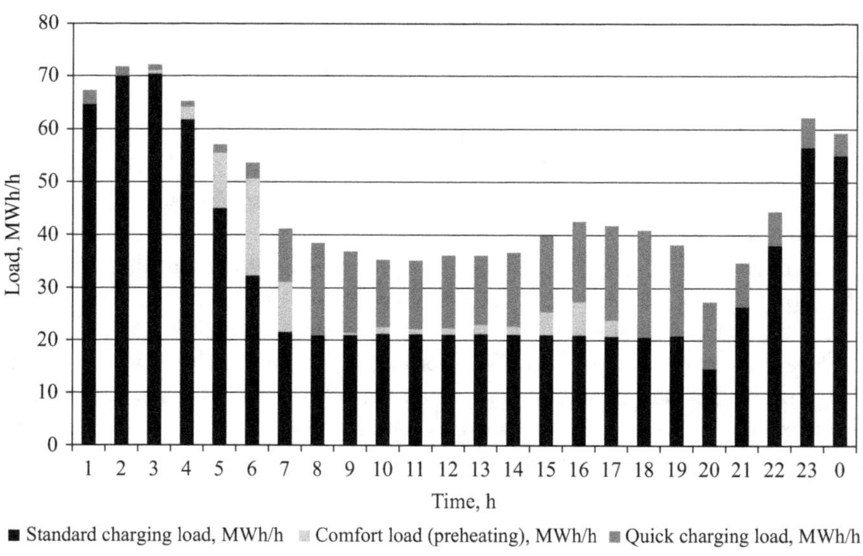

Figure 4.7 Hourly distribution of the load required by an EV fleet of 100,000 according to different load sources

uncontrolled and random, a higher load increase can be observed during peak hours between 16.00 and 18.00. This is in correspondence with the working day end and the traffic intensity changes. The coinciding increase of peak load of a power system with a peak load around 2.2 GW would be around 3%.

4.5.3.2 Industrial consumers as power generation balancing entity

In the context of the large-scale wind integration, DSR and DSM are regarded as potential sources of energy storages. Similar to the EV potential discussed in Sections 4.3.2 and 4.5.3.1, the price-based load control of industrial consumers opens an alternative source of a balancing entity for the weather-influenced generation. DSR can be implemented and preferred if it is possible to use some sort of energy storage within production line processes that consume electricity. The energy storage does not have to be a physical 'energy storage' in the literal sense, but instead an intermediate storage of a semi product, for example, in the wood industry, it is an energy storage as good as a battery or a capacitor. It enables shifting of the energy consumption of a production process without interrupting the main production line.

The following example is based on the optimisation of the energy consumption of a production line according to historical and day-ahead market price information. Electricity cost *E*, which is a stochastic parameter, can be minimised for a single load under investigation according to scenarios *s* with (4.29):

$$\min_{Q_t} \sum_{s} \sum_{t} P_t^s \cdot C^{CH}(Q_t) \cdot P(s) + C^E(Q_t) \tag{4.29}$$

where P_t is the electricity price for time period t; Q_t is the production volume of the production line for a time period t; C^{CH} is the output power of the line; $P(s)$ is a probability of the scenario s; C^E is the external cost of the load shifting, for example, start-up costs, labour costs, etc.

With the assumption that market price reflects the real situation in the power system, the optimisation algorithm could be directly extrapolated for the balancing weather-influenced generation. Neither information on long-term electricity market price nor about the weather-influenced generation capacities is available for us today. Therefore, the optimisation algorithm could be implemented only with both historical data and at best day-ahead information. The optimisation of the load according to either market price information or RE production takes into account three sets of data: historical, day-ahead and intraday information. Historical data give preliminary information about the production line output, day-ahead information is used to optimise the production plan to meet the predetermined limitations of industry if needed and intraday information provides yet again a possibility to optimise the production plan further. A simplified production control principle is illustrated in Figure 4.8.

During high price periods or low RE production part, the production line is stopped and material is used from the storage. In low price or high RE production periods, the production line is started to fill up the storage. With available historical

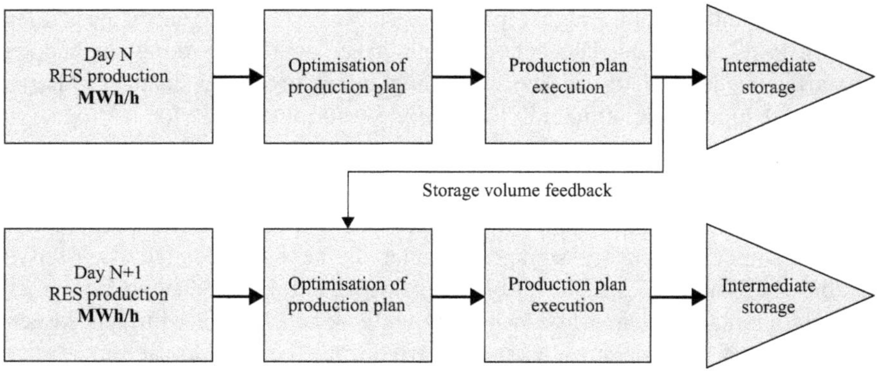

Figure 4.8 Simplified production control with day-ahead optimisation of the wood chipper's operation schedule. © 2013 IEEE. Reprinted, with permission, from [54]

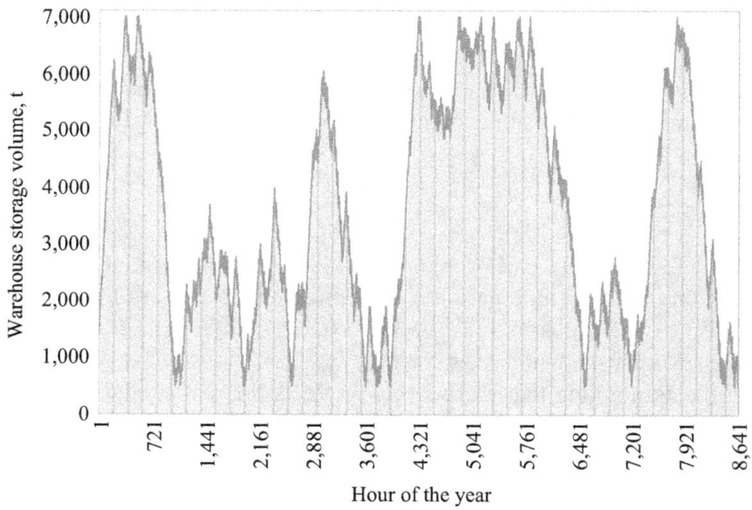

Figure 4.9 Warehouse storage volume fluctuation over a calendar year according to historical market price variations in the Nord Pool Spot area. © 2013 IEEE. Reprinted, with permission, from [54]

data manipulation in the optimisation algorithm it is possible to show the possible fluctuation of the storage capacity of the intermediate product (in this case, wood chips or sawdust), as illustrated in Figure 4.9. The optimisation algorithm can be easily adjusted to work for RE production and also in other economic sectors or fields of activities. However, it should be noted that in other fields or economic sectors, limitations may arise due to the missing storages or restrictions from the controllable devices.

4.6 Conclusion

The limited supply of fossil fuels, increasing demand for energy, and concern about the human environmental impact cannot ensure sustainability under the circumstances of existing fossil fuelled power plants. Today, wind technology is seen as the most prospective solution of alternative and sustainable sources of energy. However, its large-scale integration requires fundamental changes in the planning and operation of power systems as well as the management and attitudes towards different market and power system participants. Only synergy between consumers, producers and system operators will overcome the challenges faced with large-scale wind integration.

References

[1] *Energy and Climate Change. World Energy Outlook Special Report. International Energy Agency*. OECD/IEA, 2015.

[2] *The Road from Paris: Assessing the Implications of the Paris Agreement and Accompanying the Proposal for a Council Decision on the Signing, on Behalf of the European Union, of the Paris Agreement Adopted under the United Nations Framework Convention on Climate Change*. European Commission. Brussels, 2016.

[3] *World Energy Investment Outlook. Special Report. International Energy Agency. OECD/IEA*, 2014.

[4] *Energy Prices and Costs in Europe*. European Commission. Brussels, 2014.

[5] *World Wind Energy Association. The World Sets New Wind Installations Record: 63.7 GW New Capacity in 2015* [Online]. WWEA. Bonn, 10 February 2016. Available from http://www.wwindea.org/the-world-sets-new-wind-installations-record-637-gw-new-capacity-in-2015/ [Accessed 24 Aug 2016].

[6] Ackermann, T. *Wind Power in Power Systems*, Second Edition. John Wiley & Sons Ltd, The Atrium, Southern Gate, Chichester, West Sussex PO19 8SQ, England, 2012, 1049 pp.

[7] El-Sharkawi, M. A. *Wind Energy: An Introduction*. CRC Press, Taylor & Francis Group, 6000 Broken Sound Parkway NW, Suite 300, Boca Raton, FL 33487-2742, 2015, 355 pp.

[8] *HVDC Connection of Offshore Wind Power Plants*. CIGRE Technical Brochure No. 619. CIGRE, 2015, 92 pp.

[9] *Special Considerations for AC Collector Systems and Substations Associated with HVDC Connected Wind Power Plants*. CIGRE Technical Brochure No. 612. CIGRE, 2015, 100 pp.

[10] *European Commission's Official Webpage on Climate Policies Concerning Transportation* [Online]. Available from http://ec.europa.eu/clima/policies/transport/index_en.htm [Accessed 24 Aug 2016].

[11] Hable, M., Schwaegerl, C., Tao, L., Ettinger, A., Köberle, R., Meyer, E. P. 'Requirements on electrical power infrastructure by electric vehicles'.

Emobility – Electrical Power Train, 2010, Leipzig, 2010, pp. 1–6. doi:10.1109/EMOBILITY.2010.5668076.

[12] Drovtar, I., Rosin, A., Landsberg, M., Kilter, J. 'Large scale electric vehicle integration and its impact on the Estonian power system'. *PowerTech (POWERTECH), 2013 IEEE Grenoble*, Grenoble, 2013, pp. 1–6. doi:10.1109/PTC.2013.6652181.

[13] Turker, H., Bacha, S., Chatroux, D. 'Impact of plug-in hybrid electric vehicles (PHEVs) on the French electric grid'. *2010 IEEE PES Innovative Smart Grid Technologies Conference Europe (ISGT Europe)*, Gothenburg, 2010, pp. 1–8. doi:10.1109/ISGTEUROPE.2010.5638948.

[14] Pasaoglu, G., Fiorello, D., Martino, *et al. Driving and Parking Patterns of European Car Drivers – A Mobility Survey* [Online]. JRC Scientific and Policy Reports. 2012. Available from https://setis.ec.europa.eu/sites/default/files/reports/Driving_and_parking_patterns_of_European_car_drivers-a_mobility_survey.pdf [Accessed 24 Aug 2016].

[15] *Car Prototype Generates Electricity, and Cash* [Online]. Science Daily. 2007-12-09. Available from http://www.sciencedaily.com/releases/2007/12/071203133532.htm [Accessed 24 Aug 2016].

[16] Cleveland, Cutler, J., Morris, C. *Dictionary of Energy*. Amsterdam: Elsevier; 2006. p. 473. *ISBN 0-08-044578-0*.

[17] *Pacific Gas and Electric Company Energizes Silicon Valley With Vehicle-to-Grid Technology* [Online]. Pacific Gas & Electric. 2007-04-07. Retrieved 2009-10-02. Available from http://www.prnewswire.com/news-releases/pacific-gas-and-electric-company-energizes-silicon-valley-with-vehicle-to-grid-technology-57952787.html [Accessed 24 Aug 2016].

[18] Rodríguez-Molina, J., Martínez-Núñez, M., Martínez, J.-F., Pérez-Aguiar, W. 'Business models in the smart grid: Challenges, opportunities and proposals for prosumer profitability'. *Energies*, vol. 7, no. 9, 2014, pp. 6142–6171. doi:10.3390/en7096142.

[19] *Demand Side Response Policy Paper* [Online]. ENTSO-E, 15. September 2014. Available from https://www.entsoe.eu [Accessed 24 Aug 2016].

[20] *Estonian national Grid Code* [Online]. 2014. Available from (in Estonian) https://www.riigiteataja.ee/akt/12831412?leiaKehtiv [Accessed 24 Aug 2016].

[21] *How Can Flexible Capability Be Rewarded in the Electricity Markets of the Future? A Public Report* [Online]. Pöyry Management Consulting (UK) Ltd. February 2014. Available from http://www.poyry.com/sites/default/files/imce/files/revealing_the_value_of_flexibility_public_report_v1_0.pdf [Accessed 24 Aug 2016].

[22] Palensky, P., Dietrich, D. 'Demand side management: Demand response, intelligent energy systems, and smart loads'. *IEEE Transactions on Industrial Informatics*, vol. 7, no. 3, Aug. 2011, pp. 381–388.

[23] Paulus, M., Borggrefe, F. 'Economic potential of demand side management in an industrialized country – The case of Germany'. *10th IAEE European*

Conference, Vienna Austria; September 2009. Available from http://www.aaee.at/2009-IAEE/uploads/fullpaper_iaee09/P_488_Paulus_Moritz_31-Aug-2009,%2019:08.pdf [Accessed 24 Aug 2016].

[24] Malík, O., Havel, P. 'Analysing demand-side management potential: Situation in Europe and the Czech Republic'. *Environment and Electrical Engineering (EEEIC), 2011 10th International Conference on*, Rome, 2011, pp. 1–4. doi:10.1109/EEEIC.2011.5874781.

[25] Paulus, M., Borggrefe, F., 'The potential of demand-side management in energy-intensive industries for electricity markets in Germany'. *Applied Energy*, vol. 88, no. 2, 2011, pp. 432–441.

[26] Drovtar, I., Uuemaa, P., Rosin, A., Kilter, J., Valtin, J. 'Using demand side management in energy-intensive industries for providing balancing power – The Estonian case study'. 2013 *IEEE Power & Energy Society General Meeting*, Vancouver, BC, 2013, pp. 1–5. doi:10.1109/PESMG.2013.6672418.

[27] Ross, C. *Virtual Power Plants – Systems of the Future* [Online]. Published: July 2011. Available from http://www.ecmag.com/section/green-building/virtual-power-plants [Accessed 24 Aug 2016].

[28] Fang, X., Misra, S., Xue, G., Yang, D. 'Smart grid – The new and improved power grid: A survey'. *IEEE Communications Surveys & Tutorials*, vol. 14, no. 4, Fourth Quarter 2012, pp. 944–980. doi:10.1109/SURV.2011.101911.00087.

[29] Bayar, T. *Virtual Power Plants: A New Model for Renewables Integration* [Online]. September 30, 2013. Available from http://www.renewableenergyworld.com/articles/print/volume-16/issue-5/solar-energy/virtual-power-plants-a-new-model-for-renewables-integration.html [Accessed 24 Aug 2016].

[30] Powerstar Homepage: http://powerstar.com/virtual-power-stations/ [Accessed 24 Aug 2016].

[31] Olsina, F., Röscher, M., Larisson, C., Garcés, F. 'Short-term optimal wind power generation capacity in liberalized electricity markets'. *Energy Policy,* vol. 35, no. 2, February 2007, pp. 1257–1273, ISSN 0301-4215, http://dx.doi.org/10.1016/j.enpol.2006.03.018.

[32] MacCormack, J., Hollis, A., Zareipour, H., Rosehart, W. 'The large-scale integration of wind generation: Impacts on price, reliability and dispatchable conventional suppliers'. *Energy Policy*, vol. 38, no. 7, July 2010, pp. 3837–3846, ISSN 0301-4215, http://dx.doi.org/10.1016/j.enpol.2010.03.004.

[33] Lamont, A. D. 'Assessing the long-term system value of intermittent electric generation technologies'. *Energy Economics*, vol. 30, no. 3, May 2008, pp. 1208–1231, ISSN 0140-9883, http://dx.doi.org/10.1016/j.eneco.2007.02.007.

[34] *Scenario Outlook & Adequacy Forecast* 2015 [Online]. ENTSO-E, June 2015. Available from https://www.entsoe.eu/Documents/SDC%20documents/SOAF/150630_SOAF_2015_publication_wcover.pdf [Accessed 24 Aug 2016].

[35] Urdal, H., Martinez Villanueva, S., Kilter, J., Jahn, J., Sprooten, J., Baranauskas, A. 'Future system challenges in Europe. Contributions to solutions from connection network codes'. *2016 CIGRE USNC International*

Colloquium on Evolution of Power System Planning to Support Connection of Generation, Distributed Resources and Alternative Technologies. Philadelphia, USA, 02–03.11.2016, 1–8 pp.

[36] Urdal, H., Ierna, R., Zhu, J., Ivanov, C., Dahresobh, A., Rostom, D. 'System strength considerations in a converter dominated power system'. *IET Renewable Power Generation*, vol. 9, no. 1, 2015, pp. 10–17.

[37] Yu, M., Roscoe, A. J., Booth, C. D., *et al.* 'Use of an inertia-less virtual synchronous machine within future power networks with high penetration of converters'. *19th Power Systems Computation Conference (PSCC 2016)*; Genova, Italy, 20–24.06.2016. doi:10.1109/PSCC.2016.7540926.

[38] Urdal, H., Haesen, E., Minciuna, I. M., *et al.* 'National implementation challenges and support by ENTSO-E for European connection network codes'. *14th Wind Integration Workshop*; Brussels, Belgium, October 2015.

[39] Buchhagen, C., Rauscher, C., Menze, A., Jung, J. 'BorWin1 – First experiences with harmonic interactions in converter dominated grids'. *International ETG Congress 2015*, Bonn, Germany, 17–18.11.2015.

[40] Drovtar, I. Master Thesis: Large Scale Wind Integration Impact on Conventional Power Plant Economics on the Baltic Region's Electricity Market. *Tallinn University of Technology* 2011. 140 p.

[41] Ea Energy Analyses. Wind Power in Estonia. Final Report [Online]. 2010. 46 p. Available from http://elering.ee/public/Infokeskus/Uuringud/Tuuleenergia_Eestis.pdf [Accessed 24 Aug 2016].

[42] Kirschen, D., Strbac, G. *Fundamentals of Power System Economics*. John Wiley & Sons Ltd, Chichester, West Sussex ; UK 2004 (2010), 284 p.

[43] Wangensteen, I. *Power System Economics – The Nordic Electricity Market*. Tapir Academic Press. Trondheim 2007; 354 p.

[44] Drovtar, I., Landsberg, M., Kilter, J., Rosin, A. 'Impacts of large scale wind integration on the Baltic region's thermal power plant economics and electricity market in 2025'. *Power Electronics, Electrical Drives, Automation and Motion (SPEEDAM), 2012 International Symposium on*, Sorrento, 2012, pp. 684–689. doi:10.1109/SPEEDAM.2012.6264591.

[45] Melsas, R., Rosin, A., Drovtar, I. 'Wind park cost efficiency increase through direct cooperation with demand side response provider'. *2016 57th International Scientific Conference on Power and Electrical Engineering of Riga Technical University (RTUCON)*, Riga, 2016, pp. 1–5. DOI: 10.1109/RTUCON.2016.7763119

[46] Milligan, M., Donohoo, P., Lew, D., *et al.* 'Operating reserves and wind power integration: An international comparison'. *The Ninth Annual International Workshop on Large-Scale Integration of Wind Power into Power Systems as well as on Transmission Networks for Offshore Wind Power Plants Conference*; Québec, Canada; October 18–19, 2010.

[47] Raudsaar, U., Drovtar, I., Rosin, A. 'Overview – Pumped-hydro energy storage for balancing wind energy forecast errors'. *Electric Power Quality and Supply Reliability Conference (PQ)*; Rakvere, 11–13 June 2014, pp. 133–138. doi:10.1109/PQ.2014.6866797.

[48] Nord Pool A S. *Day-ahead marked Elspot* [Online]. 30 March 2016. Available from http://www.nordpoolspot.com/TAS/Day-ahead-market-Elspot/ [Accessed 24 Aug 2016].

[49] Ter-Gazarian, A. *Energy Storage for Power Systems*. P. Peregrinus, Wiltshire, 1994, pp. 19–31.

[50] Negnevitsky, M., Kurbatsky, V., Tomin, N., Panasetsky, D. '*Intelligent approach for preventing large-scale emergencies in electric power systems*'. IEEE PowerTech 2013; Grenoble, France, 16–20 June 2013. IEEE, 2013.

[51] International Energy Agency. *Hybrid & Electric Vehicle Implementing Agreement* [Online]. Available from http://www.ieahev.org/ [Accessed 24 Aug 2016].

[52] Lanati, F., Benini, M., Gelmini, A. 'Impact of the penetration of electric vehicles on the Italian power system: A 2030 scenario'. *Power and Energy Society General Meeting, 2011 IEEE*, pp. 1–8, 24–29 July 2011.

[53] Vee, M. *Master Thesis: EVs in Estonian Conditions (In Estonian: 'Elektriautod Eesti tingimustes')*. Department of Power Engineering, Tallinn University of Technology, Tallinn, 2012.

[54] Uuemaa, P., Kilter, J., Valtin, J., Drovtar, I., Rosin, A., Puusepp, A. 'Cost-effective optimization of load shifting in the industry by using intermediate storages'. *IEEE PES ISGT Europe 2013*, Lyngby, 2013, pp. 1–5. doi:10.1109/ISGTEurope.2013.6695404.

Chapter 5

Grid integration of large-scale PV plants: dealing with power fluctuations

Javier Marcos Álvarez[1], Íñigo de la Parra Laita[1], Luis Marroyo Palomo[1], Eduardo Lorenzo Pigueiras[2] and Miguel García Solano[1]

5.1 Introduction

The power grid comprises three distinct activities: generation, transmission and distribution. Given the fact that electricity cannot be stored, there needs to be a constant balance between demand and generation, making it necessary to forecast energy consumption and to maintain sufficient power generation reserves in order to be in a position to address possible contingencies or changes in either power generation or demand. With regard to the European electricity system, the organisation responsible for this activity is the ENTSO-E, the European Network of Transmission System Operators, represents 41 electricity transmission system operators (TSOs) from 34 countries across Europe [1]. The geographical area covered by ENTSO-E's member TSOs is divided into seven different synchronous areas (groups of countries that are connected via their respective power systems) and the system frequency, 50 Hz, is synchronous within each area. Its value is the direct consequence of the balance between the active powers generated and consumed. Frequency is an essential parameter in the quality of the power supply and must remain within a very tight range around the reference frequency, 50 $\pm$ 1 Hz. Outside this range, there is a risk that the system could collapse. If the power imbalance is significant, then the frequency moves out of the allowable range, which in turn leads to the automatic disconnection of other generating equipment, causing a chain reaction that could collapse the system. In order to avoid these situations and to guarantee stability, there are three different regulation systems available to the system operator: primary, secondary and tertiary regulation.

[1]Dpto. Ingeniería Eléctrica y Electrónica, Institute of Smart Cities, Universidad Pública de Navarra, Campus Arrosadia, 31006 Pamplona, Spain.
[2]Instituto de Energía Solar, ETSIS Telecomunicación, Polytechnic University of Madrid, 28031 Madrid, Spain

Primary regulation automatically acts through the control units of the various synchronous generators, compensating any balance deviations, as mentioned above, and which may occur throughout the synchronised area. This regulation restores, in just a few seconds, the balance between the power generated and the power consumed, although the frequency is stabilised at a value that may be different from its rated value of 50 Hz. Furthermore, if the primary reserve is insufficient to compensate the generation deficit in a regulation area, then the interconnected adjoining systems can supply the energy required, given the fact that all the regulation areas or blocks pertaining to the interconnected system jointly participate in the primary regulation process. However, the result of this action will be a deviation from the energy exchange values pre-agreed between the different areas. It is therefore necessary to have a secondary regulation which, on one hand, restores the energy balance and exchanges programme between each regulation area or block and, on the other, restores the system frequency to its rated value. At the same time, it ensures that the primary regulation reserve is once again fully available. All these actions are automatically and continuously performed by the secondary regulation, also in response to smaller deviations, which are inevitable in the normal operation of the system or in response to greater discrepancies between production and consumption. Finally, tertiary regulation controls the operating points of the generators involved in the secondary regulation. This is generally activated under the system operator's setpoint, directed at restoring the secondary regulation reserve used and distributing the energy between the various generators in the best possible way, based on cost considerations. This makes it possible to restore the reserve and to adjust the generation–demand balance for periods of 1 h or less.

Whilst the primary regulation responds almost instantaneously and reaches full availability in less than 30 s, the secondary regulation starts to act approximately after these 30 s and is not fully functional until almost 10–15 min later. Therefore, the 10-min time window represents a boundary to the system operator's reaction capacity, given the fact that in the interval from 30 s to 10 min, the operator has no room for manoeuvre whatsoever, as the primary regulation is at maximum capacity and the secondary regulation is not yet fully available. Therefore, any system imbalance of less than 10 min directly leads to a grid frequency fluctuation. Thus, any fluctuation in either the power generated or consumed, that is able to take the band frequency out of the 50 ± 1 Hz range in less than 10 min, would be a serious problem to system stability.

In this scenario, the intermittent nature of irradiance, primarily due to the variability of cloud cover, can cause significant fluctuations in the power delivered by large-scale grid-connected photovoltaic (PV) plants. As the PV generation rate increases in the electricity system as a whole, these fluctuations may have a negative impact on the quality and reliability of the electricity supply. As seen above, for time windows of less than 10 min, these fluctuations have a direct impact on the grid frequency, yet without the utility operator having sufficient response capacity to correct such deviations.

This chapter will study the dynamics of irradiance and the power delivered to the grid by PV plants in time frames of less than 10 min, paying particular attention to the influence of PV plant size and geographic dispersion on magnitude and duration. We are now going to present some general models which make it possible to simulate the power fluctuations of either a power plant or a group of power plants, based solely on irradiance measurements. Finally, we shall study the control strategies required to compensate these fluctuations, based on storage systems installed in the PV plants, analysing the energy and power requirements to compensate these variations.

5.2 The photovoltaic observatory

The study of the power fluctuations requires monitoring the power fed into the grid by the PV plants, in megawatts, with sufficient precision and resolution in order to correctly analyse the dynamics. The first database used in this chapter is from the Moura PV plant, located in Amareleja, southern Portugal (Figure 5.1). This plant covers a surface area of 250 Ha and has a total power output of 45.6 MWp, divided into 2,520 sun trackers with a rated output of 17.7–18.8 kWp. The solar trackers have a vertical axis, azimuth tracking system, tilted at an angle of 45° in relation to the ground. The corresponding inverter power P_n is 38.5 MW, and the ground cover ratio (GCR) is 0.162. The plant is divided into 70 units, each comprising 35 solar trackers connected to a DC/AC 550 kW inverter. The minimum and maximum distances between the units are 220 m and 2.5 km, respectively. Thanks to a 5-s interval monitoring, the power outputs of all 70 inverters could be recorded simultaneously. As a result of this work, it is possible to consider different sections of the plant, with power outputs ranging from 0.55 kW to 11.5 MW, as well as the plant as a whole.

The other experimental database used in this study is sourced from 6 PV plants located in Spain, specifically in Navarra, scattered in an area of 1,100 km^2, located at distances ranging from 6 to 60 km apart (Figure 5.2). The power outputs of the plants range from 1 to 9.5 MWp, making a total of 19 MWp (Table 5.1). In addition to this, data from two sections (correspondingly, 48 and 143 kWp) of Milagro PV plant have also been registered. Yet again, the plants feature vertical axis azimuth trackers, with the generator, tilted 45°. A seventh plant located at Socuéllamos (Castilla La Mancha, 345 km distance from the nearest PV plant of the previous experiment), 2.6 MWp (rated power of 1.975 MW) spread over 18 Ha is also monitored. The power delivered to the grid is recorded every second

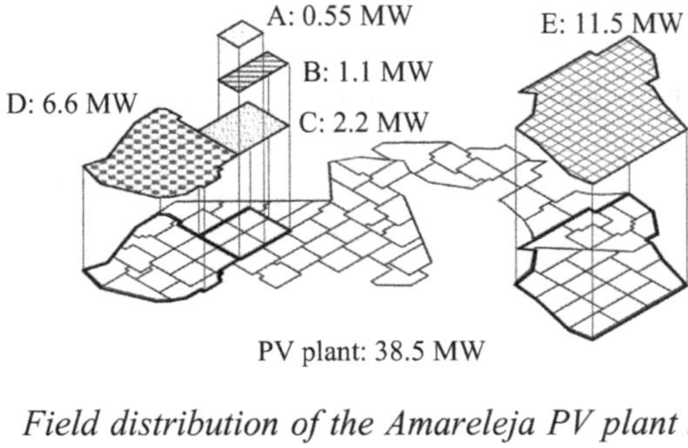

Figure 5.1 Field distribution of the Amareleja PV plant sections [2]

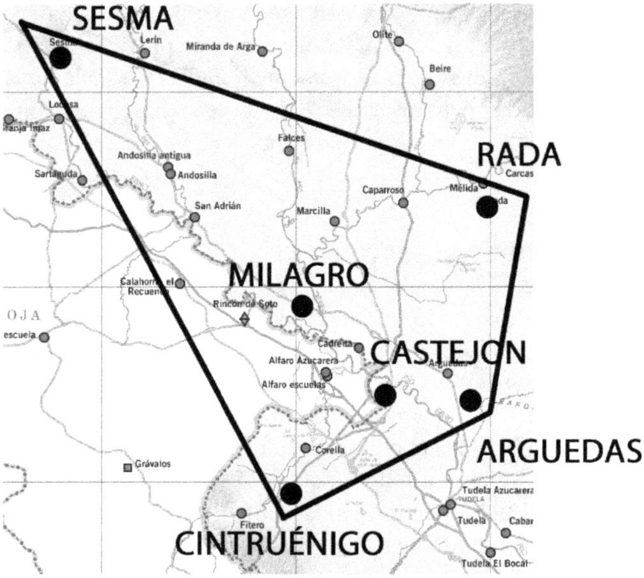

Figure 5.2 Geographic location of the PV plants under study [3]

Table 5.1 PV plants characteristics [3]

PV plant	Peak power (kWp)	Rated power, P_n (kW)	Area, S (Ha)
Arguedas	958	775	4.1
Sesma	990	800	4.2
Cintruénigo	1438	1155	6.4
Rada	1780	1400	8.7
Castejón	2640	2000	11.8
Milagro	9500	7243	52
Milagro section 1	48	35	0.21
Milagro section 2	143	100	0.63
Socuéllamos	2600	1975	18
Total	**20097**	**15348**	–

through a grid analyser and programmable logic controller (PLC). A global positioning system (GPS) guarantees synchronised data recording every second.

Thanks to the fact that our study includes plants with different power outputs and a synchronised data recording system, it has been possible to analyse the influence of the plant size and geographic dispersion on the magnitude and duration of the fluctuations.

5.3 Irradiance and power output fluctuations in large PV plants

5.3.1 At a PV plant level

The magnitude of a power fluctuation, $\Delta P_{\Delta t}(t)$ at instant t, is defined here as the difference between two power measurements separated by time interval Δt and normalised in relation to rated power P_n (Table 5.1) of the plant studied, that is shown in the following equation:

$$\Delta P_{\Delta t}(t) = \frac{[P(t+\Delta t) - P(t)]}{P_n} \times 100\ [\%] \tag{5.1}$$

This definition is also applicable to irradiance fluctuations $\Delta G_{\Delta t}(t)$, yet normalised through $G^* = 1{,}000$ W/m^2. It should be pointed out that Δt must be a multiple of the signal sampling time (1 s in our case), and that, by moving a time window with amplitude Δt over the time series, this definition makes it possible to obtain a fluctuation value for each instant of the original time series. As shown in (5.1), fluctuation is a relative value, in percentage terms in our case, somehow indicating the power output variation in a given time interval in relation to the total installed power, comprising relevant data for the operator.

Detailed below are a series of power measurements recorded at a PV plant, such as those shown in the example in Figure 5.3(a) [both irradiance and power have been normalised, $g(t)$ and $p(t)$]. To calculate the evolution of the power fluctuations associated with this series of measurements, it is sufficient to gradually move a window of amplitude Δt across the entire series and calculate the corresponding fluctuations. It should be pointed out that this window is moved across all the measurements taken. That is, second by second, the width of window Δt can be any multiple of the measurement resolution, and can also be 1 s. Therefore, Figure 5.3(b) and (c) shows the evolution of the fluctuations for irradiance and the power generated, shown in Figure 5.3, for $\Delta t = 60$ s and 600 s. In both cases, it can clearly be seen how the power fluctuations are smoothed in relation to the irradiance fluctuations. Likewise, and as was to be expected, the greater the time interval Δt, the greater the fluctuations.

5.3.1.1 Irradiance fluctuations

Figure 5.4 shows the distributions of the irradiance fluctuations observed at the Cintruénigo site in the course of 1 year and for Δt observation intervals of 1, 30, 60 and 600 s, specifically as a percentage in accordance with magnitude. These fluctuations were calculated on the daily irradiance measurements, from dawn to dusk, and the distributions are normalised so that the total area enclosed by each curve is equal to 100. This characteristic is not in line with the visual impression of the figure, given the fact that the relative frequency axis is on a logarithmic scale. Likewise, the distributions can be seen to be clearly symmetrical, which was to be expected, given the fact that cloud passage causes a decrease in irradiance, when

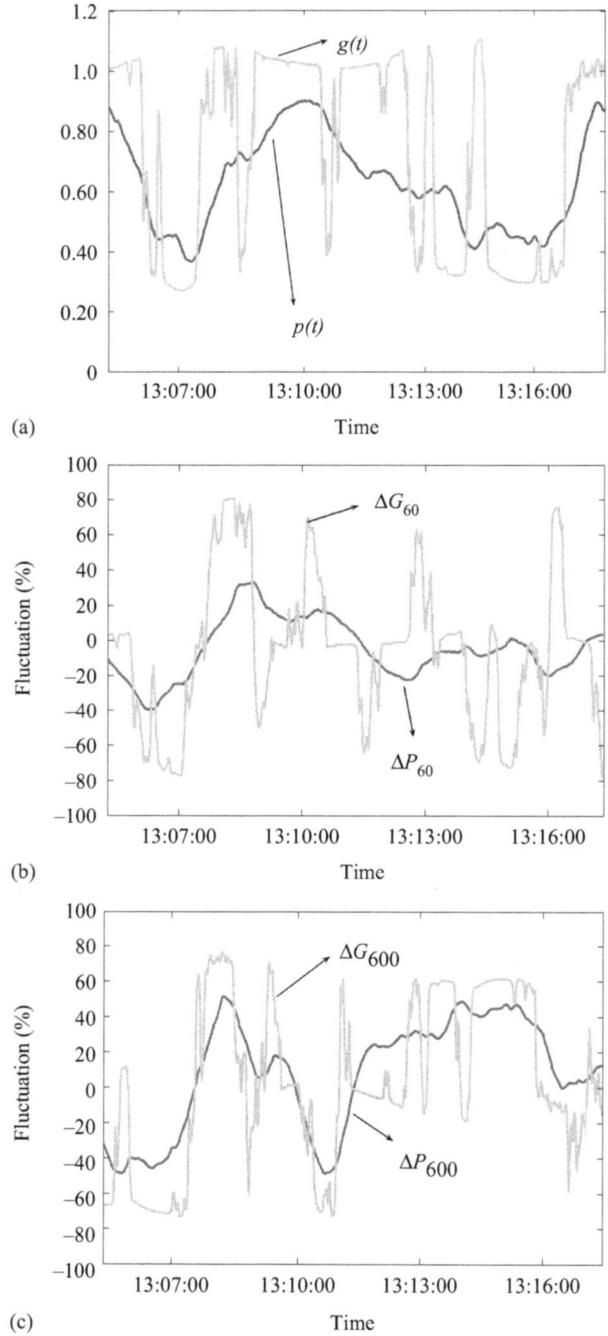

Figure 5.3 *(a) Evolution of the normalised irradiance* g(t) *and power output* $p(t)$ *recorded at the Milagro plant (9.5 MWp) over a 15-min period on 12 August 2008. The lower part of the figure shows the evolution of the irradiance and power fluctuations corresponding to the (a) signals for* Δt *observation intervals (b) 60 s and (c) 600 s. There is an evident smoothing of the power fluctuations in relation to the irradiance fluctuations [3]*

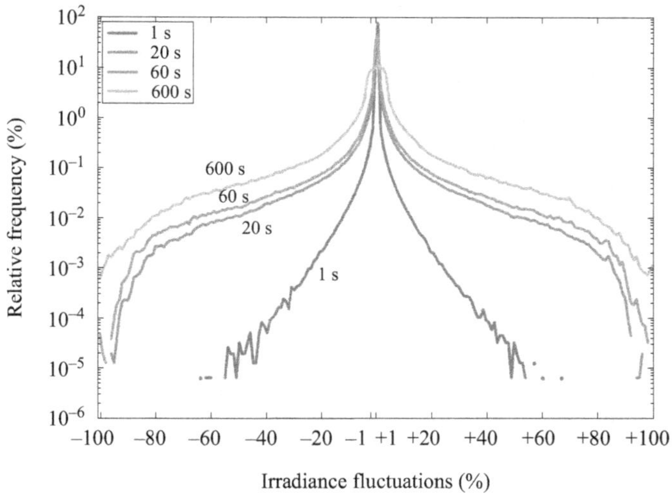

Figure 5.4 Distribution of the irradiance fluctuations recorded at the Cintruénigo site over an entire year for $\Delta t = 1$, 30, 60 and 600 s. Note that the Y-axis is logarithmically scaled [3]

Table 5.2 Influence of observation interval Δt on the frequency of occurrence of irradiance fluctuations recorded in the course of 1 year at the Cintruénigo site. Note that the occurrence of significant fluctuations is a relatively rare phenomenon, although it does increase with Δt [3]

$\Delta G_{\Delta t}$ (%)	Δt (s)			
	1 s	**30 s**	**60 s**	**600 s**
$0\% \le \Delta G \le 3\%$	98.9	92.5	87.5	59.3
$3\% \le \Delta G \le 10\%$	0.9	4.5	7.4	28.2
$10\% \le \Delta G \le 50\%$	0.1	2.7	4.4	11.2
$50\% \le \Delta G \le 100\%$	2×10^{-4}	0.3	0.7	1.3
$0\% \le \Delta G \le 100\%$	100	100	100	100

the shade of the cloud enters the study area, and then an increase once the shade has gone. In addition, as could already be observed in Figure 5.3, there is a positive correlation between the magnitude of the fluctuation and observation window Δt. Table 5.2 shows the values for the relative frequency of occurrence for given fluctuation intervals. It should be noted that the occurrence of significant fluctuations of, say, more than 3% in 1 s is a relatively rare phenomenon and only occurs for 1% of the time. In fact, this is why the frequency axis is shown in logarithmic scale in Figure 5.4. However, for 600 s, the relative frequency of fluctuations of more than 3% is close to 41%.

This study was extended to cover the remaining sites located in southern Navarra and similar results were obtained regardless of the location considered.

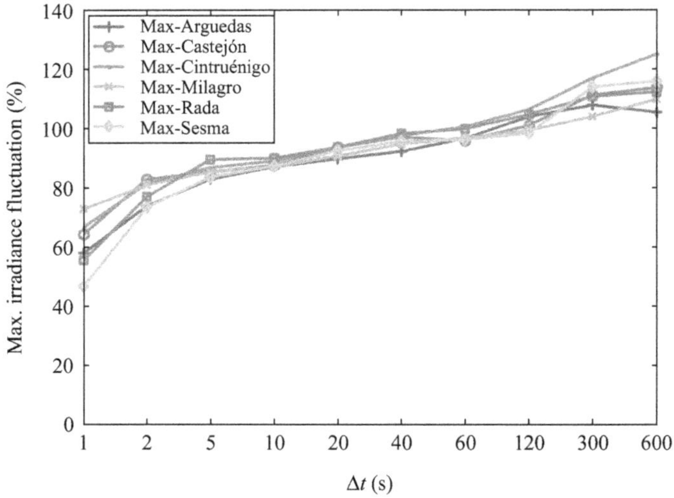

Figure 5.5 Maximum irradiance fluctuations recorded, according to site, since the start of the monitoring, based on observation interval Δt

It can therefore be said that, for the conditions of this experiment, the occurrence of irradiance fluctuations is independent of the site considered. Consequently, as the irradiance fluctuations are the primary cause of power fluctuations, it is possible to extend this result and affirm that the power fluctuations will not be affected by the geographical location of the PV plant in southern Navarra.

Although the distributions above have already provided a glimpse of the maximum value that can be reached by the fluctuations, this is now going to be discussed in greater detail. Figure 5.5 shows the maximum irradiance fluctuations based on observation interval Δt, during 1 year and at all the sites considered. All the values are positive, given the fact that the maximum value was calculated on the absolute values of the fluctuations since; in principle, the grid fluctuation compensation capacity is symmetric. It is clear that the irradiance dynamic is very rapid: the maximum fluctuations reach 50%–75% in 1 s and 100% in 60 s. Fluctuations of more than 100% are due to the reflection of the white cloud edges, providing irradiance values of more than G^*. Yet again, there are no relevant differences between the various sites, supporting the idea that, for the conditions of the PV observatory located in Navarra, the performance of the irradiance fluctuations is independent of geographic location.

5.3.1.2 Power fluctuations

The section above shows how the intermittency of the cloud cover causes rapid irradiance fluctuations which, moreover, increases in magnitude and frequency of occurrence as the observation time increases (or Δt). However, irradiance is measured using a calibrated cell which can be considered to be practically located at a specific point, in comparison with the ground cover of a PV plant. Therefore, a priori, it could be thought that the power fluctuations will be less than the irradiance

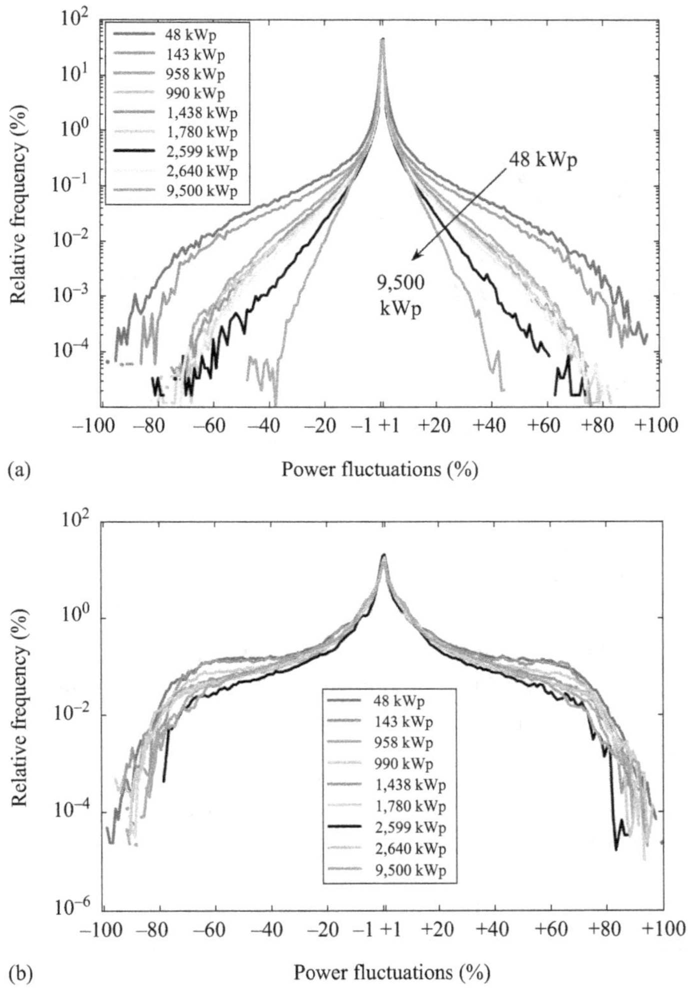

Figure 5.6 Distribution of the power fluctuations recorded in the course of 1 year at each of the PV plants for (a) $\Delta t = 1$ s, (b) $\Delta t = 20$ s and (c) $\Delta t = 600$ s. Those plants with the greatest power output show a lower occurrence of fluctuations for small Δt [3]

fluctuations and, moreover, they will be inversely proportional to the size of the plant. This, for a start, is intuitive, given the fact that the larger the surface area, the longer it takes to be completely shaded. On the other hand, for the same plant surface area, it seems logical to think that the greater the value of Δt, the greater the occurrence of fluctuations. These hypotheses are studied below, using the same tools discussed in the section above for irradiance fluctuations.

Figure 5.6 shows the annual distributions of the power fluctuations recorded at each PV plant for (b) $\Delta t = 20$ s and (c) $\Delta t = 600$ s during 1 year. As was to be expected, there is a smoothing effect on the occurrence of fluctuations and

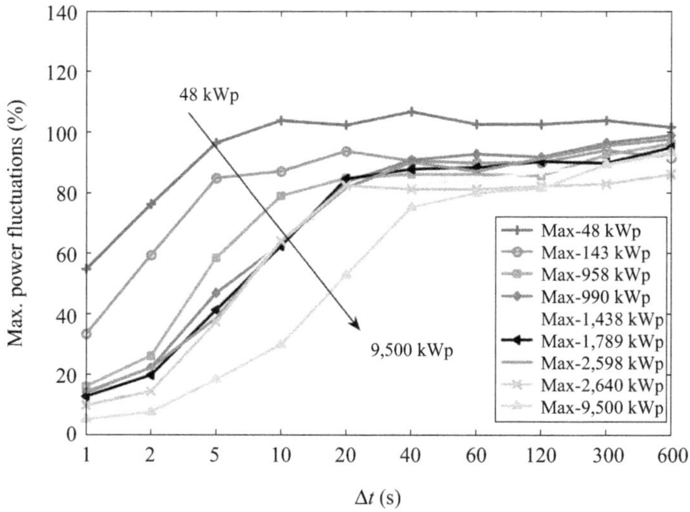

Figure 5.7 Maximum power fluctuations recorded, according to site, in relation to time interval Δt [3]

their magnitude. This becomes more evident as the size of the PV plant increases. However, it should be noted that this fluctuation smoothing gradually decreases as the time interval for observing the fluctuations increases. In fact, for $\Delta t = 600$ s, there are hardly any differences between the fluctuations observed at the various plants.

Figure 5.7 shows the maximum power fluctuation recorded, as an absolute value, at each of the plants compared to observation interval Δt. Yet again, the increased smoothing effect is noticeable as the plant size increases. For $\Delta t = 1$ s, the differences are more than significant: the 143 kWp plant has fluctuations of more than 30%, compared to 10% for a 990 kWp plant and the 5% of the 9,500 kWp at Milagro. However, these differences are clearly reduced as the Δt increases. For $\Delta t = 40$ s, the 990 kWp plant has the same fluctuations as the 143 kWp plant, yet, from 300 s onwards, the maximum fluctuations are very similar for all plants. In fact, the fluctuation values reached at these high Δt are close to 80%–90%, which practically means moving from a completely clear sky (diffuse and direct radiation) to completely cloudy (only diffuse radiation). In other words, for these Δt, the shade moves quickly enough to completely cover a plant occupying a surface area of 52 Ha, as is the case of the 9,500 kWp Milagro plant.

It is therefore clear that there is a significant relationship between the magnitude of the fluctuations and the size of the PV plant. This invites us to attempt to propose a mathematical description in order to scale up the results of this specific experiment to a more general situation. Here we have addressed this task on the 99th percentile of $\Delta P_{\Delta t}$, 99th($\Delta P_{\Delta t}$), or a value that leaves 99% of the total fluctuations under it. From the operator's point of view, the relevant variable is the maximum fluctuation. However, the values measured could respond to very specific situations which, in turn, could negatively affect the validity of this analysis.

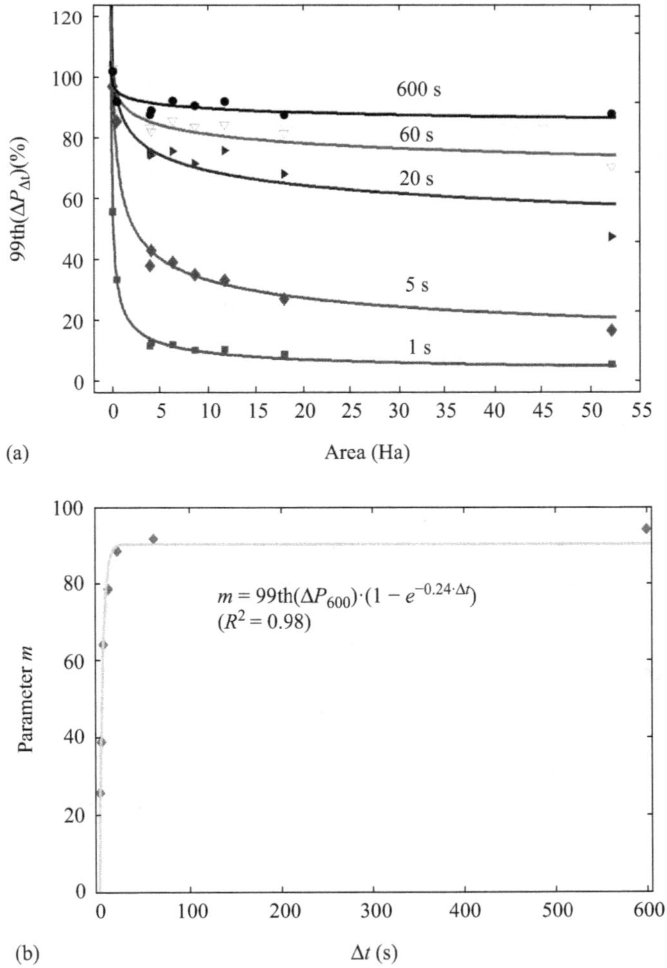

Figure 5.8 (a) 99th ($\Delta P_{\Delta t}$) based on the area of each plant, with Δt as a parameter. (b) Estimated relationship ($R^2 = 0.98$) between parameter m and observation interval Δt [3]

Figure 5.8(a) shows the value of the 99th($\Delta P_{\Delta t}$) based on the area of each plant, with Δt as a parameter.

These observations correctly fit an exponential function of the form, (5.2):

$$99\text{th}(\Delta P_{\Delta t}) = m \cdot S^{-c}, \qquad m, c > 0 \tag{5.2}$$

with S in hectares and where m and c are two parameters dependent on Δt. It is summarised by the values of these two parameters, the result of a least square adjustment, together with R^2 of the regression. It is interesting to point out that for $\Delta t = 1$ s, c is very close to 0.5. This is equivalent to saying that, for small

Table 5.3 Values of parameters m and c, estimated by regression, for the expression proposed in (5.2). Likewise, the goodness of fit is shown, represented by value R^2 [3]

Δt (s)	m	c	R^2
1	25.55	0.49	0.99
0.5	64.09	0.29	0.96
20	88.2	0.11	0.89
60	91.6	0.05	0.86
600	94.07	0.02	0.61

monitoring times, the smoothing effect of the area is form $1/\sqrt{S}$. This relationship is easily explained, considering the fact that shade moves in a single dimension whilst the power output is related to ground cover, in other words, to two dimensions. However, for higher Δt (up to 600 s in our case), the smoothing effect is practically negligible ($c = 0.02$), which implies that the power fluctuations are not affected by the surface area and, for this reason, regression coefficient R^2 decreases for large Δt. These results are consistent with the intuitive fact that 600 s is sufficient time for moving shade to completely cover the PV plants studied in this paper. This is represented by the m values of Table 5.3, based on Δt. It could be interpreted that as the dependence in relation to the surface area is concentrated in c, then m describes the variability of the irradiance, specifically at the 99th($\Delta P_{\Delta t}$), based on the observation interval. Moreover, assuming that the value of m for $\Delta t = 600$ s (94.07%) it is precisely the value of the 99th percentile at this same interval 99th (ΔP_{600}), through a regression analysis ($R^2 = 0.98$), we obtain an expression that provides the value of m for any Δt, Figure 5.9(b)

$$m = 99th(\Delta P_{600}) \cdot (1 - e^{-0.24 \cdot \Delta t}), \tag{5.3}$$

Substituting (5.3) in (5.2), we obtain (5.4)

$$99\text{th}(\Delta P_{\Delta t}) = 99th(\Delta P_{600}) \cdot (1 - e^{-0.24 \cdot \Delta t}) \cdot S^{-c} \tag{5.4}$$

This empirical expression provides the 99th percentile value for the power fluctuations for any surface area and observation interval. Noting that, since the start of the monitoring period, the maximum fluctuation has never been greater than the 99th($\Delta P_{\Delta t}$) in more than 9%, then (5.4) makes it possible to estimate, with a certain safety margin, the maximum fluctuation that could occur in periods of less than 10 min.

5.3.2 Power fluctuations at a PV plant group level

The study of the fluctuations of a group makes it necessary to clarify what is understood to be a fluctuation in the power delivered to the grid by a group of PV plants. In line with the previous chapter, the magnitude of the power fluctuation of

N plants at instant t, $\Delta P_{\Delta t,N}(t)$, is understood to be the difference between two samples of normalised power delivered by group p_N, separated by sampling time Δt, which is shown in the following equation:

$$\Delta P_{\Delta t,N}(t) = [p_N(t + \Delta t) - p_N(t)] \times 100 \quad [\%] \tag{5.5}$$

Strictly speaking, $p_N(t)$ is the result of first adding the power delivered by each ith plant and then dividing by the total of the rated power of the PV plant group, which is shown in the following equation:

$$p_N(t) = \frac{\sum_{i=1}^{N} P^i(t)}{\sum_{i=1}^{N} P_n^i} \quad [\%] \tag{5.6}$$

where P_n^i is the ith plant rated power. However, account should be taken of the fact that, when the power output of one of the plants is considerably higher than that of the others (as is the case of the Milagro plant), then the performance of the group is strongly affected by this and conceals the influence of geographic dispersion. This drawback can partially be resolved by using a different expression to calculate the normalised power, as shown in the following equation:

$$p_N(t) = \frac{1}{N}\sum_{i=1}^{N} \frac{P^i(t)}{P_n^i} \quad [\%] \tag{5.7}$$

This equation first proposes normalising the power outputs measured at each plant, then subsequently adding them together and, finally, dividing them by the number of plants in the group. In this way, the fluctuations for all the plants are equally weighted in the analysis. Admittedly, it is possible to argue that this method does not completely eliminate all the influence of the plant size, given the fact that the corresponding smoothing is intrinsic to the same and impossible to eliminate. This infers the desirability of conducting the study with irradiance measurements. However, as will be seen later on in this chapter, the smoothing of the power fluctuations due to geographic dispersion is far more significant than that associated with the size of the plant, thereby minimising the potential drawback of dealing with power measurements. It is our understanding that, the derived benefit is far higher due to its greater representativeness.

By way of example, Figure 5.9(a) shows the normalised power output p_1 for single PV plant in Cintruénigo ($P_n^1 = 1.155$ MW), and that corresponding to the group of six plants p_6 located in Navarra, ($P_n^6 = 13.373$ MW), on the 2 February 2009, from 12.30 to 13.00 h. The smoothing effect of the group can easily be seen. Figure 5.9(b) and (c) shows the fluctuations corresponding to these measurements for $\Delta t = 20$ s and 600 s, respectively, and which highlight this smoothing effect still further, if at all possible.

Figure 5.10 shows the normalised distributions for the fluctuations recorded in 1 year for $\Delta t = 60$ s and 600 s corresponding to a single plant (Arguedas, $P_n^1 = 0.755$ MW), to a group of three plants (Castejón, Milagro and Sesma, $P_n^3 = 10.043$ MW) and to another group of six plants (all located in the south of

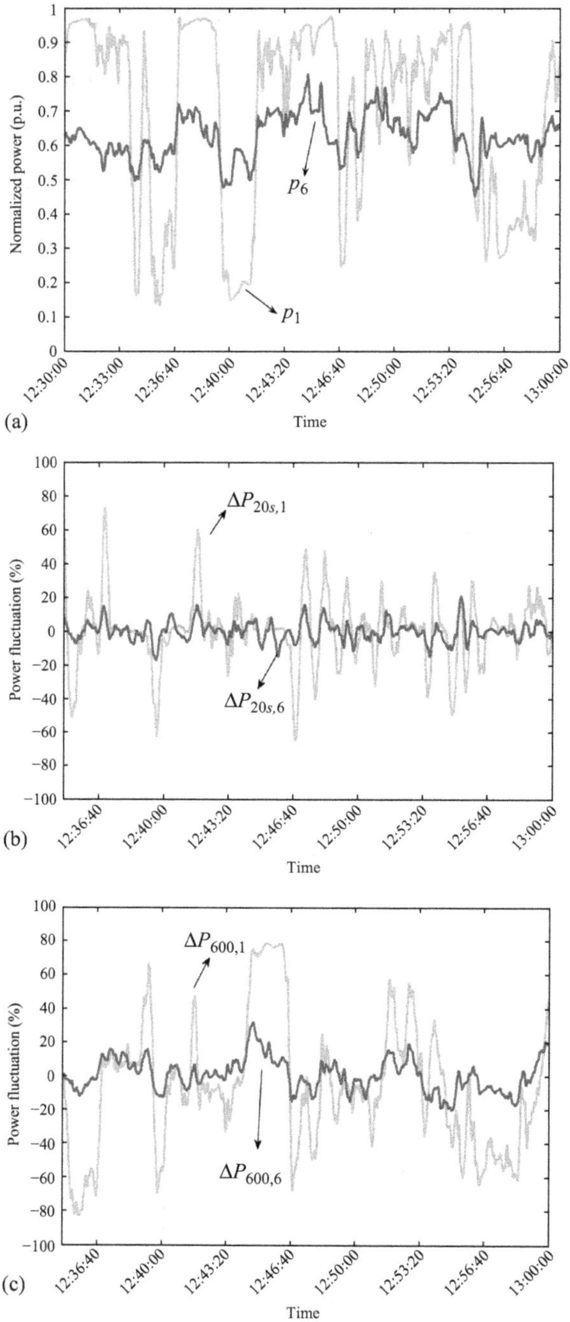

Figure 5.9 *(a) Normalised power output p_1 at the PV plant at Cintruénigo ($P_n^1 = 1.155$ MW) and that corresponding to the group of 6 plants located in Navarra, p_6, ($P_n^6 = 13.373$ MW). The lower part of the figure shows the performance of the power fluctuations corresponding to the (a) signals for Δt observation intervals (b) 20 s and (c) 600 s. The smoothing effect is shown in both cases [4]*

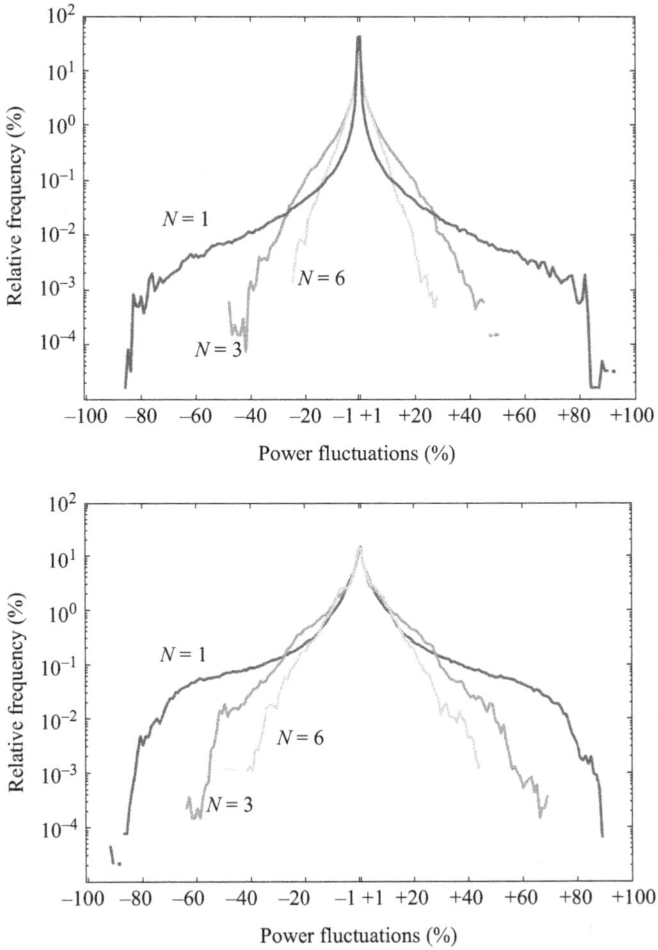

Figure 5.10 Distribution of the power fluctuations $\Delta P_{\Delta t,N}$ recorded in the course of 1 year at the PV plant of Arguedas, for the group of (a) three plants (Castejón, Milagro and Sesma) and (b) six plants in Southern Navarra, for Δt is equal to 60 s (a) and 600 s (b) [4]

Navarra $P_n^6 = 13.373$ MW) It can be clearly seen how the magnitude and frequency of the fluctuations decrease in line with the number of plants grouped together N, and it is important to stress that the smoothing effect also applies to the long observation intervals ($\Delta t = 600$ s) contrary to what has been observed for area smoothing.

Figure 5.11(a) shows the maximum power fluctuations observed for all potential PV plant combinations in Southern Navarra (from N is equal from 1 to 6, where N is equal to 1 corresponds to the PV plant at Arguedas, $P_n^1 = 0.755$ MW) The smoothing effect due to geographic dispersion is evident: for Δt is equal to 1 s,

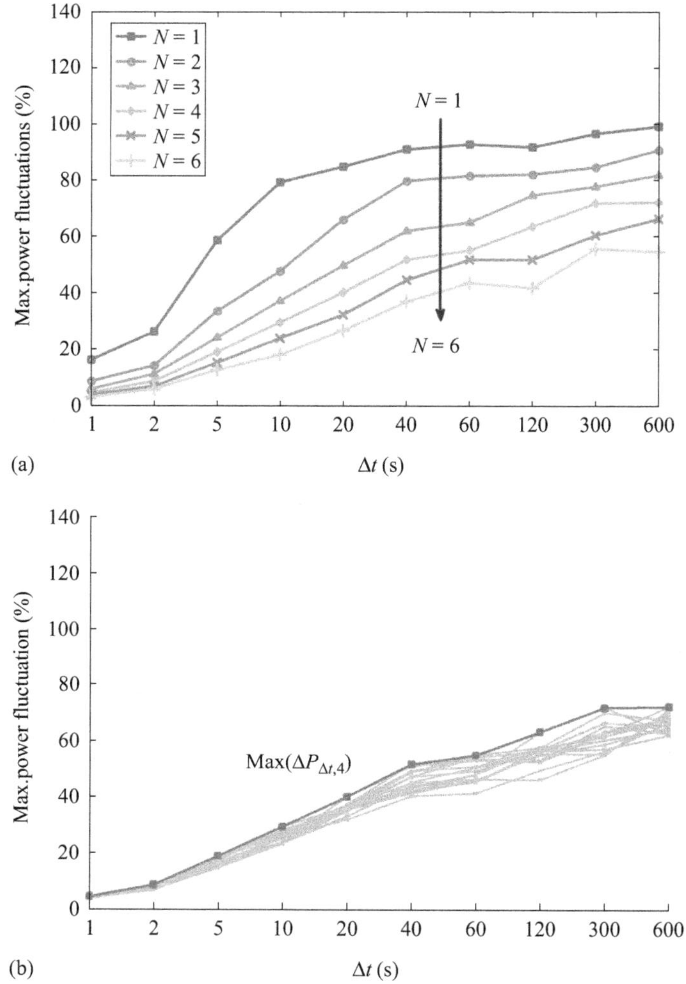

Figure 5.11 (a) Maximum power fluctuations Max($\Delta P_{\Delta t,N}$) recorded throughout 1 year for all possible combinations of $N = 1 \ldots 6$ PV plants (located in Navarra). (b) Maximum power fluctuations recorded in 1 year for all the 15 possible combinations of $N = 4$ [4]

the maximum fluctuation is reduced from 16.1% to 3% and for Δt is equal to 600 s, it is reduced from 99.2% to 54.4%. On the other hand, Figure 5.11(b) represents the maximum fluctuation recorded at each of the 15 potential combinations of four PV plants. As can be seen, no combination offers a greater smoothing effect than another. And although here, we have only shown the case for N is equal to four, this is repeated for any N under consideration.

It is therefore possible to draw the same conclusions as those discussed in the section above: the smoothing effect due to geographic dispersion continues to take

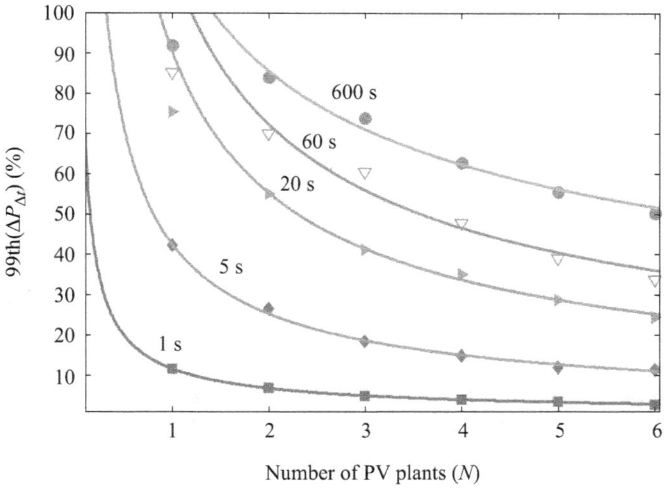

Figure 5.12 99th Percentile of the fluctuation observed for all the potential N plant combinations, 99th($\Delta P_{\Delta t,N}$), for Δt is equal to 1, 5, 20, 60 and 600 s [4]

place even for observation time intervals of 10 min and, for a combination of N number of plants, the location of each PV plant in relation to the others within the group has no effect on the magnitude of the smoothing effect. Therefore, everything appears to indicate that the smoothing effect phenomenon solely depends on the number of aggregated sites. However, it should be emphasised that this idea presupposes the condition that the power fluctuations are independent of each other. It is easy to understand that there is an inverse relationship between the distance between two plants and the correlation of their fluctuations, in other words, the greater the distance the lower the correlation. At one extreme, if two plants are adjoining, to such an extent that, viewed from the air, they could be considered to be a single plant, then the fluctuations will be at the same rate, and there will be no significant smoothing effect as far as geographic dispersion is concerned. Evidently, there will still be a smoothing effect related to the PV plant footprint which, as seen above, for large observation intervals, is of minor importance. Therefore, everything appears to indicate that the distance of 6 km, being the shortest distance between any plant and our observatories (located at Arguedas and Castejón) is sufficient to guarantee the non-correlation of plant power fluctuations for time intervals of less than 10 min or, to put it another way, to guarantee smoothing by geographic dispersion.

Figure 5.12 shows the 99th percentile for the fluctuations for all possible combinations of up to 6 PV plants FV, 99th($\Delta P_{\Delta t,N}$) with Δt as a parameter. The form of the results suggests the approximation by a geometric function of the form, (5.8):

$$99\text{th}\left(\Delta P_{\Delta t,N}\right) = 99\text{th}\left(\Delta P_{\Delta t,1}\right) \cdot N^{-a}, \quad a > 0 \tag{5.8}$$

Table 5.4 Estimated 'a' parameter and goodness of fit coefficient for empirical equation (5.8) [4]

Δt (s)	***a***	***R*²**
1	0.77	0.99
5	0.75	0.99
20	0.71	0.99
60	0.63	0.98
600	0.46	0.98

Table 5.5 Estimated 'a' and 'c' parameter for empirical equation (5.9) [4]

Δt (s)	***c***	***A***
1	0.49	0.77
5	0.29	0.75
20	0.11	0.71
60	0.05	0.63
600	0.02	0.46

Table 5.4 shows the values of parameter a and coefficient R^2 resulting from the regression.

Equation (5.8) can be combined with (5.4) to estimate the 99th percentile for the fluctuation of a group of N plants 99th($\Delta P_{\Delta t,N}$) with footprint S and for an observation interval Δt of less than 10 min, (5.9):

$$99\text{th}(\Delta P_{\Delta t,N}) = 99\text{th}(\Delta P_{600,1}) \cdot (1 - e^{-0.24 \cdot \Delta t}) \cdot S^{-c} \cdot N^{-a} \quad a, c > 0 \tag{5.9}$$

This equation describes the smoothing of the fluctuations resulting from the plant footprint and also geographic dispersion. Whenever coefficients a are greater than c (Table 5.5), the smoothing due to geographic dispersion is greater than that due to the plant footprint. By making the appropriate transformations, these results are consistent with some theoretical studies on fluctuation variance, as seen in [5]. For Δt of a few seconds and plants that are very close together (a situation termed *Crowded Region* in [5]), the footprint smoothing is proportional to $\sqrt{S}$; ($c = 0.49$ in our case) whilst the smoothing due to geographic dispersion is proportional to $1/N$ ($a = 0.77$ in our case). For 10-min time intervals and PV systems at a sufficient distance apart, a condition termed *Spacious Region* in [5], the smoothing effect achieved is form $\sqrt{N}$ ($a = 0.46$ in our case).

In an endeavour to show how N and S affect power fluctuations, an exercise is detailed below to compare the variability of distributed versus centralised PV generation. Given a scenario in which the aim is to install 100 MW of PV power,

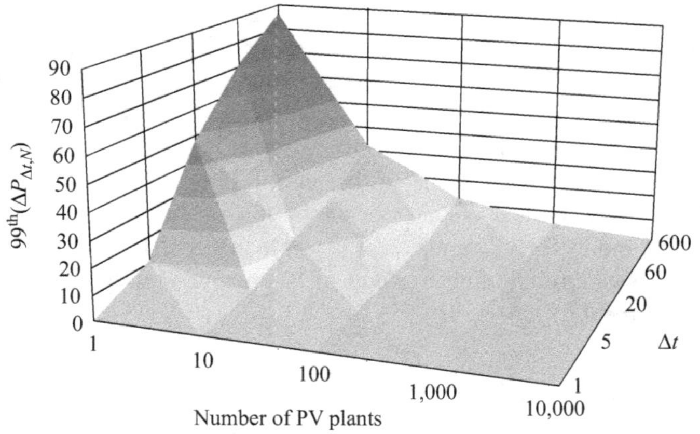

Figure 5.13 99th Percentile of the fluctuations to be expected from a 100 MW PV generation, based on the number of independent plants in which it is distributed, N, and observation interval Δt [4]

the grid operator must decide on the degree of geographic dispersion of the PV generation (size of each plant P_n and number of plants N), so that 100 MW $= N{\cdot}P_n$. Assuming that all the plants have the same construction parameters, then Figure 5.13 shows the fluctuations corresponding to a power output of 100 MW distributed between N similar power plants and with a distance apart of at least 6 km, after applying (5.9). It is possible to see how the fluctuations of 100 1-MW plants are approximately 10% of the fluctuation of a single 1 MW PV plant. In the case of 50 2-MW plants, then the fluctuation are 20% at 600 s. Typically, conventional electric generators have ramp rates of 25%–40% at 600 s [6]; therefore, the need for greater dispersion is questionable.

5.4 Simulating power fluctuations at PV plants

Throughout this chapter, it has been possible to see how, in practical terms, there are two natural mechanisms that smooth out the power fluctuations. First, there is a smoothing effect due to size: the fluctuations from incident irradiance on a PV plant are smoothed out in proportion to $\sqrt{S}$, where S is the square root of the surface area. Therefore, the greater the size of the plant, the smaller the fluctuations for the same time window. The second mechanism is geographic dispersion for the same installed inverter power P_n. The fluctuations are smaller if this power is divided between N different power plants, which are at a sufficient distance apart in order to ensure that, in small periods of time, the fluctuations are not correlated. Specifically, in time windows of just a few minutes, the fluctuations are smoothed out in proportion to the square root of the number of combined plants $\sqrt{N}$.

Using both mechanisms, we are now going to propose a simple model to simulate the power variability generated by a group of geographically dispersed PV plants, solely based on the irradiance measurements of a single site. Specifically, the model comprises two different transfer functions: the first being a low pass filter to convert the measured irradiance time series into power generated by a PV plant, based solely on a knowledge of the plant footprint. And, second, a transfer function that solely applies dispersion smoothing to the number of plants considered, N. The model therefore makes it possible to simulate power fluctuations in any electricity system including a group of PV plants, simply based on the irradiance measurements.

5.4.1 PV plant model

The periodic nature of the solar resource opens the door to use the Fourier series to analyse the irradiance and power signals. The following figure shows the Discrete Fourier Transform (DFT) applied to the normalised irradiance signal $g(t)$ recorded at the tracker plane over the entire year, at the site of Milagro (Figure 5.14). The transform was calculated using a Fast Fourier Transform algorithm. In general, for a time series of measurements for a particular variable, $x = x1, \ldots, xN$, the value of each point in the frequency domain, X_k, is calculated using the following equation:

$$X_k = \sum_{j=1}^{N} x_j \ \omega_N^{(j-1)(k-1)} \tag{5.10}$$

where N is the length of the signal and ω_N is calculated as follows:

$$\omega_N = e^{(-2\pi \ i)/N} \tag{5.11}$$

Following the recommendations proposed by other authors [7], and in order to reduce the variance in coefficients X_k, the original signal has been divided into a number of segments (32). Therefore, coefficients X_k are calculated for each segment and are then averaged for each frequency. Despite the fact that the Nyquist theorem states that the DFT transform can be calculated up to a maximum frequency that is double that of the sampling frequency (0.5 Hz in our case), here it is restricted to 0.05 Hz, which is sufficient for our objectives. It is possible to fit the linear region of the spectrum with sufficient accuracy to a function of the type $f^{-0.7}$, a trend which is consistent with the observations made by other authors [7]. It is also possible to observe the typical variability of the solar resource, with a 24-h harmonic (1.15×10^{-5} Hz).

This same analysis has been applied to other pre-normalised annual power output series recorded at a number of PV plants. Figure 5.14(a) also shows the power output spectrum calculated for the Milagro (9.5 MWp) and Sesma (0.99 MWp) plants, and the irradiance spectrum shown above. In this figure, it is easy to identify two different linear areas that correctly fit the functions of form $f^{-0.7}$ and $f^{-1.7}$, where the cut-off frequency of both defines cut-off frequency f_c.

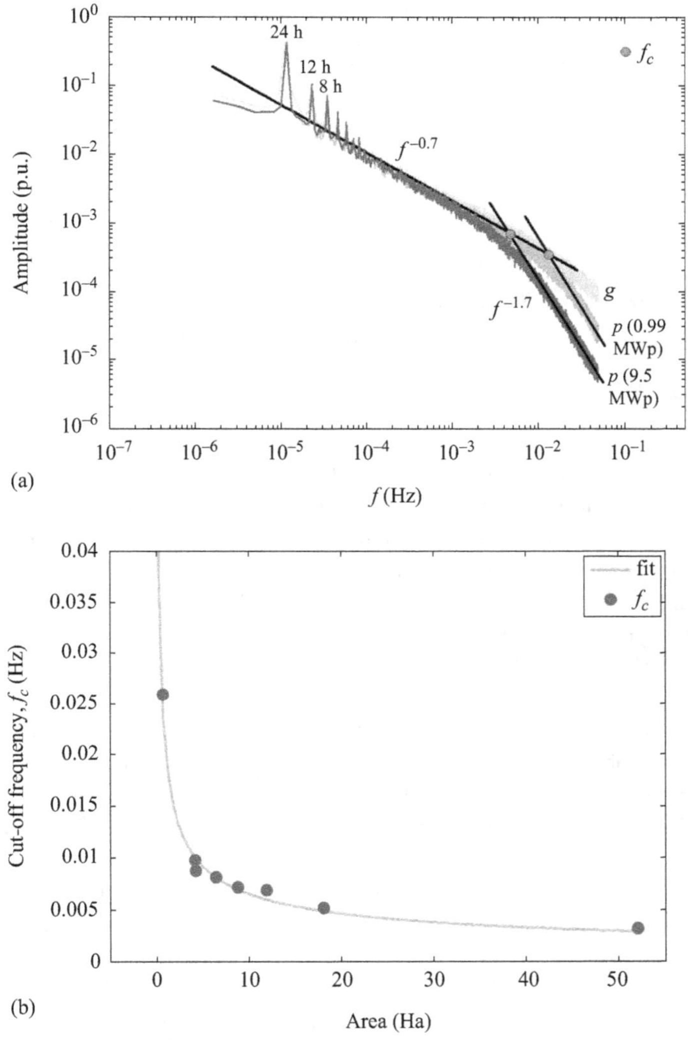

Figure 5.14 (a) Spectrum of the irradiance g(t) recorded at Milagro site, outpower at Sesma p(t) (0.99 MWp; 0.8 MW) and Milagro (9.5 MWp; 7.243 MW) during 1 year. The linear region for the larger frequencies of the power spectrums can be well fitted by a function of the form $f^{-1.7}$. (b) Cut-off frequency, f_c, as a function of the PV plant area, S. The tendency can be properly fitted ($R^2 = 0.99$) by the function $f_c = a \cdot S^b$, where $a = 0.020$ and $b = -0.499$ [8]

In other words, from the point of view of the fluctuations, a PV plant can be seen as a first-order low-pass filter for the irradiance signal. By conducting this same exercise for all the plants located in Navarra, with an extremely wide range of power outputs and footprints, different cut-off frequencies have been obtained for

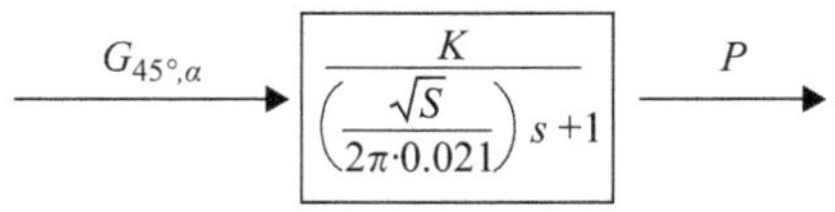

Figure 5.15 Proposed transfer function for a PV plant with an area equal to S. The value of the pole t, has been replaced by the expression $\tau = (2\pi f_c)^{-1}$ [8]

all these plants. Figure 5.14(b) shows these cut-off frequencies in relation to the plant footprint. This curve can be well fitted ($R^2 = 0.99$) by a function of the type

$$f_c = a \cdot S^b \tag{5.12}$$

where $a = 0.02$ and $b = -0.5$, with the cut-off frequency in Hertz and S in hectares. In other words, it can once again be seen that the smoothing effect of the fluctuations is inversely proportional to $\sqrt{S}$. This result is readily understandable, taking into account the fact that shade typically moves in a single direction whilst the power output is related to the surface area (two dimensions).

The validity of this equation was questioned using another PV plant located in a different region, Socuéllamos (Castilla La Mancha, at a distance of 345 km from the closest PV plant to the other PV plants in the experiment) and occupying a surface area of 18 Ha. The cut-off frequency, following the Fourier analysis of the 1-year data, is $f_c = 0.0051$ Hz, practically similar to the frequency provided by (5.3). It should be emphasised that the Socuéllamos PV plant has a GCR of 0.15, which is considerably smaller than that of the other plants (0.18). Looking at the f_c values, it can be said that the real factor responsible for smoothing the fluctuation is the surface area, rather than the installed power.

With all this data, we are now in a position to simulate the power delivered to the grid by a PV plant based on irradiance measurements, using the transfer function proposed in Figure 5.15. It should be noted that parameter K corresponds to ratio P_n/G^* [m^2] for the PV plant in question, and the pole of function τ has been replaced with expression $\tau = (2\pi f_c)^{-1}$. By way of example, Figure 5.16 shows the result of this simulation for 2 h at the Milagro PV plant (52 Ha). The similarity between both signals can be clearly observed.

5.4.2 Model of a group of PV plants

The same analysis in the frequency domain was applied to the normalised power signals of PV plant combinations, recorded throughout 1 year. Figure 5.17(a) shows the power output spectrum for the Milagro p_1 PV plant, the Milagro and Castejón p_2 group and the Arguedas, Castejon, Milagro, Cintruenigo and Rada, p_5 group. The normalised power at moment t of a group of power plants N was calculated according to (5.7). The spectra are very similar for very low frequencies, yet again coinciding with the daily variability cycle of the solar resource. However, from approximately frequencies of around 4×10^{-4} Hz, it is possible to see how the

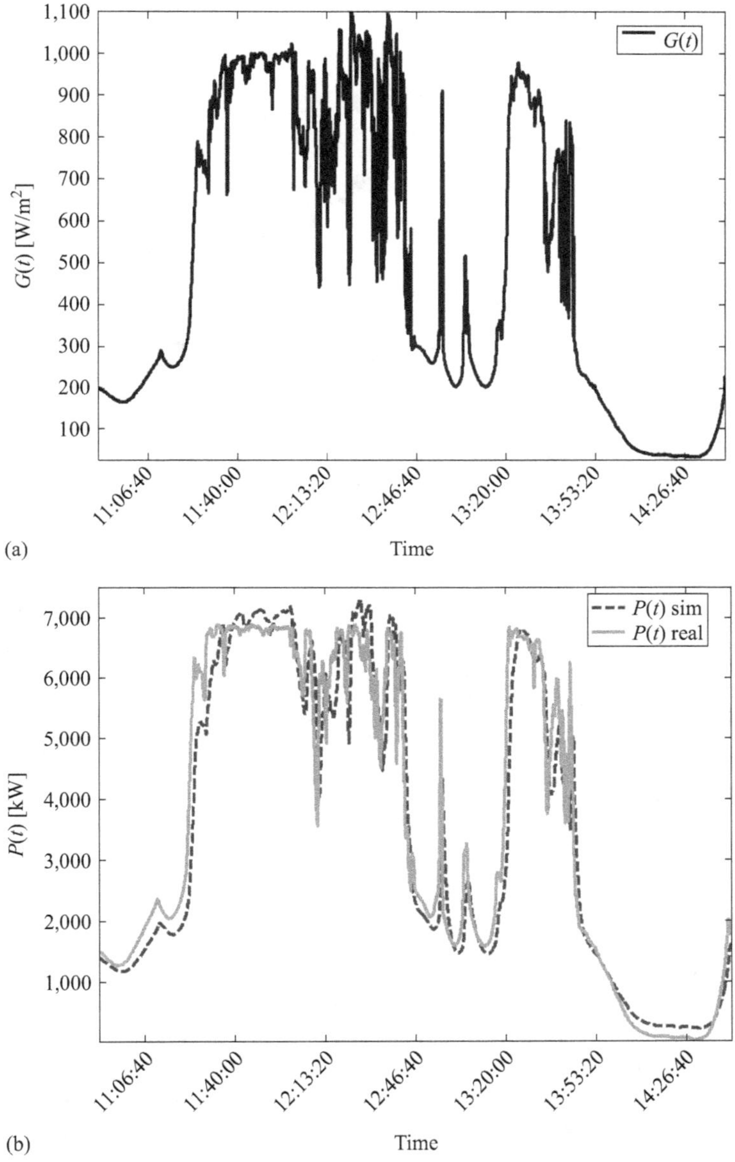

Figure 5.16 Irradiance G (45°,a) recorded from 12.30 to 15.00 h on 25 October at Milagro (52 Ha), (b) output real power, P_{real} (continue line) recorded during the same period at Milagro and output power simulated, P_{sim} (dashed line) by the transfer function showed in Figure 5.6 [8]

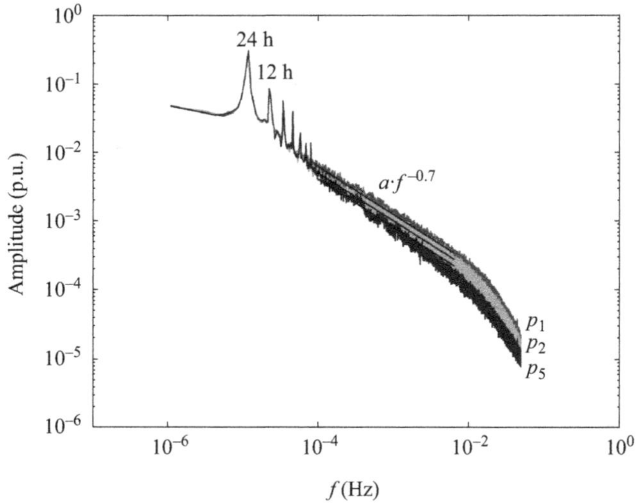

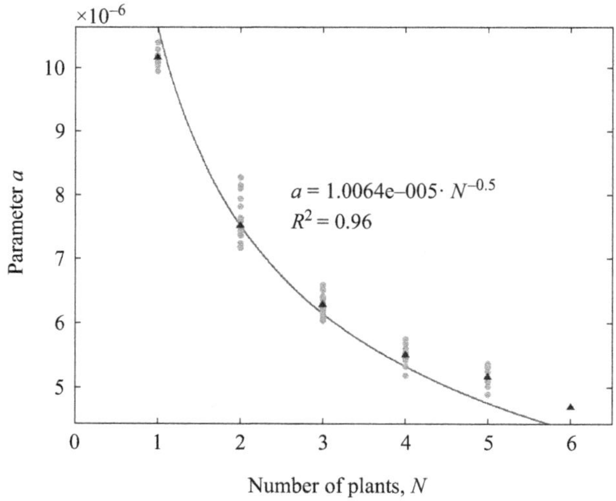

Figure 5.17 (a) Spectrum of the normalised power signals, p_N, recorded in the course of a full year for N = 1, 2 and 5 (p_1, p_2 and p_5, respectively). (b) Parameter a as a function of N. The trend can be properly fitted ($R^2 = 0.97$) by function $a = m \cdot N^b$, where $m = 1.0E^{-5}$ and $b = -0.41$ [9]

signals are smoothed out and, to a large extent, in line with the number of plants taken into account. This exercise has been extended to all the potential combinations of N power plants in Navarra, $N = 1.6$. Figure 5.17(b) shows value a in relation to N. The curve fits very well ($R^2 = 0.96$) to an expression of style a:

$$a = m \cdot N^{-0.5} \tag{5.13}$$

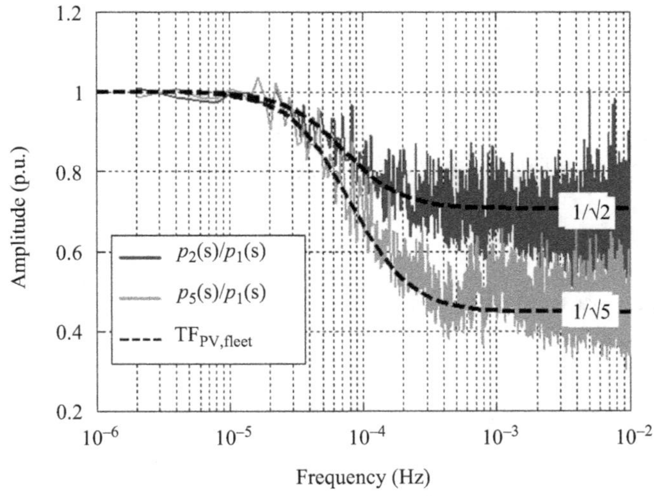

Figure 5.18 Spectra resulting from ratio p_2/p_1 (dark grey) and p_5/p_1 (light grey). The superimposed dashed line corresponds to the Bode diagram of the model proposed in (5.16). It can be seen how the high frequencies tend to smooth out due to $\sqrt{N}$ [9]

where $m = 1.007 \times 10^{-5}$. It can yet again be seen that the smoothing of the various spectra is proportional to $\sqrt{N}$, being consistent with the fluctuations analysis considered above. The reason for the fit not being quite so good is due to the presence of different sized PV plants and, therefore, the high frequencies are smoothed differently.

Given a normalised power signal from a single power plant, p_1, the aim is to obtain the power signal that would correspond to a group of N power plants, p_N. In this way, the transfer function sought $TF_{PV,fleet}$, is as follows:

$$TF_{PV,fleet} = \frac{p_N}{p_1} \tag{5.14}$$

For example, for the case of $N = 2$ and $N = 5$, the experimental transfer function $TF_{PV,fleet}$ is obtained from the quotient between p_2/p_1 and p_5/p_1, respectively, as shown in Figure 5.18. Taking into account the fact that the high frequency smoothing is proportional to $\sqrt{N}$, then the Laplace domain transfer function must be present in the following equation:

$$TF_{PV,fleet} = \frac{p_N(s)}{p_1(s)} = \frac{\left(K/\sqrt{N}\right)s + 1}{Ks + 1} \tag{5.15}$$

In our case, a good fit is obtained with $K = 2{,}400$, (5.16). The Bode diagram of the proposed function is shown also in Figure 5.18, showing the good fit in relation to the experimental transfer functions.

$$TF_{PV,fleet} = \frac{p_N(s)}{p_1(s)} = \frac{\left(2{,}400/\sqrt{N}\right)s + 1}{2{,}400\ s + 1} \tag{5.16}$$

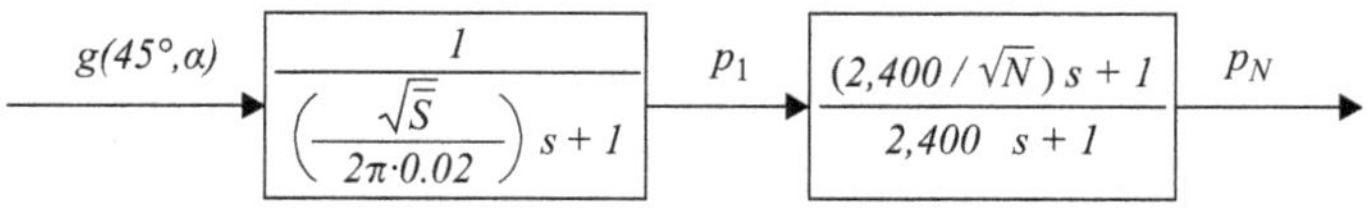

Figure 5.19 Proposed transfer function for a PV fleet, consisting of N PV plants with a mean size equal to $\overline{S}$ [9]

Then, given a irradiance measurement time series, the power that would be produced by a group of PV plants can now be obtained by simulation (e.g., with software such as MATLAB® Simulink™), by combining the first model corresponding to a PV plant with that of a group of plants, that is Figure 5.19. We would emphasise the simplicity of the model, given the fact that the only input parameters required are the number of plants in group N and the footprint of the same, S. The actual value of S in a scenario in which there are a number of PV plants with different footprints, is debatable. However, as shown above in this same chapter, the smoothing effect due to dispersion is much more effective than the smoothing effect due to plant footprint. For this reason, some satisfactory results can be obtained if S is considered to be the mean value of all the PV plants, $\overline{S}$ (15.0 Ha in our case). Figure 5.20(b) shows the evolution of the power generated on a specific day for the group of 6 PV plants, $p_{6,\text{real}}(t)$ compared to the results obtained by simulation $p_{6,\text{sim}}(t)$, applying the model proposed in Figure 5.19 to the irradiance measured at Milagro, Figure 5.20(a), with $\overline{S} = 15$ Ha y $N = 6$. The profiles of $p_{6,\text{real}}(t)$ and $p_{6,\text{sim}}(t)$ are very similar except for the beginning and end of the day.

To quantify the goodness of fit of the model, a year of radiation data $g(t)$ were simulated, measured every second, and the daily aggregate ramp rate (DARR) was calculated. The power profile variability is important because it is directly related to the spinning reserve levels required to compensate fluctuations. Some authors have proposed the (DARR) [10] to quantify this variability, a parameter which also makes it possible to categorise the days according to their level of fluctuation, based on mean minute-to-minute variability, in other words, (5.17):

$$\text{DARR} = \sum_{t=1}^{1440} \frac{|x(t) - x(t-1)|}{X_n} \tag{5.17}$$

where $x(t)$ is a time series of measurements averaged every minute and X_n the nominal value of x. Therefore, if $x(t)$ is equal to $p_{N,\text{sim}}(t)$ we obtain $\text{DARR}_{N,\text{sim}}$ and if $x(t)$ is equal to $p_{N,\text{real}}(t)$, we obtain $\text{DARR}_{N,\text{real}}$. And, as both signals are normalised, X_n is equal to 1. Figure 5.21 shows the 365 values of $\text{DARR}_{6,\text{real}}$ compared to simulated values $\text{DARR}_{6,\text{sim}}$. The regression coefficients obtained confirm the goodness of fit of the model (RMSE $< 2\%$).

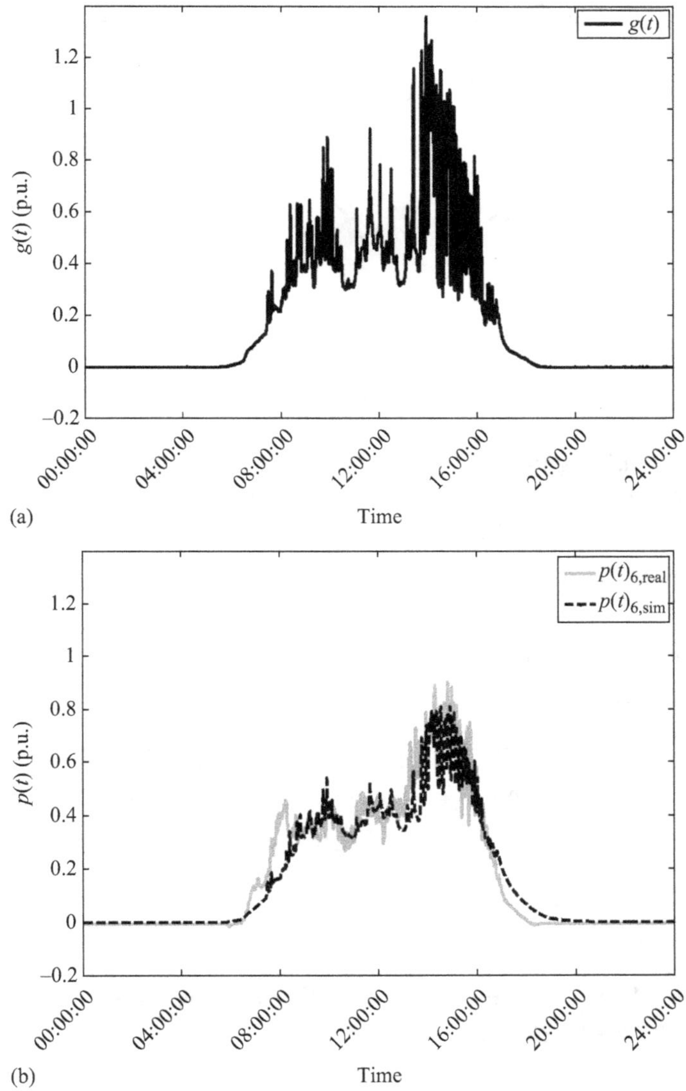

Figure 5.20 *(a) Irradiance G (45°,α) recorded at Milagro (52 Ha), (b) output real power, P_{real}, (continue line) recorded during the same period at Milagro and output power simulated, P_{sim} (dashed line) by the transfer function showed in Figure 5.19 [9]*

This regression analysis was repeated for different combinations of plant groups $N = 2, \ldots, 5$ (Table 5.6). The small improvement obtained in the regression residuals suggests that the model precision improves in line with an increase in the number of plants considered. This model has been validated for different climate zones, with satisfactory results, as discussed more extensively in [9].

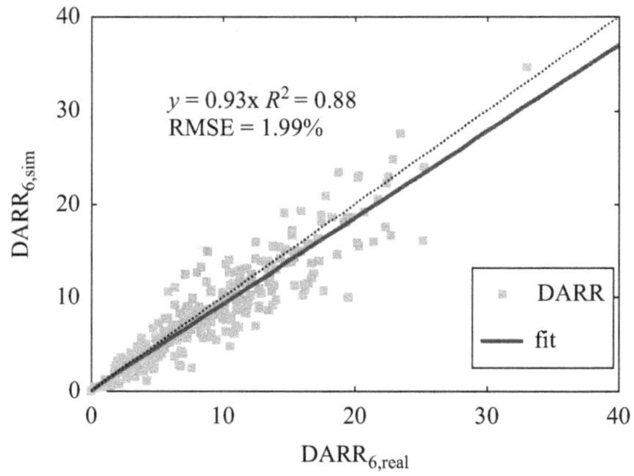

Figure 5.21 Daily aggregate ramp rate simulated values for the fleet of six plants $DARR_{6,sim}$ compared to the real values, $DARR_{6,real}$ for 1 year data (365 values) [9]

Table 5.6 RMSE error evolution as a function of N [9]

N	RMSE Max($\Delta P_{1\min,N}$)	RMSE $DARR_N$
2	7.1	2.75
3	7.0	2.36
4	5.9	2.35
5	5.2	2.15
6	4.6	1.99

5.5 Smoothing power output fluctuations by using energy storage systems

As has been seen, the magnitude of fluctuations in the power delivered to the grid can negatively affect the quality and reliability of the electricity supply. This variability is particularly significant in relatively small grids such as islands with high levels of PV penetration, given the fact that smoothing by geographic dispersion may be intrinsically limited. It was precisely an island grid operator, The Puerto Rico Electric Power Authority, that recently opened the door to the regulation of the power delivered by PV plants, by imposing a maximum fluctuation of 10%/min in relation to the plant rated capacity [11]. And, as has been seen up to now, a power plant has considerably higher fluctuations, of up to 70%–90% per minute.

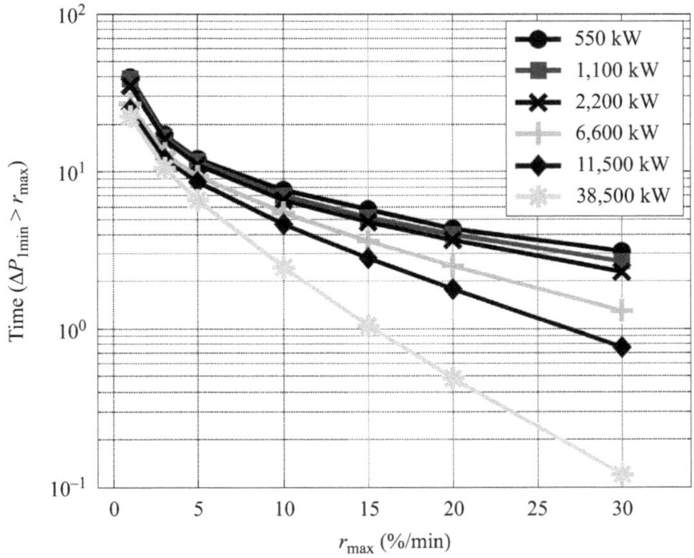

Figure 5.22 Frequency over 1 year of PV power fluctuations calculated in 1-min time window, $\Delta P_{1min}(t)$, is superior to a given ramp $r_{\max}$ (%/min). The frequency value is given in relative terms to the total production time (4,380 h) [2]

Specifically, it is possible to compare a time series of fluctuations $\Delta P_{\Delta t}(t)$ with a maximum ramp rate per minute limitation, $r_{\max}$, and count the number of times that the fluctuations exceed a given ramp rate ($abs\ [\Delta P_{\Delta t}(t)] > r_{\max}$). This analysis was performed for different sections of the PV plant at Amareleja and for different values of r in the course of a year (Figure 5.22). This plant was used in order to make the frequency of occurrence of fluctuations completely independent of geographic location, and thereby solely observe dependency of fluctuations on plant size. As was to be expected, the fluctuation frequency of occurrence decreases with $r_{\max}$ and with P_n. For $r_{\max} = 1$%/min and $P_n = 550$ kW, the power fluctuations exceed the ramp rate for 40% of the time. For this same ramp rate, if we increase the plant size to $P_n = 38.5$ MW, then the time that the maximum ramp rate is exceeded is reduced by up to 23% whilst, for a much less demanding ramp, $r_{\max} =$ 30%/min, these values drop to up to 3% and 0.1%, respectively. These examples show that the enforcement of maximum ramp rates on the power generated (typically around 10%/min) makes it necessary to install some type of energy storage system (ESS), even for large plants.

The cost per power unit is an important function of their capacity, and too high a cost is prohibitive for industrial and commercial acceptance. It is therefore essential to establish a method to optimise the size and operation of such ESS in

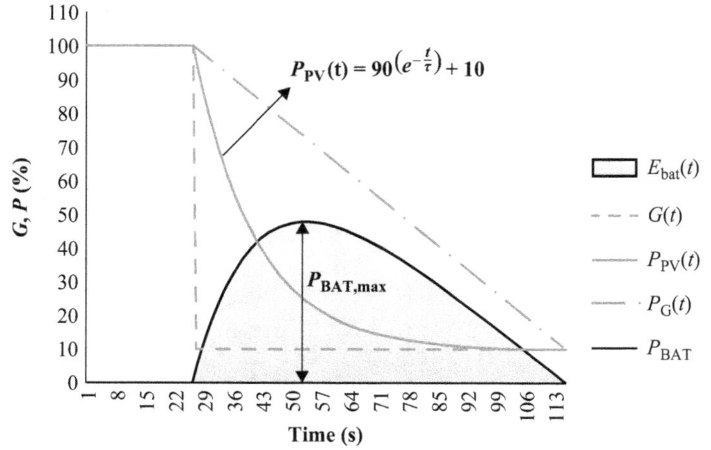

Figure 5.23 Worst fluctuation model. The solid grey line represents the $P_{PV}(t)$ response to an irradiance fluctuation (dashed line) and the dashed and dotted grey line is the power injected to the grid P_G with a ramp-rate control. The difference between P_G and P_{PV} is P_{BAT} (solid black line), the maximum difference corresponds to $P_{BAT,max}$ and the defined integral of P_{BAT} corresponds to $E_{BAT,max}$ [2]

order to meet the application constraints. Detailed below is a method to calculate the minimum energy C_{BAT} and power required $P_{BAT,max}$, for any plant size and maximum allowable ramp rate. This method, called the *worst fluctuation model*, is based on the worst fluctuation that can take place at a PV plant and is a function of the shortest measurement for the PV plant perimeter.

5.5.1 The worst fluctuation model

The study of the successive fluctuations occurring at a PV plant makes it possible to calculate the worst fluctuation that could occur as an exponential drop (Figure 5.23) from the plant rated capacity P_n up to 0.1 P_n (or an exponential rise from 0.1 P_n up to P_n) which means that the beam irradiance disappears and only the diffuse light remains. The time constant τ (s) is empirically correlated to the shortest PV plant perimeter dimension, l (m), according to an expression of the type:

$$\tau = a \cdot l + b \tag{5.18}$$

with $a = 0.042$ [s/m] y $b = -0.5$ s.

Therefore, the energy requirements $E_{BAT,max}$ for different maximum ramp rate values are determined by the area between P_G (dashed and dotted grey line) and P_{PV} (solid grey line) and can be described as follows

$$E_{BAT,max} \approx \frac{0.9P_n}{3{,}600}\left[\frac{90}{2 \cdot r_{max}} - \tau\right] \tag{5.19}$$

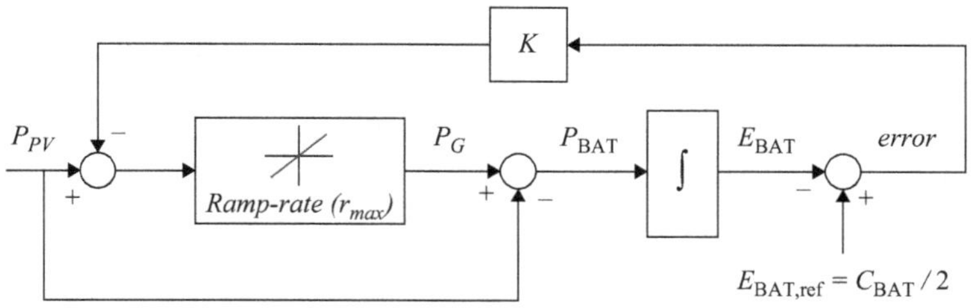

Figure 5.24 Ramp-rate control. Notice that the SOC control action is also smoothed by the ramp-limiter in order to guarantee that power fluctuations are always below r_{max} [2]

where P_n is expressed in kW, $r_{\max}$ in %/s, τ in s and $E_{\text{BAT,max}}$ in kWh. Moreover, the maximum power $P_{\text{BAT,max}}$ is determined by

$$P_{\text{BAT,max}}(t) = \frac{P_n}{100}\left[90 - \tau \cdot r_{\max}\left(1 + \ln\frac{90}{\tau \cdot r_{\max}}\right)\right] \tag{5.20}$$

Once the energy and power requirements to offset the worst PV plant fluctuation have been determined, it is then necessary to operate with the right strategy.

5.5.2 Conventional ramp-rate control

As an initial approach, the implementation of a ramp-rate strategy to smooth out fluctuations considering a maximum allowable ramp rate value of the power injected into the grid, $r_{\max}$ [%/min], is shown in Figure 5.24. The power generated, $P_{\text{PV}}(t)$, passes through a ramp rate limiter that is only activated if $P_{\text{PV}}(t)$ dynamics exceed the ramp-rate limitation, that is shown in the following equation:

$$\frac{\left|\Delta P_{\text{PV,1 min}}(t)\right|}{P_n} \cdot 100 > r_{\max} \tag{5.21}$$

where $r_{\max}$ is expressed in %/min. The ramp-rate limiter output is the power that needs to be delivered to the grid, $P_G(t)$. This power is similar, except that $P_G(t)$ has the ramp rates limited. The power through battery, $P_{\text{BAT}}(t)$ is the difference between $P_G(t)$ and $P_{\text{PV}}(t)$ and the energy through battery $E_{\text{BAT}}(t)$ is obtained simply by integrating $P_{\text{BAT}}(t)$. Moreover, a reference value for the energy stored in battery ($E_{\text{BAT,ref}}$) needs to be established in order to absorb the positive and negative fluctuations alike. It should be noted that, as the sign of the first fluctuation is unknown, a battery that is double the capacity of the one given in (5.19) is necessary in order to absorb the upward and downward fluctuations alike ($C_{\text{BAT}} = 2\,E_{\text{BAT,max}}$). For this reason, we have implemented a state of charge (SOC) control to ensure that SOC = 50%: continually, $E_{\text{BAT}}(t)$ is compared to $E_{\text{BAT,ref}} = C_{\text{BAT}}/2$ (SOC = 50%), and the error is multiplied by constant k to obtain the necessary power for the SOC control. This power is added/subtracted from $P_G(t)$. It should be

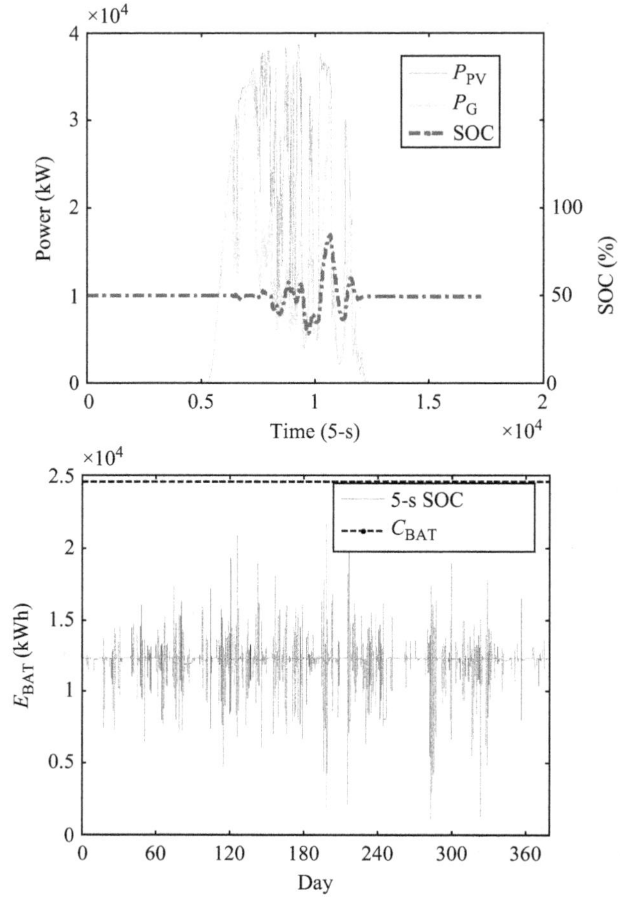

Figure 5.25 (a) Ramp-rate control for a specific day. (b) Annual effective 5-s SOC profile for the classic ramp-rate control strategy [12]

noted that, in order to ensure that condition r_{max} is met at all times, the control action must be implemented prior to the ramp-rate limiter.

To assess this strategy, $k = 6$, being a value which allows a good trade-off between speed and system stability. To give an example, Figure 5.25 shows the response of the control on the 21 November 2012 with extreme fluctuations measured at $P_n = 38.5$ MW Moura PV plant for $r_{max} = 2\%/\text{min}$. For this particular case, and as has been explained above, twice the energy is required for the worst case fluctuation [see (5.19)], the minimum theoretical effective battery needed is therefore $C_{BAT} = 24{,}558$ kWh. The battery requirements for this specific day are the result of the corresponding energy values and it can be seen that these are under the capacity limits: the effective capacity used ($C_{BAT,used}$) is $C_{BAT,used} = E_{BAT,max} - E_{BAT,min} = 14{,}041$ kWh, which corresponds to a $SOC_{max} = 85.07\%$ and

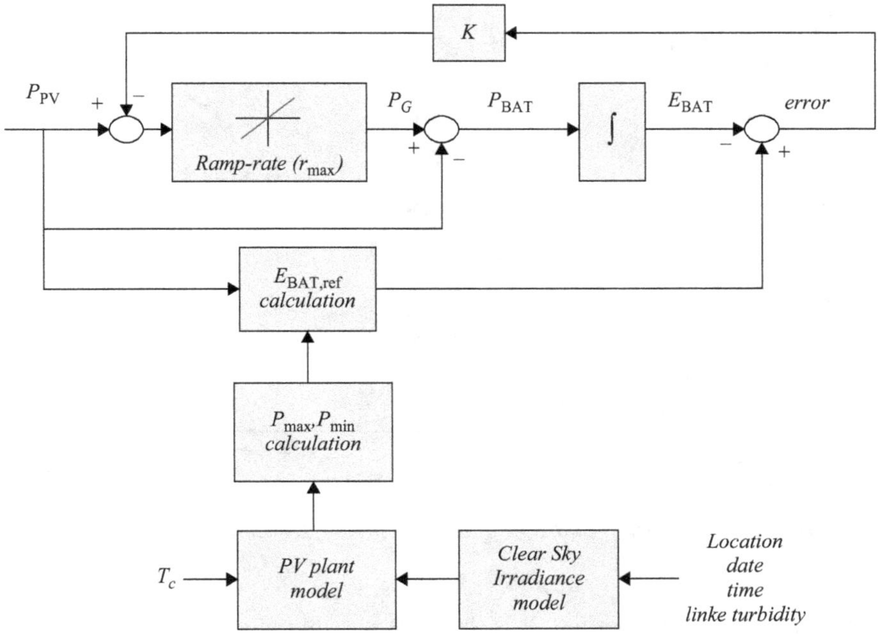

Figure 5.26 Ramp-rate control based on the PV plant model. Notice that the SOC control action is also smoothed by the ramp-limiter in order to guarantee that power fluctuations are always below r_{max} [12]

$SOC_{min} = 27.90\%$. However, throughout the whole year, is close to the theoretical value (<5%), as it can be checked in Figure 5.25(b).

5.5.3 Power Ramp-Rate control based on the PV power plant model

The great advantage of ramp-rate control is that it only acts when the fluctuation exceeds the maximum allowable ramp-rate value, thereby implying lower cycling degradation in the ESS [13]. However, the main drawback of the latter control is that, as the sign of the first fluctuation is unknown, a double capacity battery was required to absorb both the upward and downward fluctuation, setting the SOC reference at 50%. Considering the fact that this increased capacity involves high PV plant overheads, it therefore follows that a second control would be extremely useful in order to avoid the need to double the storage system capacity. In this way, this new control would offer the advantage of a lower use of the ESS capacity. A possible strategy for correctly operating with the minimum requirements necessary to function properly with the worst fluctuation model is to control the battery SOC based on the PV plant instantaneous power and its production limits. Figure 5.26 shows a block diagram of the control strategy.

The instantaneous power of a PV plant $P_{PV}(t)$, for some given irradiance values, $G(t)$, and cell temperature, $T_c(t)$, may be easily estimated from a parametric

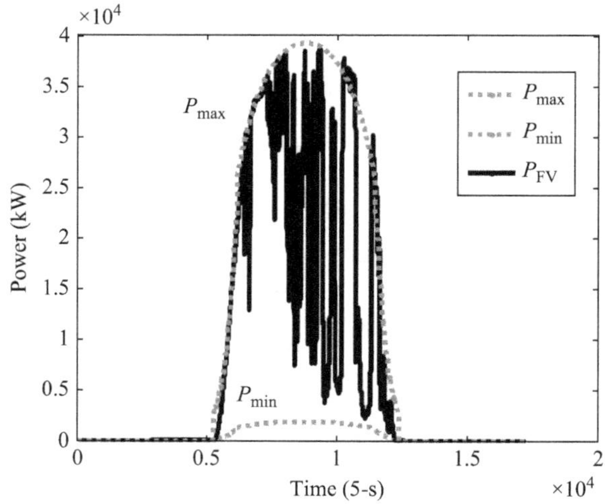

Figure 5.27 P_{max} and P_{min} limits for the 21 November 2012 [12]

model of the plant in question. It is therefore possible to estimate the plant production limits at all times. These are the maximum power that can be delivered by the plant in clear sky conditions, $P_{PV,max}(t)$, and the power that the plant would deliver on a completely cloudy day on which only diffuse radiation would reach the panels. $P_{PV,min}(t)$.

In order to calculate $P_{PV,max}(t)$, the first step is to calculate the clear sky irradiance and then apply a parametric model to move from irradiance to power output. The determination of the clear sky irradiance that reaches the ground has been a key parameter in the field of solar radiation modelling and evaluation. In this way, many empirical and physical models are available in the literature (details can be found in [12]). Given the fact that high accuracy is not essential for modelling clear sky days for the goodness of the control, for the purpose of our study, the model selection criteria must be based on implementation simplicity and input parameter availability (Linke turbidity or aerosol optical depth) alike. We therefore implemented the clear sky model of the European Solar Radiation Atlas which only requires the Linke turbidity as an input parameter [14]. Monthly values of the Linke turbidity factor are sufficient for the purpose of our study and can be obtained from [15]. Once the clear sky irradiance has been obtained, simply by implementing a parametric model, it is possible to calculate the maximum power to be delivered by the PV plant on a clear day, $P_{PV,max}(t)$. By way of example, Figure 5.27 shows these limits and the real PV power, $P_{PV}(t)$, for the 21 November 2012.

Once the PV plant production limits are known, it is relatively simple to implement a SOC control that takes these limits into account and changes the storage system reference based on the weather conditions of the day under consideration. In this respect, it is possible to set the storage system SOC close to 100% on a clear day and operate the PV plant at full power given the fact that the only

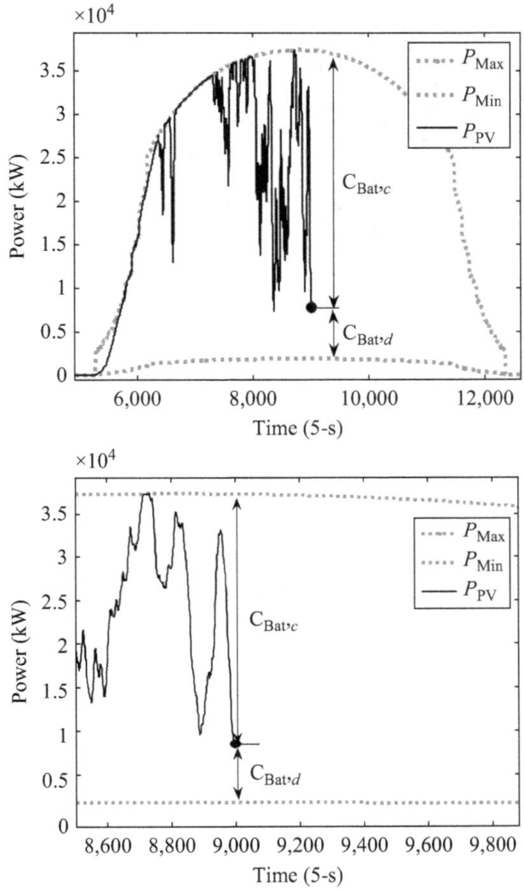

Figure 5.28 SOC control calculation during an instant on 21 November 2012 [12]

event that could occur would be a negative fluctuation which would be smoothed out by the energy provided by the storage system. On the contrary, on a completely cloudy day and with the PV plant only operating with diffuse irradiance, then the storage system SOC would be too low to absorb the only possible positive fluctuation that could occur. Any other intermediate production state would imply setting a reference SOC that would be a function of the instantaneous weather conditions. To evaluate this strategy, we are going to work with $k = 6$. Figure 5.28 shows the capacity required at a time close to midday on the 21 November 2012 in order to correctly absorb any potential upward and downward fluctuations alike. Note that, on the one hand, the available capacity required for discharging in the event of the maximum possible downward fluctuation ($C_{BAT,d}$) is the stored energy that needs to be fed into the grid in order to smooth out this fluctuation. On the other hand, the capacity required to charge in the event of the maximum upward

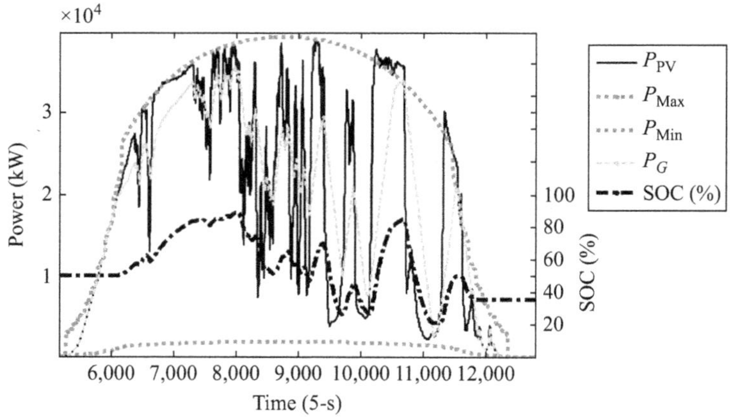

Figure 5.29 SOC control during 21 November 2012 [12]

fluctuation ($C_{\mathrm{BAT},c}$) is the difference required in the ESS to absorb this fluctuation. $C_{\mathrm{BAT},d}$ and $C_{\mathrm{BAT},c}$ requirements are determined by adapting (5.19) to the following expressions:

$$C_{\mathrm{BAT},d}(t) = \frac{(P_{\mathrm{PV}}(t) - P_{\min}(t))}{3{,}600}\left[\frac{(P_{\mathrm{PV}}(t) - P_{\min}(t)) * 100/P^*}{2 \cdot r_{\max}} - \tau\right] \tag{5.22}$$

$$C_{\mathrm{BAT},c}(t) = \frac{(P_{\max}(t) - P_{\mathrm{PV}}(t))}{3{,}600}\left[\frac{(P_{\max}(t) - P_{\mathrm{PV}}(t)) * 100/P^*}{2 \cdot r_{\max}} - \tau\right] \tag{5.23}$$

It is worth noting that there are many points in which the sum $C_{\mathrm{BAT},d} + C_{\mathrm{BAT},c} < C_{\mathrm{BAT,ramp}}$, which means that it is not just a single point that meets the ramp limitation but an entire area. If the SOC is within this area, then no modification will be necessary, making it possible to reduce the ESS cycling degradation should this be necessary. This results in the following control logic, (5.24):

$$\begin{aligned}
&\text{if } E_{\mathrm{BAT}}(t) < C_{\mathrm{BAT},d}(t)\\
&\quad E_{\mathrm{BAT,ref}}(t) = C_{\mathrm{BAT},d}(t)\\
&\text{else if } C_{\mathrm{BAT}} - E_{\mathrm{BAT}}(t) < C_{\mathrm{BAT},c}(t)\\
&\quad E_{\mathrm{BAT,ref}}(t) = C_{\mathrm{BAT},c}(t)\\
&\text{else}\\
&\quad E_{\mathrm{BAT,ref}}(t) = E_{\mathrm{BAT,ref}}(t-1)\\
&\text{end}
\end{aligned} \tag{5.24}$$

By way of example, Figure 5.29 shows the performance of the proposed SOC control during the fluctuations on the 21 November 2012. Despite the fluctuations recorded in the course of the day, the strategy worked successfully. The SOC of the ESS was continually adjusted to meet the ramp rate limitation implemented and also any possible fluctuations that could take place. Furthermore, as discussed

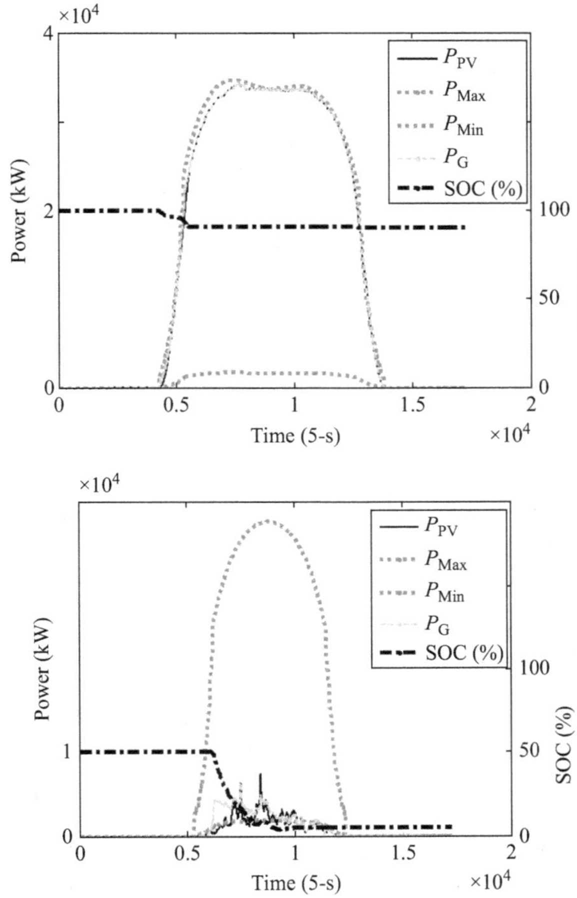

Figure 5.30 SOC control during (a) a clear day (26 August 2012) and (b) a cloudy day (23 November 2012) [12]

above, at the beginning and end of the day when $E_{BAT}(t)$ values meet the ramp rate limitation, the SOC remains constant in accordance with (5.24). In addition, according to (5.19), the minimum theoretical effective battery required is C_{BAT} = 12,279 kWh. On this particular day, SOC_{max} = 88.84% and SOC_{min} = 21.43%. Thus, the effective capacity used on this day was $C_{BAT,used} = E_{BAT,max} - E_{BAT,min}$ = 8,276 kWh. Yet again, this strategy allows C_{BAT} to be half the C_{BAT} required in conventional ramp-rate control.

Furthermore, Figure 5.30 shows the performance of the SOC on a clear day (26 August 2012) and a cloudy day (23 November 2012). As was to be expected, on the clear day when $P_{PV}(t)$ is close to the $P_{PV,max}(t)$ limit, once the SOC level of the ESS can provide enough energy to smooth out the only potential downward fluctuation, it remains constant. On the other hand, on a cloudy day, when the SOC

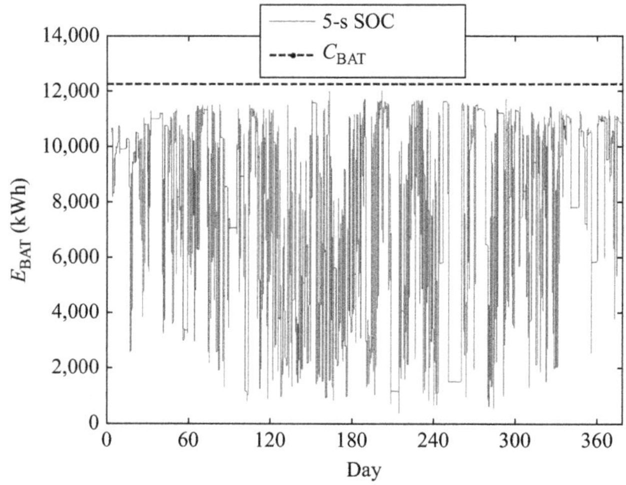

Figure 5.31 Annual effective 5-s SOC profile for the strategy based on the PV plant model. Note that the capacity required in classic ramp-rate control is double the capacity required in this one [12]

level of the ESS is low enough to absorb the energy required to smooth out the only potential upward fluctuation, it also remains constant. It goes without saying that, less use of the ESS leads to smaller losses and lesser ESS cycling degradation.

Figure 5.31 shows the effective SOC evolution for the whole year and for the last strategy proposed. Once more, $C_{BAT} = E_{BAT,max} - E_{BAT,min}$ throughout the whole year is near to the theoretical value. In addition, the required capacity is half the required in conventional ramp-rate control. Finally, the SOC evolution shows that the ESS is much more used because it is continuously adjusting throughout the year depending on the climate conditions (there is not a SOC reference value).

Hence, this second strategy provides a most efficient SOC control in terms of capacity, based on the actual power delivered by the PV plant and the production limits. Although a parametric model of the PV plant is needed, it is easy to implement and does not require great accuracy. Regardless of PV plant size, the strategies can operate correctly with the minimum theoretical effective capacity based on the worst fluctuation model in order to comply with a maximum allowable ramp-rate given by a grid code regulation.

Finally, it is important to highlight that the worst fluctuation model can be also applied to a group of PV plants, in order to size the energy necessities of a centralised ESS connected at the distribution level. Taking advantage of the smoothing effect of geographical dispersion on PV power, the centralised solution lets to obtain similar savings up to 50% in both capacity and power requirements for strictest values of r_{max} instead of a distributed one [16].

References

[1] European Network of Transmission System Operators, (ENTSO-E). *Annual Report 2015.* [online]. ENTSOE report. Available at: https://www.entsoe.eu/Documents/Publications/ENTSO-E%20general%20publications/ENTSO-E_AR15_FINAL.pdf?Web=1 [Accessed 26 Jan. 2017].

[2] Marcos, J., Storkël, O., Marroyo, L., Garcia, M., and Lorenzo, E. (2014). Storage requirements for PV power ramp-rate control. *Solar Energy*, 99, 28–35.

[3] Marcos, J., Marroyo, L., Lorenzo, E., Alvira, D., and Izco, E. (2011). Power output fluctuations in large scale PV plants: one year observations with one second resolution and a derived analytic model. *Progress in Photovoltaics: Research and Applications*, 19(2), 218–227.

[4] Marcos, J., Marroyo, L., Lorenzo, E., and García, M. (2012). Smoothing of PV power fluctuations by geographical dispersion. *Progress in Photovoltaics: Research and Applications*, 20(2), 226–237.

[5] Hoff, T.E. and Perez, R. (2010). Quantifying PV power output variability. *Solar Energy*, 84, 1782–1793.

[6] Rahman, S., and Bouzguenda, M. (1994). A model to determine the degree of penetration and energy cost of large scale utility interactive photovoltaic systems. *IEEE Transactions on Energy Conversion*, 9(2), 224–230.

[7] Curtright, A. and Apt, J. (2008). The character of power output from utility scale photovoltaic systems. *Progress in Photovoltaics: Research and Applications*, 16, 241–247.

[8] Marcos, J., Marroyo, L., Lorenzo, E., Alvira, D., and Izco, E. (2011). From irradiance to output power fluctuations: the PV plant as a low pass filter. *Progress in Photovoltaics: Research and Applications*, 19(5), 505–510.

[9] Marcos, J., de la Parra, Í., García, M., and Marroyo, L. (2016). Simulating the variability of dispersed large PV plants. *Progress in Photovoltaics: Research and Applications*, 24, 680–691. doi: 10.1002/pip.2719.

[10] Haaren, R., van Haaren, R., Morjaria, M., and Fthenakis, V. (2014). Empirical assessment of short-term variability from utility-scale solar PV plants. *Progress in Photovoltaics: Research and Applications*, 22, 548–559.

[11] Puerto Rico Electric Power Authority , (PREPA). Minimum Technical Requirements for Photovoltaic Generation (PV) Projects [online]. Available at: http://www.fpsadvisorygroup.com/rso_request_for_quals/PREPA_Appendix_E_PV_Minimum_Technical_Requirements.pdf. [Accessed Nov. 2014].

[12] De la Parra, I., Marcos, J., García, M., and Marroyo, L. (2015). Control strategies to use the minimum energy storage requirement for PV power ramp-rate control. *Solar Energy*, 111, 332–343.

[13] Marcos, J., de la Parra, I., García, M., and Marroyo, L. (2014). Control strategies to smooth short-term power fluctuations in large photovoltaic plants using battery storage systems. *Energies*, 7(10), 6593–6619.

[14] Rigollier, C., Bauer, O., Wald, L. (2000). On the clear sky model of the ESRA – European solar radiation atlas – with respect to the Heliosat method. *Solar Energy*, 68, 33–48.

[15] SODA, Solar energy services for professionals. Available from: http://www.soda-pro.com/ [Accessed Jan. 2016].

[16] De la Parra, I., Marcos, J., García, M., and Marroyo, L. (2015). Storage requirements for PV power ramp-rate control in a PV fleet. *Solar Energy*, 118, 426–440.

Chapter 6

Towards the extensive use of renewable energy resources: needs, conditions and enabling technologies

Isabel M. Moreno-Garcia[1], *Rafael Real-Calvo*[1], *Victor Pallares-Lopez*[1], *Miguel J. Gonzalez-Redondo*[1] *and Isabel Santiago*[1]

6.1 Introduction

Traditional electricity grids are currently involved in a process of change towards the so-called smart grids, with the aim of making the system able to face the continued growth of energy demand in a cleaner, more sustainable, efficient, economical, reliable and secure way. In this new concept of network, the planning, management and operation of the electricity grids must be redefined [1]. They must integrate in an intelligent way all of their components, for example generators, grid operators and end-users, whose behaviour and actions should be monitored and optimally controlled in real time.

Large-scale integration of renewable and low carbon energy sources, like wind and photovoltaic (PV) systems, are among the new members of these future intelligent grids, in addition to the active consumers, with possible energy self-generation (prosumers) and demand management and storage systems. These types of energy sources are typically available in a decentralised way as distributed energy resources (DERs) [1], so the traditional role of the central generation, transmission and distribution is being transformed by adding the aggregation of distributed resources, which results in a microgrid architecture [2]. Moreover, the fluctuating, unpredictable and unmanageable character of renewable generation systems, subject to intermittent atmospheric conditions, increases the difficulty of ensuring the reliable and stable management of the electricity systems. Smart grids will support these variable renewable energy sources by providing operators with real-time system information that enables them to manage generation, demand and power quality (PQ) [3].

This new infrastructure requires, in all of the components of the networks, the integration of modern technologies such as advanced metering infrastructure and

[1]Universidad de Córdoba, Córdoba, Spain

diagnosis, sophisticated information and communication architectures or advance automation and control techniques. This will allow the incorporation, in a complex architecture, of a great amount of intelligent and interconnected components that will exchange energy and information with a bidirectional flow [1]. In such a way, the future Smart Grid will be constituted as a system of systems, where all of its components will independently operate, but they will be integrated, working in a cooperative mode in order to achieve a higher performance for the entire system [4]. The Smart Grid will also be equipped with self-healing, self-organising and self-configuring capabilities, to balance supply and demand by providing the PQ for a range of needs [3].

Although some technological achievements are already available, this challenge still requires the wide research and development of technical, financial and regulatory solutions to overcome the multiple and significant barriers still existing. And in this context, some of the needs, conditions and enabling technologies that will allow for the extensive use of renewable energy resources in the future electrical networks are explained and analysed in this chapter.

The implementation of all of this infrastructure must be carried out on different levels in the power system infrastructure, from generation to end-users, incorporating in all of its components a series of intelligent electronic devices (IED), with a high integration and multifunctional character, in order to monitor and manage their operation [1].

On the one hand, there is a clear tendency within the energy sector to achieve complete control of the operation of renewable plants. The objective is to have a better knowledge of the energy they generate, to optimise the operation of their components and improve their maintenance, to control the quality of the injected signal to the network, and to operate, following a network manager, in an efficient, competitive and reliable way. To this end, a real-time monitoring and management system for such plants becomes absolutely necessary [5]. An example of this will be presented in this chapter.

On the other hand, in the new smart grids, renewable DERs are connected directly to any point in the medium- and low-voltage power grid. This supposes an additional factor of disturbance of the network, fomenting the appearance of problems of PQ, stability, loss of reliability or presence of bi-directional flows, among others. In recent years, a number of technical problems related to the interaction of these systems with the power grid have been partially overcome. The development and implementation of IEDs that control the interface with the network of this type of renewable generation plants has been promoted [6,7]. Part of the solution to moderate the impact of these distributed installations is located in their own equipment, as it can be in the interface of power converter [8]: that is, the inverter in the distributed PV systems. Driven by the 2011 LV–MV directives from Germany [9], inverters can contain intelligent functions, such as reactive power control, to help with network integration. By the end of 2015, more than 50% of inverters were able to perform these functions. This amount could increase with the innovation of the energy storage incorporated in the converter [10] and with the vertiginous advance of the microgrids [11].

Originally, in the presence of any disturbance, the distributed generation (DG) had to be disconnected as quickly as possible to return the system to its original status. However, this position has changed over time, to the extent that there has been an understanding of the benefits that can be obtained with renewable generation. Companies have realised that distributed power systems bring economic and strategic advantages to their own networks. Power enterprises, to maximise their profits, are inclined to operate power systems very close to their limits, increasing the likelihood of service interruptions. Moreover, an electrical system operating under more forced conditions increases the impact of suffering other problems such as weak connections, unforeseeable and unexpected events, failures undetected in protection systems or human errors that, in the case of a failure of a part of the network system, may bring it into instability or even 'blackout'. In underdeveloped countries, power systems are even less reliable, and they often suffer from a large imbalance between demand and generation, with inadequate transmission capacity and large distances between load centres and generation units. Thus, if a contingency occurs in these systems, due to their lack of sophistication, the risk is greater of suffering more serious consequences, such as a complete power fault.

To address these problems, on the one hand, it is necessary to introduce new mechanisms for the prevention, detection and localisation of possible anomalies that can trigger supply problems with serious consequences, to reduce the times of interruption of supply or even for automatic self-healing. This will be possible through a constant monitoring of the entire network and an increase in information processing capacity [12,13]. But also, the use of renewable generation arises as support for the electrical system in case of an event of even a blackout, either from the point of view of support for the supply in the affected area, or as an active part of the strategy to avoid the collapse of the network. For this, operators will have to know more precisely the state of the service areas in order not only to manage the system more efficiently, but also to establish safe methods of network stability to stand up to failures and to avoid the spread of problems such as the dreaded 'Trigger Effect' [14].

It is therefore necessary, in addition to plant level and network interconnection, to carry out real-time monitoring and display of power system components over large geographic areas, which will help operators to understand and optimise the behaviour and performance of all these network components. Numerous systems will have to coexist to control both the power generation equipment, as previously mentioned, and the monitoring and control equipment for transformation substations and distribution networks, intelligent systems for efficiency and devices specialised in an effective response to the demand. This wide-area monitoring will require advanced technologies including wide-area situational awareness, wide-area monitoring systems and wide-area adaptive protection, control and automation, with advanced capture and processing capabilities, in order to generate data to inform decision making, mitigate wide-area disturbances and improve transmission capacity and reliability [3].

In this area, one of the key challenges is to coordinate all of these systems or integrated IEDs across the network through the appropriate communication channels. Specifically, standard IEC 61850 [15] seeks to integrate all of the interoperability and communication needs among all these systems that are responsible for the supervision, management and control of electrical energy. This standard defines the real-time synchronisation and information transfer needs. Some international committees are working on the process of integrating the standard of synchrophasors or phasor measurement units (PMUs) for critical infrastructures [16]. The integration is defined in section IEC 61850-90-5 of the standard [17]. These synchrophasors, which are governed by the IEEE standards C37.118.1 [18] and C37.118.2 [17], and whose voltage and current values must be transferred to a central data processing system, allow the determination of the stability of the electrical network in an extensive geographical area with microsecond accuracy. Therefore, synchronisation is crucial for the surveillance, protection, control and state estimation in power systems.

The first parts of the IEC 61850 standard were published between 2003 and 2005 and were focussed on the scope of the substations. The standard is based on the definition of a hierarchical model of data that models all of the elements present in the installation, both the physical devices and the functions performed by them, with the so-called logical nodes (LNs). It supports semantic definitions for DER [19,20]. It is possible that other semantic interpretations are needed for the new capabilities: processing information nodes with complex structures [complex event processing (CEP)] [19]. In this case, new LNs can be defined to cover these shortcomings. As an example, in this chapter, we list some of the Logical Devices (LDs) and LNs associated with a PV plant. They are included in section IEC 61850-7-420 [21], which is specifically developed for renewable installations.

Taking into account all the above, it is clear that the model of communication chosen is of great importance [22]. A model is required that provides support for automation and distributed monitoring with decentralised management and that supports optimised functions for critical events. Special emphasis is dedicated to minimising the time required for the transmission of information with full synchronisation guarantees.

Moreover, for distributed synchronisation, it is necessary to keep in mind the relation between the precision of the synchronism and the distance between thenodes of the system, so that to design any system, the limitations imposed by this aspect must be taken into account. As the transmission distance increases, it is more difficult to share the signal between systems to maintain synchronisation because the distance between nodes degrades the clock signal. Therefore, from a certain limit it is necessary to use another method to synchronise the clocks. A GPS located in each point does not suffer degradation with distance and will allow all the points to take a very precise common signal as a reference.

Alternatively, IEEE 1588 protocol must be adopted [23], whose analysis is also included in this chapter. Instead of directly sharing a clock signal, these devices periodically exchange information and adjust their local clock based on the clock that acts as a master. This process of synchronising clocks in a distributed form requires a continuous process. A watch is essentially composed of two parts, an oscillator and an accumulator. In theory, if two clocks are identical and their oscillators work at the same frequency, they should behave with the same accuracy. In practice, accuracy is limited by changes brought about over time and by changes in temperature. The accuracy of the synchronism in the electric sector depends on the type of task entrusted to the system [24].

Definitely, there is a scenario that demands an effort in the development of systems for the management of the electricity networks, at the level of all of their members, and multiple efforts must be added to develop, mature and deploy all these systems in order to achieve a large-scale integration of renewable resources and to increase the 'smartness' of the grid.

6.2 Measurement and assessment of the renewable generation

Within the framework of DER, PV energy is a highlight for having experienced one of the greatest technological developments and offering the greatest possibilities for competing in the market. Its numerous advantages, such as simple installation, high reliability, zero fuel costs, very low maintenance costs, and lack of noise due to the absence of moving parts [25] have resulted in a high growth rate. In this way, PV consolidates itself, after hydro and wind energy, as the third most important type of renewable energy in terms of installed capacity worldwide [26]. Opportunities for improvement in this sector are based on the increase in the level of installed capacity and the need for developments of technological advances that allow for optimisation of the production of electricity, in order to reduce generation costs and to be able to manage the integration of PV plants into the future context of smart grids [1,27].

Initially, the problem of PV variability was easily accommodated when these sources turned out to be a very small percentage of the power supplied to the grid. But as more large-scale solar parks (Multi-MW Solar PV Parks) have been built and integrated into the grid, variability is becoming a significant problem. This problem could be overcome with the use of storage systems, wherein the management, in real time, of the power flows involved in the electrical system is needed to ensure coverage of demand. The use of storage systems, one of the most widely accepted solutions to ensure the coverage of electricity demand [28], requires a detailed knowledge of the facility production mode in order to optimise the investment of economic resources in such systems, which in the near future will be an inevitable part of DER [29–32].

Therefore, to address the high penetration of PV power, future smart grids need to implement efficient monitoring and control technologies to improve their operational performance. Information regarding energy flowing in and out of the solar park must be fully monitored. That is to say, for a PV plant to be dispatchable, it is vital that the facility is perfectly characterised, providing at all times a precise knowledge of the operation mode of all the components that integrate it [33]. This knowledge allows the following:

- Comparing at each set time the production of the facility with the previous forecast made by a prediction system, determining the deviations of the actual production compared to the expected. This will allow, on the one hand, evaluation of the prediction system, and, on the other hand, once the system has been evaluated, detection of the presence of failures in the facility.

- Characterisation of the actual margins in which each component of the facility operates, to optimise production and reduce as far as possible the different causes of losses, detecting possible design failures or correct selection of system components. Given that the power generated by DERs is continually increasing worldwide, a small percentage of improvement in their performance would lead to a significant net improvement in production, and therefore a reduction in the net cost of these facilities. This advance is necessary to improve their competitiveness in the face of traditional modes of energy generation.
- Obtaining a more complete diagnosis of the energy signal injected to the grid than those provided by inverters, including PQ analysis in real time (on time). This is possible by increasing the sampling frequency in the measures to accomplish the PQ standard: specifically, the IEC 61000-4-30 [34] standard, which describes the measurement methods for each type of parameter at the grid site. This standard is detailed in Section 6.2.2.
- Carrying out a predictive maintenance of the same, implementing techniques for the diagnosis, supervision and detection of failures in those facilities. This will also result in a greater number of hours of operation and in the increase of the long-term stability of the components, allowing an improvement of the productive processes and, consequently, an increase in their profitability.
- And finally, having the operation under control would allow modification of the delivery regime of the energy produced by the facility to the grid, based on the specifications that are marked at any moment by the system operator, or based on the production planned and previously agreed with the control centre and management centre, so that its management condition can be positively valued.

6.2.1 Use of a PV monitoring system on time in a grid-connected PV park

For several years, the interest for the analysis of the requirements of generation prediction programmes to combat the deficiency in the management has increased [31,33]. In this line, a system for monitoring the behaviour and mode of operation of a PV plant is presented in this section. This system focusses on providing a solution to the management of the electric power supply commanded from an operations centre.

To advance in this issue, an intelligent control system, which incorporates an advanced sensor and communications network, has been deployed. The system requirements were defined based on the work developed with IED [35–37], distributed synchronisation techniques for instrumentation systems [38–40] and the integration of communication models in smart grid [41–43]. Specifically, a monitoring system has been developed that records, with very high sample frequencies, environmental and electricity production measures, based on standard IEC 61724: 1998 [44], which will be shortly substituted by a new version IEC 61724-1, and PQ measures [34]. From these measures, it is possible to completely monitor the performance of all components, quantify different loss mechanisms (in wiring, degradation of components, etc.) and detect faults or deviations from optimal production [45], based on Technical Specification IEC 61724-3: 2016 [46]. Also, these measures can be integrated as LNs in the IEC 61850 standard dedicated to communication networks and systems for power utility automation. The modelling of some elements that are usually present in a DER will be explained later in Section 6.4.4.

The architecture of the monitoring system was defined on the basis of the characteristics of the 6-MW PV plant in which it has been implanted and tested. Figure 6.1 provides more details about the proposed system, showing the configuration scheme for monitoring in real time the large-scale PV plant connected to the grid. The structure of the developed system comprises sensors, programmable controllers, and a monitoring and supervision system, along with the different relationships between them. The system implemented has the novelty of measuring PQ of the electrical signals generated by the inverters, the inclusion of high synchronisation performance in the data capture, and the management of real-time data transmission according to the industrial communication standard IEC 61850 [47].

The developed system uses the IEEE 1588 standard, precision time protocol (PTP) and Gigabit-Ethernet switch with boundary clock (PTP active) interconnected as a synchronisation method. With PTP, a precision of nanoseconds can be achieved and distances of up to 10 km can be saved, as is already being studied in critical infrastructures such as CERN with the White Rabbit project [48]. This will allow the correct establishment of the comparison between the parameters measured in various inverters located in large PV parks. Therefore, the proposed system offers a technological solution for the integrated management of both the Multi-Mega and dispersed parks, providing accurate and real-time information on the state of operation of the plants, complying reliably and securely with the requirements of the power system operators. A more detailed explanation about the synchronisation stability with PTP appears in Section 6.4.5.

In addition, the system has other features that make it an autonomous monitoring system for a PV plant or a wind park because it is equipped with a very advanced user interface that integrates a level of analysis and processing adaptable to other plants with different configurations. In addition, a complete detection and specific alarm mechanism can be established. Ultimately, the developed platform provides a system of real-time monitoring of the performance of all components, a quantification of the Key Performance Indicators and the different mechanisms of losses of the plant, and the detection of the presence of faults and deviations from a production optimal. An example of this application is shown in Figure 6.2.

The top bar shows the menu options which enable the detail panel of the devices or measures groups: 'System', 'Graphs', 'Production', 'PQ', 'WSN', 'Weather station' and 'Diagram'. The figure illustrates the monitoring system in which the option 'Graphs' is selected. On this screen, a record of the production of the four inverters is shown.

At the top of the screen, the measurements of the active power generated by inverter are shown. At the bottom of the screen, the variation values (increments or decrements) of these powers are shown every 10 s. Both types of graphs are shown in percentages.

On the right side, there is a menu dedicated to visualising other temporal scales in the increments of the active powers. Specifically, the monitoring system processes

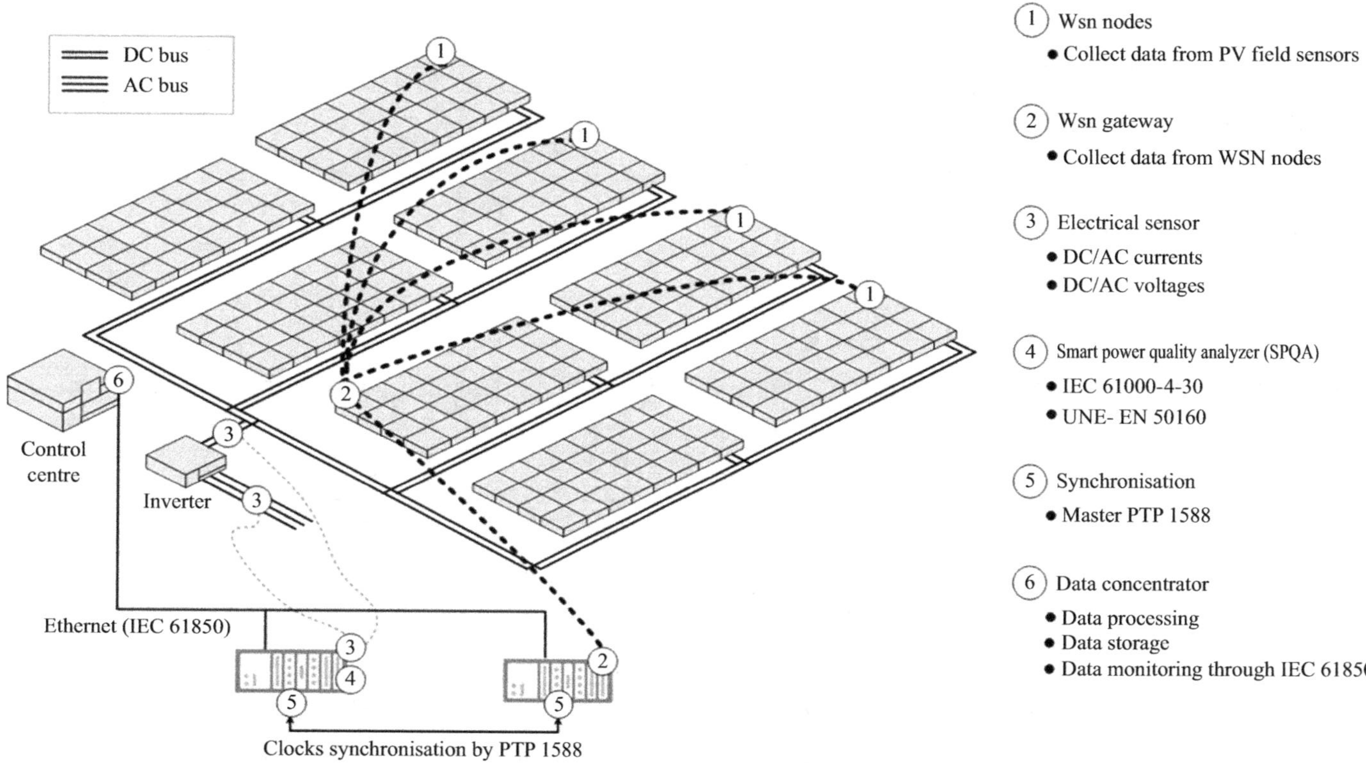

Figure 6.1 Proposed scheme for the real-time monitoring of grid-connected PV plant

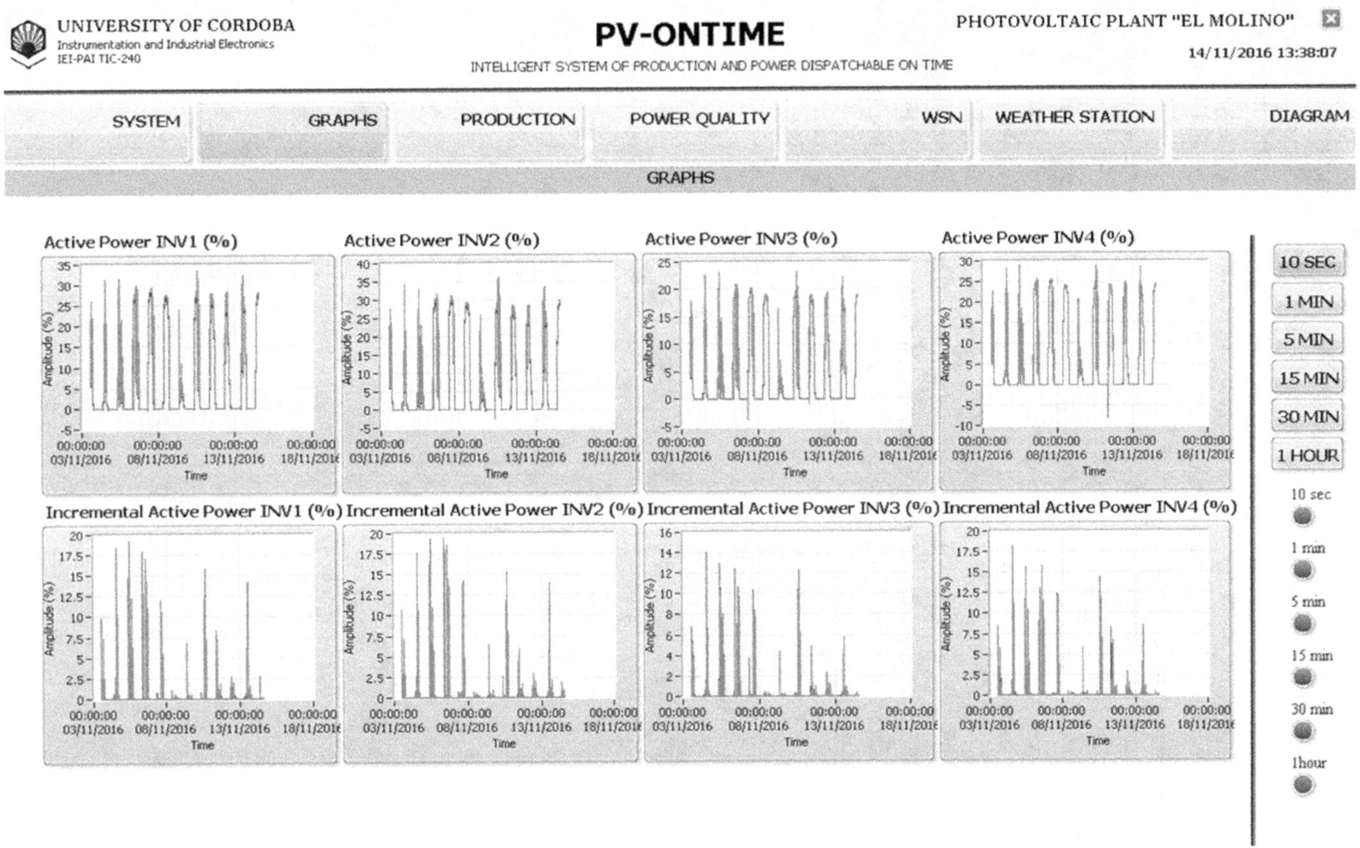

Figure 6.2 Monitoring application for on-time supervising of a PV plant

this operation for 10 s, 1 min, 5 min, 15 min, 30 min and 1 h. On this screen, a new record is represented by a visualisation icon (LED indicator) by each temporal scale. This screen is very useful for viewing the correlation between inverters.

One of the main novelties of the system is its ability to control in real time the PQ of the signal injected to the grid [49], as well as the influence of the installation on the power system. With the measurement of these parameters, it is possible to detect and record the occurrence of electrical disturbances. In a global way, with this system it is possible to centralise the monitoring and control of the equipment in the plant itself and process the data for the operation and maintenance personnel.

Among the different PQ parameters evaluated by this system, an experimental test has been carried out concerning the THD voltage (THD_V), comparing it with the irradiance and production of the inverters. For this aim, PQ parameters at the output of an inverter, the active power (P_a) and 9th voltage (V_{9th}) harmonic, have been compared with the solar irradiance (G_i). The research was based on data recorded in the database of the developed monitoring system for three cases: a clear day (Figure 6.3), a cloudy day (Figure 6.4) and a day with sunny spells (Figure 6.5). The results depict that there is no effect of solar radiation in the PQ of the signal produced by the inverter.

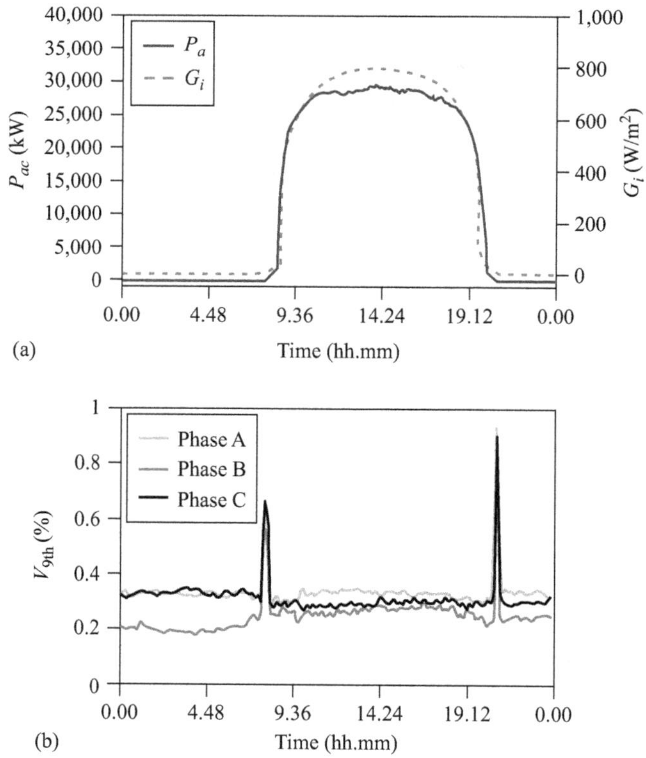

Figure 6.3 Clear day

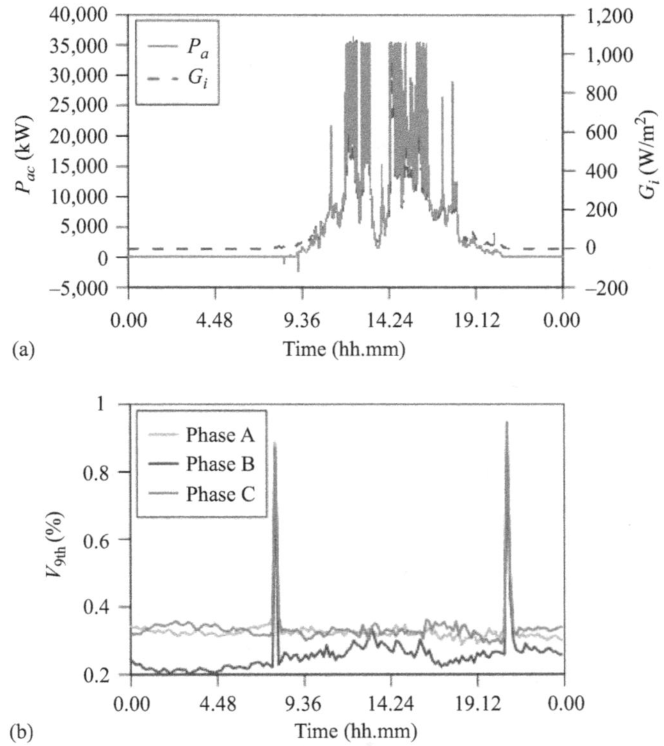

Figure 6.4 Cloudy day

The upper graphs, Figures 6.3(a), 6.4(a) and 6.5(a), show the active power and the solar irradiance. For the three cases studied, the active power P_a produced by the inverter is obviously linearly dependent with the solar irradiance G_i. On the contrary, the harmonic distortion, shown in the lower graphs, Figures 6.3(b), 6.4(b) and 6.5(b) are not affected by the solar irradiance G_i. Specifically, these graphs show the $V_{9\text{th}}$ registered at the output of the inverter. Each horizontal line of these graphs corresponds to one channel or phase. As shown in these graphs, a high harmonic distortion is produced when the inverter starts and ends to produce energy. The spikes on the $V_{9\text{th}}$ in the three phases are easily visible. Although these instants match with sunrise and sunset, low solar irradiance conditions, the $V_{9\text{th}}$ is not influenced by the solar irradiance G_i. An obvious example is shown in Figure 6.4, which shows a cloudy day, where it can be seen that the fluctuations in the P_a [Figure 6.4(a)] do not generate harmonics at the output of the inverter [Figure 6.4(a)].

6.2.2 Temporal requirements in the measurement of parameters to control the power quality of the generated signal

Given the requirements in terms of PQ and safety that involve the integration of the DER, controlling the PQ of signals generated by them becomes absolutely necessary.

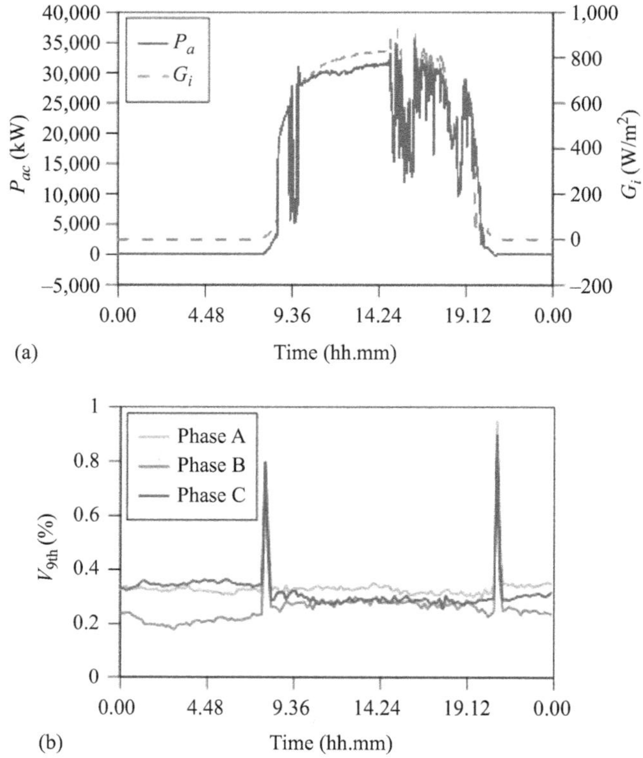

Figure 6.5 Sunny spell day

A breakthrough in this regard is fundamental for companies in the electric power business to obtain greater energy efficiency and a decrease in supply cut-off times.

The IEDs integrated in the power system equipment can provide alternative solutions to the problems in the integration interface of renewable generation with the grid. State-of-the-art equipment already has these intelligent integrated functions, but these devices still represent a small percentage of installed equipment. The real problem arises when applying these new functionalities in conventional equipment to update it to the new needs and requirements of the power system. The continuous increase of the integration and flexibility of the embedded systems, and the reduction of costs in the design, as in production, suppose significant advantages for the use of IEDs in the interface of the power system.

Monitoring parameters complying with the requirements of the international regulations allow for carrying out a control of the PQ at the interconnection of the DER with the power system. The IEC 61000-4-30 standard defines the methodology for measuring the PQ parameters in the 50- and 60-Hz electrical systems and interprets the resulting values. This standard describes the measurement methods for each type of parameter at the grid site, so that by following its guidelines reliable, reproducible and comparable results can be obtained regardless of the instrument used and its surrounding conditions. Also, it is important to note that IEC 61000-4-30 defines functional methods of measurement but does not specify how they should be carried out, thus leaving open the implementation of such methods. Nor does it define the ranges of values between which the measurement of parameters must be performed, so that these should be chosen on the basis of specific regulations, especially in the case of security and protection applications. In Section 6.3.1, a comprehensive analysis of the application standards at interconnection is performed.

Consequently, with regard to the time intervals for the measurement of the electrical parameters, the reference is standard IEC 61000-4-30. For each parameter measured, the IEC 61000-4-30 standard contemplates two classes of measurement methods: Class A and Class S methods. Class A methods are used when precise measurements are necessary for high-quality equipment and instrumentation. Class A requirements imply that measurements of a parameter made with two different instruments, when measuring the same signals, will produce consistent results within the specified uncertainty range. These Class A requirements will impose demanding characteristics of bandwidth and sampling frequency, adequate to ensure measurements of each parameter with the uncertainty specified in the standard. Class S methods are more relaxed; they can be used for applications where high precision is not required, such as statistical studies and anomaly investigation. In this case, the standard allows the designer to choose the method of measurement, depending on the specific application.

6.2.2.1 Temporary measurement intervals

Four-time intervals or windows of observation for the methods of Class A are raised

1. *Basic interval:* Must be 10 periods for a network of 50 Hz or 12 periods for a network of 60 Hz (200 ms). With this window, the amplitude of the voltage, harmonics, interharmonics and voltage unbalance are measured.

   The following three intervals will be constructed in aggregate form, depending on the duration and the temporal nature of the electrical phenomenon whose parameter is measured.
2. *3-s interval:* the interval data of 150/180 periods must be added after 15 intervals of 10/12 periods.
3. *10-min interval:* added from 200 intervals of 3 s.
4. *2-h interval:* added from 12 intervals of 10 min.

So, standard IEC 61000-4-30 proposes a basic measurement interval and three aggregations of measure time; these are calculated by the square root of the arithmetic mean of the squares of the values of input.

When the measurement parameters correspond to electrical phenomena, which, due to their transient nature, must be observed at the cycle level, the temporary windows treated above are not applicable. Usually, the likelihood of these

Table 6.1 Intervals and aggregation algorithms in IEC 61000-4-30

IEC 61000-4-30	
Measuring intervals (Class A)	**Aggregation algorithm**
One cycle (refreshed every 1/2 cycle)	$U_{\mathrm{ef}(1/2)} = \sqrt{\frac{1}{T}\int_T u^2(t)\,dt}$
10/12 cycles (50/60 Hz)(200 ms)	$U_{\mathrm{RMS}(200\ \mathrm{ms})} = \sqrt{\frac{1}{200\ \mathrm{ms}}\int_{200\ \mathrm{ms}} u^2(t)\,dt}$
150/180 cycles (50/60 Hz)(3 s)	$U_{\mathrm{RMS}(3\ \mathrm{s})} = \sqrt{\frac{1}{15}\sum_{i=1}^{15} U^2_{\mathrm{RMS}(200\ \mathrm{ms})}}$
10 min (200 × 3 s)	$U_{\mathrm{RMS}(10\ \mathrm{min})} = \sqrt{\frac{1}{200}\sum_{i=1}^{200} U^2_{\mathrm{RMS}(3\ \mathrm{s})}}$
2 h (12 × 10 min)	$U_{\mathrm{RMS}(2\ \mathrm{h})} = \sqrt{\frac{1}{12}\sum_{i=1}^{12} U^2_{\mathrm{RMS}(10\ \mathrm{min})}}$

PQ events' occurrence is evaluated on the basis of the RMS change across windows. Despite its simplicity, the RMS-based method is effective in detecting amplitude-related distortions. According to EN-61000-4-30, this RMS value should be measured over one cycle, commencing at a fundamental zero crossing, and refreshed each half cycle ($U_{\mathrm{ef}(1/2)}$).

For class *S*, the manufacturer must indicate the method, number and duration of aggregation time intervals.

Table 6.1 schematically shows the relationship between intervals and aggregation algorithms.

6.2.2.2 Time stamping

During certain non-stationary events, the measurement algorithms could cause doubtful values in more than one parameter. Thus, the same event can be reflected in several parameters and can be interpreted erroneously as distinct events. The temporary markup avoids counting a given event more than once in different parameters [50].

All parameters calculated at the cycle level will be temporarily marked, such as voltage sags, temporary over voltages and interruptions. In addition, if a value is marked and registered during a certain time interval, the added values that include it must be stamped and registered as well. Table 6.2 shows the measurement intervals for the different electrical parameters that the standard involves.

Table 6.2 Measurement of parameters according to IEC 61000-4-30

IEC 61000-4-30		
Magnitude	**Averaging time**	**Parameter**
Frequency	10 s	–
Voltage (RMS)	200 ms	$V_{\mathrm{RMS\ (200\ ms)}}$
Flicker	10 min	Short-term (P_{st})
	2 h	Long-term (P_{lt})
Sags	One cycle	$U_{\mathrm{ef(1/2)}}$
Over voltages	One cycle	$U_{\mathrm{ef(1/2)}}$
Transients	Not defined	
Unbalance	200 ms	$V_{\mathrm{RMS\ (200\ ms)}}$
Voltage harmonics	200 ms	C_{ng}
Voltage interharmonics	200 ms	$C_{n-200\ \mathrm{ms}}$
Signal transmission	10/12 cycles (200 ms)*	–
Rapid voltage changes	Not defined	
Under- and over-deviation	200 ms	$V_{\mathrm{RMS\ (200\ ms)}}$

*In this parameter, the standard aggregation algorithm does not apply.

6.3 The interconnection between renewable generation and the electricity grid

The intelligent control of the interconnection interface between the DG and the electrical system is one of the key points for efficient network management (Figure 6.6). The new scenario of DG, with a high penetration of renewable generation and the possibility of the user injecting power into the grid, results in two-way flows of power and the medium-term need to control these flows and the entire state of the network. This new context has also led to the appearance of new regulatory changes that refer to the behaviour of inverters in PV generation, which must control the injection or the absorption of reactive power according to the needs of the electric service.

The IEDs integrated in the equipment of the renewable generation are those that must provide a series of characteristics oriented to the intelligent control of the interconnection with the electric network. Fundamentally, the following could be highlighted:

- Temporary response suitable for monitoring parameters and for planned protection actions, using the normative context in force.
- Real-time interaction between the equipment and system management remote entities, as well as with different entities within the distribution area [51,52].
- Temporal determinism in the communications and temporal labelling of the data, so that the state of the interconnection interface can be precisely established [53,54].
- Multifunctionality, since these IEDs will have to act in different contexts, covering functions of PQ, protection or both simultaneously [55].
- Modularity, since they must be able to integrate into a very heterogeneous set of equipment [56].

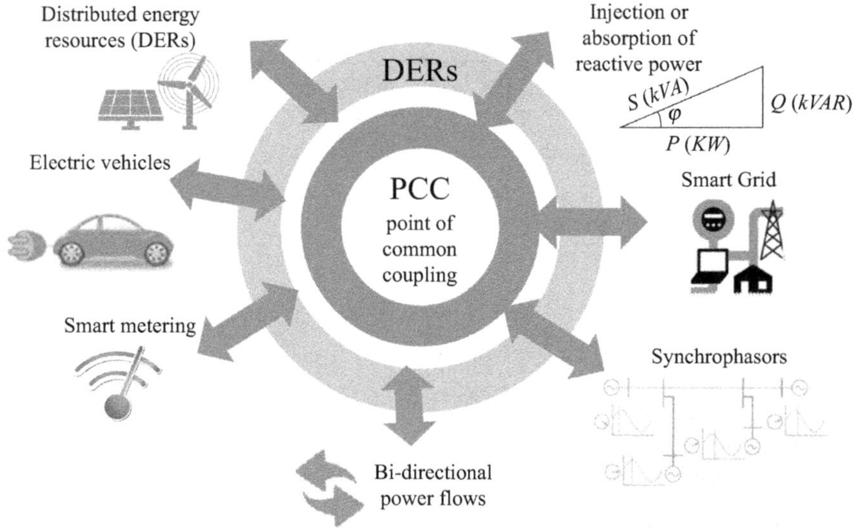

Figure 6.6 Interconnection scenario between distributed renewable generation and smart grid

The latest equipment already has these intelligent functions integrated, but these devices still represent a small percentage of installed equipment. The real problem arises when applying these new functionalities in the conventional equipment to update it to the new needs and requirements of the electrical system. The continuous increase in integration and flexibility in embedded systems, and the reduction of costs in both design and production, represent significant advantages for the use of these devices in the interconnection interface.

6.3.1 Temporary requirements for protections

In order to meet the protection functionality in the IEDs, the response times of the equipment must be taken into account when it is necessary to generate a safety action, in case the operating limits of certain electrical parameters are exceeded. Traditionally, the most common protection action has been to disconnect the DG of the network, to preserve the integrity of the electrical system, but in recent times this approach has been replaced by continuing to inject energy with characteristics that will contribute to the stability of the system. In any case, these actions must be performed at specific times, for which the main international standards must be taken into account.

This section presents a description of the temporary requirements for protection actions based on disconnection, leaving for the next section the active management of the interconnection. For this purpose, the main international standards have been compared, such as IEEE 1547 [57], IEEE 929 [58], IEC 61727 [59], IEC 62116 [60] and VDE 0126-1-1 [61].

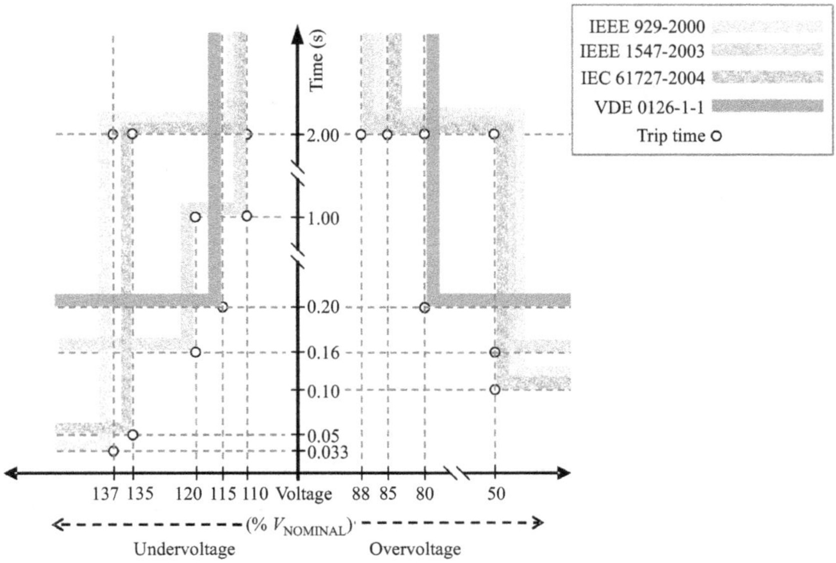

Figure 6.7 Comparison between standards for out-of-range voltage response

6.3.1.1 Voltage operational limits

As can be seen in Figure 6.7, the IEC 61727 and IEEE 929 standards are very similar with respect to operating ranges and disconnection times. In these standards, when the operation is outside the normal range, it is necessary to disconnect the equipment in less than 2 s. The following disconnection points are at a relatively large distance in percentage of the nominal working voltage, so they will require low disconnection times (less than or equal to 0.10 s). IEEE 1547 tries not to reach such high percentages of the nominal voltage, looking for a compromise solution for the surges, so that the times of disconnection can be increased somewhat. Finally, the VDE 0126-1-1 standard aims to provide a simpler solution in the control of the device, with two unique breakpoints, for over voltages and under voltages, respectively.

6.3.1.2 Frequency operational limits

Regarding frequency variations, the operating window and disconnection times vary depending on the regulations, as reflected in Figure 6.8. Both IEEE 929 and IEEE 1547 are American standards that operate at the nominal frequency of 60 Hz with the same operating ranges, although IEEE 1547 slightly relaxes the times of disconnection. The IEC 61727 standard is based on a generic value of nominal frequency and has an operating range of 1 Hz above and below it. The VDE 0126-1-1 standard applies to 50-Hz systems and, among the standards studied, is the most restrictive of the operating frequency range. Both IEC 61727 and VDE 0126-1-1 have disconnection times (less than 0.20 s) higher than the IEEE standards.

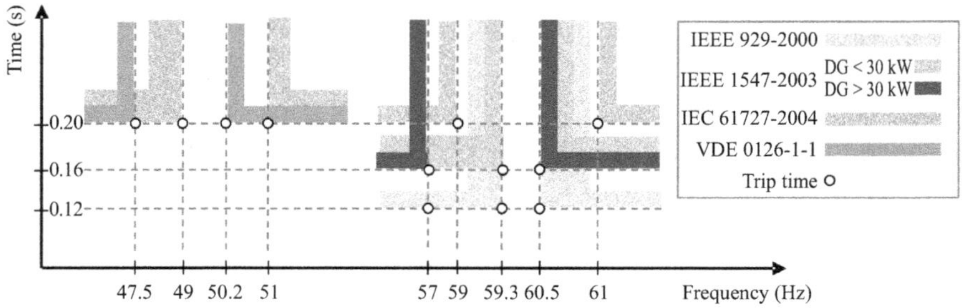

Figure 6.8 Comparison of different regulations for abnormal frequency variations

Table 6.3 Response to islanding in different standards

	Anti-islanding	
	Operation conditions	**Disconnection time (s)**
IEEE 929-2000 y UL 1741	(Match load-generation <50%) OR (load power factor <95%)	$\leq$ 10 cycles
	(Match load-generation $\geq$ 50%) OR (load power factor $\geq$ 95%) Q_{LOAD} $\leq$2.5	$\leq$2 s
IEEE 1547-2003	Any islanding condition	$\leq$2 s
IEC 62116	Any islanding condition	$\leq$2 s
VDE 0126-1-1	Single-unit operation	$\leq$5 s
	Multiple-unit operation	$\leq$5 s (Break command time $\leq$0.2 s)

6.3.1.3 Islanding

Islanding is a situation that arises when the main power grid runs out of power but DG continues to power the lines unintentionally. This scenario, if uncontrolled, is potentially dangerous from the point of view of the safety of people and equipment. So DG systems must be able to detect it and activate the protection systems in a given time to take the appropriate actions, which traditionally implies deactivating the DG of the electric grid [62,63].

It can be seen in Table 6.3 that in the standards there is a general consensus of assigning a response time of less than 2 s to disconnect the equipment from an island situation. However, the VDE 0126-1-1 standard relaxes the switch-off times to less than 5 s but also considers operating situations with multiple devices in which it defines a time below 0.2 s to generate the command that disconnects all inverters. This consideration is very useful, for example, when several very close inverters are simultaneously operating.

Table 6.4 Response in recovery of normal network conditions in different standards

	Response to utility recovery
IEEE 929-2000	Voltage and frequency within the specified ranges AND delay for 5 min
IEEE 1547-2003	No DR reconnection shall take place until the area EPS voltage is within Range B of ANSI C84.1-1995, Table 6.1 AND frequency range of 59.3 to 60.5 Hz AND delay for 5 min (or fixed delay of 5 min)
IEC 61727	Voltage and frequency within the specified ranges AND delay for 20 s to 5 min
VDE 0126-1-1	Voltage and frequency within the specified ranges AND delay for 30 s (≤ 5 s, for short interruption[1])

[1]A short interruption is characterised by the upper and lower limits.

6.3.1.4 Response to recovery of normal power grid conditions

Once the protections have acted and disconnected the equipment from the mains, it is important to define the reconnection procedure. This procedure causes a delay in the activation of the equipment, so that it is possible to verify a series of basic and quality parameters before reconnecting it to the power grid. Table 6.4 summarises these conditions. We see that VDE 0126-1-1 is the standard that requires less delay and also provides consideration for a 'short' delay for 'fast' interruptions.

6.3.2 The active management of the interconnection

The technological development related to the renewable generation has meant a great integration of these in the electrical systems. This has led to the fact that renewable generation, mainly PV and wind, has been introduced in the energy policies of many states to boost the countries' own energy resources and reduce energy dependence from abroad. This context has given rise to a greater complexity in the management of the electrical system, especially in the distribution areas, where the focus of a passive model of energy consumption is giving way to an active model in which bidirectional energy flows may exist. In the coming years, this trend will be consolidated, which will require the system to evolve towards a much smarter active model known as Smart Grid [64–66].

Increasing competitiveness in the economic costs of renewable energy facilities is leading to an increasing penetration of renewable generation. For the particular case of PV systems, the low cost of panels is increasing the proliferation of these systems, especially among home users. As PV generation is intermittent by its very nature, the high penetration of PV generation that is currently occurring can lead to grid instability if the generation systems themselves do not adequately regulate the interconnection with the grid. According to some authors, these effects begin to be significant in the distribution areas, where the penetration of PV energy reaches between 10% and 15% [67]. Therefore, those areas where solar energy is an important part of the grid are exposed to higher economic costs due to possible cuts and increased maintenance needs.

Given that inverters are now acquiring a greater degree of intelligence, a possible relatively low-cost solution to mitigate the voltage changes caused by fluctuating solar power generation would be for the inverters themselves to generate the active power and reactive power required to control injection or reactive power absorption at the point of interconnection according to electrical service requirements [68]. With this type of active management, the consequences of intermittent generations, such as solar or wind power, can be mitigated, avoiding possible problems in the electrical system.

Based on this approach, the main international standards governing the operation of DG are being reconsidered, since this approach is in precisely the opposite direction to the criteria that these regulations have followed up to now. Traditionally, these regulations have forced inverters to disconnect from the network at the first sign of problems, but in current electric grid scenarios this simple security action can be counterproductive, as it can cause even more instability.

As an example, the graphs in Figure 6.9 show the case of Spanish legislation [69,70]. Graph (a) shows the voltage-time curve showing the area in which the inverter must operate when a problem occurs in the electric grid. Graph (b) shows the operating area for the reactive current absorbed or generated according to the voltage at the point of interconnection.

In this new smart grid context, the control of the electrical interconnection between DERs, loads and the rest of the system becomes critical. This will require the equipment that manages the interconnection to meet the new requirements, which often implies an increase in the intelligence of their electronic systems. And it is in this scenario in which embedded systems and IEDs play an important role,

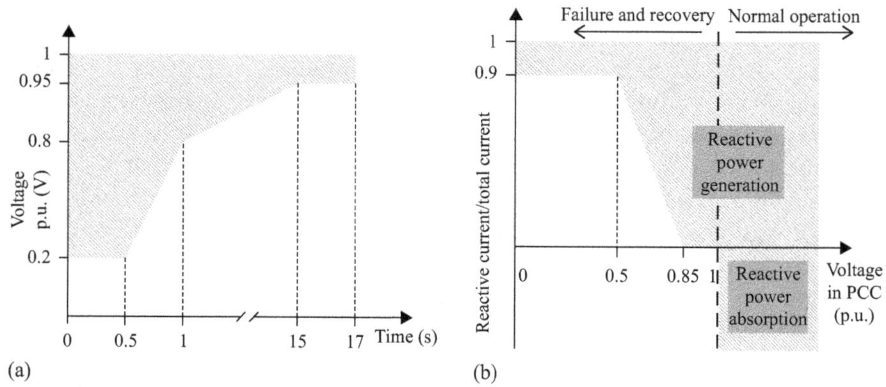

Figure 6.9 Operational requirements for the inverter in the interconnection in accordance with Spanish regulations (Operation Procedure 12.3). (a) Voltage-time curve for the operating ranges of the inverter on a fault. (b) Injecting or absorbing reactive power

either with their integration into new equipment put on the market, or to improve the equipment already installed and adapt it to the new scenarios that are coming.

6.3.3 Solutions for the interconnection with electrical grid: smart inverter

An intelligent active inverter can be developed by considering, on the one hand, the conventional systems of an inverter (power electronics and control) and, on the other hand, an IED specifically designed to provide the inverter with the specific functions to control and manage the interconnection with the electrical grid [71,72]. Therefore, the IED is responsible for continuously analysing the operational context of the inverter and the grid, indicating the most suitable operating conditions for the state of the service at all times. This implies that the inverter, in addition to fulfilling its role as a DC/AC converter, can manage energy accumulation when there is a lack of demand in the electrical system, or inject reactive energy to compensate for any event that degrades the quality of the electric service. This makes it possible for the inverter to change the energy injected according to the needs of the service, according to the scenario shown in Figure 6.10. In this way, the inverters become active parts of the electric grid, achieving one of the objectives considered as research and development priorities for the European Smart Grids Platform [73].

For this purpose, the functions of monitoring electrical parameters and energy quality must be implemented, taking into account the appropriate time intervals in the process of extraction and measurement of parameters defined in IEC 64000-4-30 [34,74,75]. The system must also have protection functionality, considering the appropriate ranges of operation of different parameters in the electrical system and the response times of the protection systems against abnormal conditions, taking as a basis the most representative regulation applied to the interconnection with the electrical system. These needs require a flexible architecture in the design of the IED that brings intelligence to the inverter, so that all the subsystems can interact with each other, and the equipment can communicate with other external entities. An example of an open and flexible architecture, capable of integrating the different functional blocks that make up the system, is described in Figure 6.11.

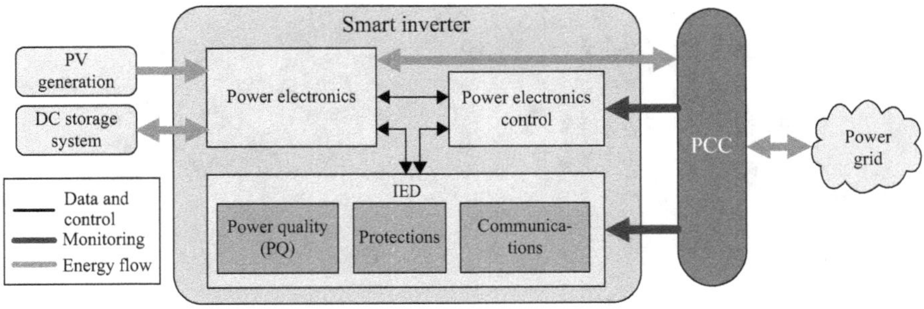

Figure 6.10 Context of operation of a smart inverter

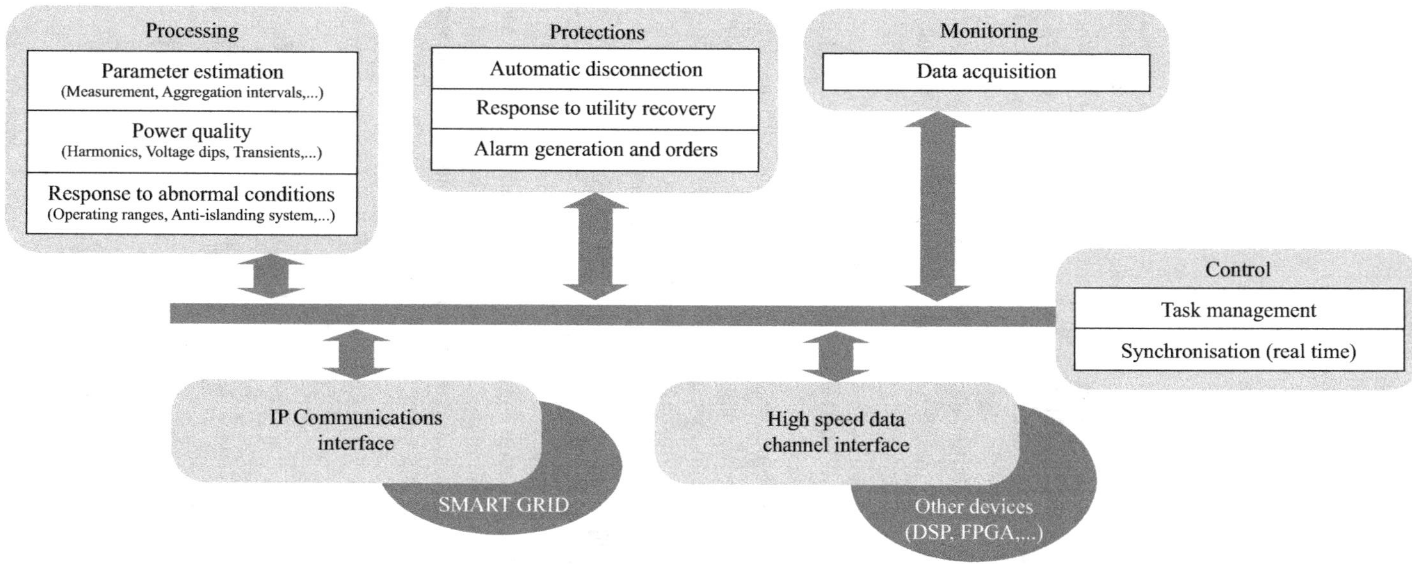

Figure 6.11 Multifunctional architecture to integrate an IED into a smart inverter

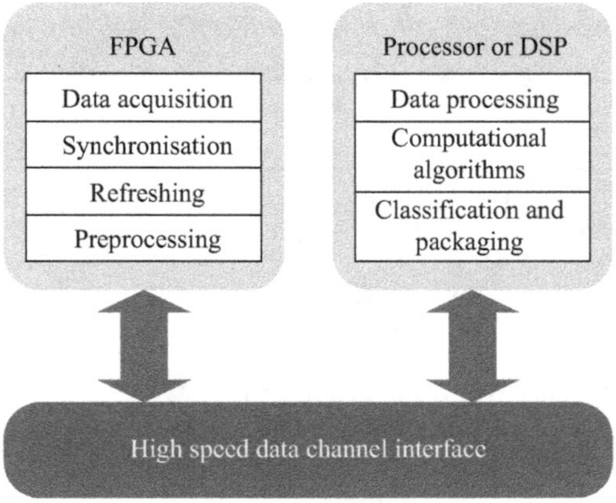

Figure 6.12 Strategy for the distribution of tasks between processor or DSP and FPGA

In order to achieve a high level of functionality, it may be necessary to use several specific parallel processing devices that share the tasks of the different subsystems in order to avoid the appearance of bottlenecks [76]. It is possible to propose the integration of two or more DSPs in the same IED; however, this procedure can be complex to implement and also expensive. An interesting and versatile solution consists of integrating a processor and FPGA (Field Programmable Gate Array) in order to distribute the tasks that have to be executed. In this sense, the strategy to achieve optimal efficiency is to assign to the FPGA those tasks that depend heavily on hardware resources, such as data acquisition, synchronisation, refreshment and even some degree of preprocessing, and on the other hand, assign to the DSP or to the processor those tasks of heavy computational load, such as data processing and calculation and decision algorithms. Figure 6.12 describes such a strategy. FPGA-based compromise solutions are now available, with these devices being increasingly integrated into platforms and embedded systems alongside processors or DSP. In these cases, high-speed busses such as SPI or PCI Express must be used for communication between different processing devices.

The added features in the smart active inverter are due to the coordination between the IED, which provides the new functionalities, and the power systems of the inverter. To do this, a series of information data should be exchanged, as detailed in Table 6.5 for guidance.

6.4 Wide-area network: data model with the IEC 61850 standard for smart grid

When talking about an intelligent wide-area network, an efficient network with systems that cooperate from the points of power generation to the consumer is proposed, regardless of the distance between the management systems. The aim is to rapidly and predictably detect the occurrence of anomalies that can trigger

Table 6.5 Data exchange between IED and power electronics in a smart inverter

Description of information	Transmitted (TX) or received (RX) by IED
Start and stop order	TX
State of the power grid	TX
State of the inverter	TX/RX
Type of event that is occurring in the power grid	TX
Order of maximum power (in kW) that must be generated by the inverter	TX
Active power (in kW) generated by the inverter	RX
Limitation on the active power generation	RX
Order of reactive power (in kVAR) that must be generated by the inverter	TX
Reactive power (in kVAR) generated by the inverter	RX
Limitation on the reactive power generation	RX
Second harmonic value that must be generated by the inverter	TX
Third harmonic value that must be generated by the inverter	TX
Fifth harmonic value that must be generated by the inverter	TX
Seventh harmonic value that must be generated by the inverter	TX
Eleventh harmonic value that must be generated by the inverter	TX
Thirteenth harmonic value that must be generated by the inverter	TX
Limitation on harmonic generation	RX
Values of unbalance between phases in the inverter	TX/RX
Limitation on generating unbalanced currents	RX

supply problems with serious consequences [12]. It does so with the ability to reconfigure according to the operational needs of each system that is part of the whole, improving response times. In short, these systems are contributing to greater energy efficiency [13]. Therefore, there is a scenario that demands an effort in the development of so-called distributed management systems [77,78].

Almost all systems specially designed for critical tasks, such as control and protection systems for substations and distribution networks, are evolving with the incorporation of advanced capture and processing capabilities that allow identifying and diagnosing the origin of the anomaly. In this category the so-called Smart Power Quality Analyser (SPQA) can be included [45].

6.4.1 Integration of renewables in wide area networks

As mentioned above, the current trend in network management is to increase the share of DG systems [79] with energy sources such as solar and wind power. Greater penetration of micro-generation, as provided by small-scale PV panels, changes the traditional radial structure of the grid. This substantial change in the structure of the gird produces new problems that were previously unknown to engineers [80,81], mainly facing the network management.

The intermittent nature of wind and solar generation poses operational problems and challenges in microGrids. To ensure the continuity of electrical supply, other microGrids must be able to respond quickly to replace sources that generate energy intermittently [82]. As noted above, when penetration of renewables increases, it becomes more difficult to maintain equilibrium with increasing variability and uncertainty in the supply of electricity [83]. One of the key requirements for achieving intelligent energy distribution is the integration of intelligent devices that communicate in real time [36,84] and do so with a high degree of stability, coordination and synchronism [35]. This communication must be based on internationally recognised standards, such as IEC 61850.

However, DG, conventional or renewable, has great potential to improve the reliability of the electrical system because it can reduce the need for new transmission lines and is usually close to the customer's load. It can also be configured to meet various power demands. With careful planning and a correct choice of DG, it is possible to help stabilise the system, and even avoid major power cuts, in some cases.

In intentional islanding systems, this can play a vital role in maintaining the balance between generation and load, ensuring that each area is balanced. However, the main and known disadvantage of DG is that it may not be available in the locations that, at any given time, would need it.

6.4.2 Detection of faults in cascade and fall of the network (blackout)

A 'blackout' or failure of the electrical system is the state in which the entire system, or large areas of it, can collapse completely. This situation is usually preceded by a sequence of cascade error events, which cause the transmission lines and the generation units to fall. Any major disruption in load generation and load balancing in a massively interconnected system can lead to undesirable variations in power flows and line voltages. Sometimes, this imbalance can spread without control over an entire electrical system, causing large areas to fall.

Large network failures often start with unintended islanding, triggered by some sort of contingency of a diverse nature, when some areas are disconnected from the rest of the system by forming islands and causing a power imbalance. Then the areas affected by the network failure may grow as a result of a series of power outages and cascading faults. In a typical scenario, the disturbance spreads rapidly because the protective relays, installed at the critical nodes, disconnect the key components in an attempt to isolate the damage. Although the reasons behind these collapses may vary from case to case, there are some factors that are significant in most documented network failures. Among them it is possible to highlight the following:

- The degree of interconnection of the network.
- The existing imbalance between the generating capacity and the transmission capacity of the lines.
- The imminence of an event that implies greater demand for load and generation.
- The power flowing through the areas at a given instant.

In this context, other systems that can coexist with the SPQA are the so-called PMUs or synchrophasors governed by standards C37.118.1 [16] and C37.118.2. These meters allow for determination of the stability of the electrical network in a very wide geographic zone with an accuracy of microseconds. Therefore, the synchronisation of PMUs is key for monitoring, protection, control and state estimation in power systems. The values of the voltage and current phasors must be transferred to a central data-processing system, known as a phasor data concentrator (PDC). Each PDC collects the results of the phasor measurements of a given area. Finally, the entire power grid data are made available to a central PDC management system. Thanks to the accuracy (close to the microsecond) and the number of phasors transferred per second (from 10 to 50) from the meters, the system is able to capture and represent the dynamic behaviour of the energy network.

Currently, the PMUs allow estimating the frequency and detecting phase deviations with errors close to 1 μs. To achieve this level of accuracy, all PMUs use the pulse-per-second signal from a GPS [85,86] and stabilise their clocks locally. To maintain this quality, the synchronism subsystem has to be of high precision.

The IEEE 1588 protocol, which is the subject of analysis in this chapter, can be adopted as an alternative. Instead of directly sharing a clock signal, these devices periodically exchange information and adjust their local clocks based on the clock that acts as a master.

Finally, the IEC 61850 standard [15] seeks to integrate the needs for interoperability and communication between these systems in charge of the supervision, management and control of electric energy. This standard defines the synchronisation and information transfer needs in real time. Some international committees are working on the process of integrating the synchrophasors standard for critical infrastructure [16]. The integration is defined in part IEC 61850-90-5 of the standard [17].

6.4.3 Data model with the IEC 61850 standard

As we have specified in the previous section, the international standard IEC 61850 [76] facilitates the integration of PMUs and SPQAs through a communication architecture based on the principles of distributed interconnection and based on a semantic definition that adapts to the type of interoperability desired [87]. IEC 61850 supports semantic definitions for DER [19,20] and, therefore, can be applied to the microgrid. Other semantic interpretations may be needed for new capabilities in microgrid applications. In this case, new LNs can be defined to cover these shortcomings.

In this chapter, part 7-420 of IEC 61850 [47] is used to define the LDs and the LNs associated with a real PV installation that, in conditions of maximum efficiency, can generate until 6 MW. Section 7.2 of IEC 61850-7-420 is dedicated to LNs for PV systems. This part of the standard classifies it into the category of 'large PV system' recommending the individual monitoring of PV-String and some auxiliary systems. So a sub-array of a sector with four inverters of 100 kW each has been chosen. The focus of the study is on the measurements of the current generated by each of the 22 strings of the selected sub-array and the power and energy measurements at the output of each inverter. These measures are carried out with the deployment of a wireless monitoring system.

In real-time systems, the way in which the data are grouped in the information nodes for their transmission has a direct impact on the response time [19]. Therefore, it is important to optimise the processing time prior to the transmission of the information. This factor is decisive because it is necessary to process information nodes with CEP structures [19]. The PV plant object of this study is based on the definition of several LNs, such as those dedicated to electrical measurements (MMXU) with at least 20 floating point parameters [47].

6.4.4 IEC 61850 modelling for distributed energy resource applications

The top level of the IEC 61850 hierarchical model is occupied by the physical device and represents the visible behaviour of the device. A physical device can host one or more LDs, whose mission is to represent the information relative to a set of specific functions of the application. An LD can perform one or several functions and these functions are modelled by LNs included in the LD.

The first letter of an LN name identifies the group to which it belongs. Thus, for example, the first letter of the LNs for measurement applications is M, and the letter P is the first letter of the LNs for protection applications. The extension of IEC 61850 to other areas of the electricity sector has led to the expansion of the models and LNs originally defined in part 7–4, as well as the definition of new models and LN included in new parts of the standard.

For example, it is possible to identify the different nodes for the experimental PV installation defined above. Figure 6.13 shows a classification based on three levels. At the installation level there is a sub-array with several strings identified by the LN DPVA, the PV plates are defined with the LN DPVM and the types of inverters with the LNs ZINV.

At the equipment level, there is an SPQA which is based on the MMDC logic nodes for voltage and current measurements. The MMXU LN is used for voltage and current measurements and the MMTR logic node for the power measurements and energy. The accuracy and quality of the synchronisation is defined with the RSYN LN. It is important to note that LNs have been included for the detection of disturbances at the inverter output. Other LNs are associated to frequency variations (QFVR node), current transients (QITR), voltage transients (QVTR) and voltage events (QVVR).

Finally, LNs have been included for the meteorological measures, like temperature, with the MMET LN or measures of irradiance with the LN STMP.

As mentioned before, all of the LNs have been chosen for the exchange of information with DER as defined in part 7-420 of IEC 61850.

6.4.5 Stability with synchrophasors and synchronisation with PTP

Electricity distribution networks are working near the stability limit more often. The fundamental cause is the increase in generation systems with renewable energies and the continuous growth of the systems of electronic control of power for converting

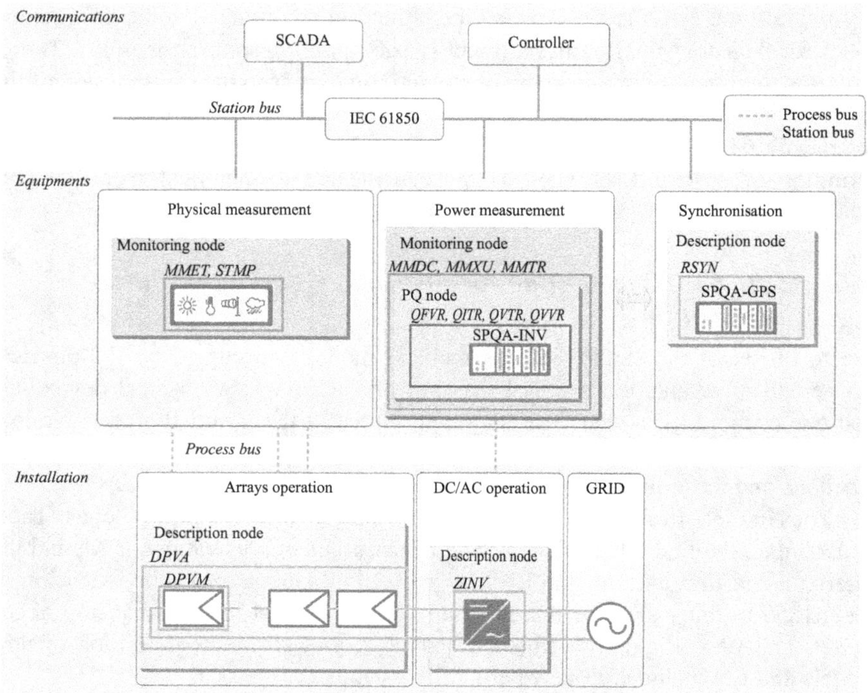

Figure 6.13 Logical nodes for the PV plant

energy. These systems are in charge of a correct transformation of the energy according to the characteristics of the load. The result is a change in the static behaviour of the distribution networks towards a more dynamic behaviour. This requires dynamic control of the network in real time with a large number of measurement systems that determine the degree of stability of each network as a whole.

PMUs meet the requirements for analysing the stability of the network in real time and are based on the C37.118-2005 standard [88] and its evolution towards IEC 61850-90-5 [89]. These two standards include all specifications for synchronisation, processing and transmission of voltage and current phasors. Communication can be established between the PMUs and a central team acting as a PDC.

Obviously, an electric grid has a very large extension. Measuring the stability of the entire network is highly complicated. The distance and the large number of measurement points complicate the definition of a measuring system with a unique synchronism. Decisions are generally taken to minimise the number of measurement points and to try moving that measurement to a large number of nearby nodes.

It is important to mention that the energy sector uses a time standard, known as IRIG-B standard, for event monitoring systems in substations. However, with receivers following the IRIG-B standard, the accuracy of the synchronisation is in the order of 1 ms, which is not sufficient to obtain accurate measurements of the energy distribution system (a tolerance of 1 ms corresponds to an uncertainty of approximately 18° in the phase). In contrast, with 1 μs of latency, it is possible to achieve accuracy around a phase angle of 0.018° for a 50 Hz system.

A suitable solution (Figure 6.14) combines GPS technology with the PTP protocol, known as the IEEE 1588 standard [23]. The technique consists of choosing a Grandmaster PTP (first level) located in a strategic geographic position to act as coordinator of several devices called master PTPs (second level) located in remote facilities. The master PTP is responsible for coordinating the equipment (slave PTP) located in the nearby local network.

PTP master devices act as PMUs [39,90] and control the stability of the electrical network with immediate actions. The same equipment can integrate a relay for temporary disconnection until the phase stabilises, or it can transmit that information to other devices in the local network with a generic object oriented substation event (GOOSE) message, which guarantees very fast transference. A remote server periodically receives phasors with manufacture message specification (MMS) messages for a centralised system to maintain global network information. This server acts as a PDC. Transfer times are no longer critical. MMS messages work with transfer times greater than those obtained with GOOSE messages.

The SPQAs, as shown in Figure 6.14, detect the potential disturbances in each installation, record them and transfer them with GOOSE messages to the local server. This equipment can execute a protection action immediately. The typical transfer times are less than 3 ms. The same local server may be receiving periodic supply quality information from the SPQA; as this information is not critical, it is feasible to receive the data with MMS messages.

The GOOSE and MMS messages are part of the IEC 61850 standard. GOOSE messages carry critical information and usually establish communication with protective equipment located on the same local network. MMS messages can be transferred to remote systems using the conventional TCP/IP protocol.

To apply this technology to both PV and wind power plants, it is necessary to develop applications that unify the IEC 61850-7-420 standards for PV installations with the IEC 61400-25 standard for wind installations.

6.4.6 Justification of the distributed synchronism through the IEEE 1588 v2 protocol

All measuring systems need a precise and stable time base. When analysing the factors that influence the instability of the oscillators in charge of maintaining the clock signal, it has been observed that this lack of stability can produce considerable error locally over the long term. This problem can be aggravated when trying to maintain a simultaneous synchronisation of different clocks for measuring systems working in a distributed way.

Continuous technological advances have led to the emergence of new techniques that have progressively improved the accuracy and stability of local sources. For the proposed application, the reference provided by the GPS system based on atomic clocks and guarantees levels of synchronism greater than 1 μs was used as the reference source.

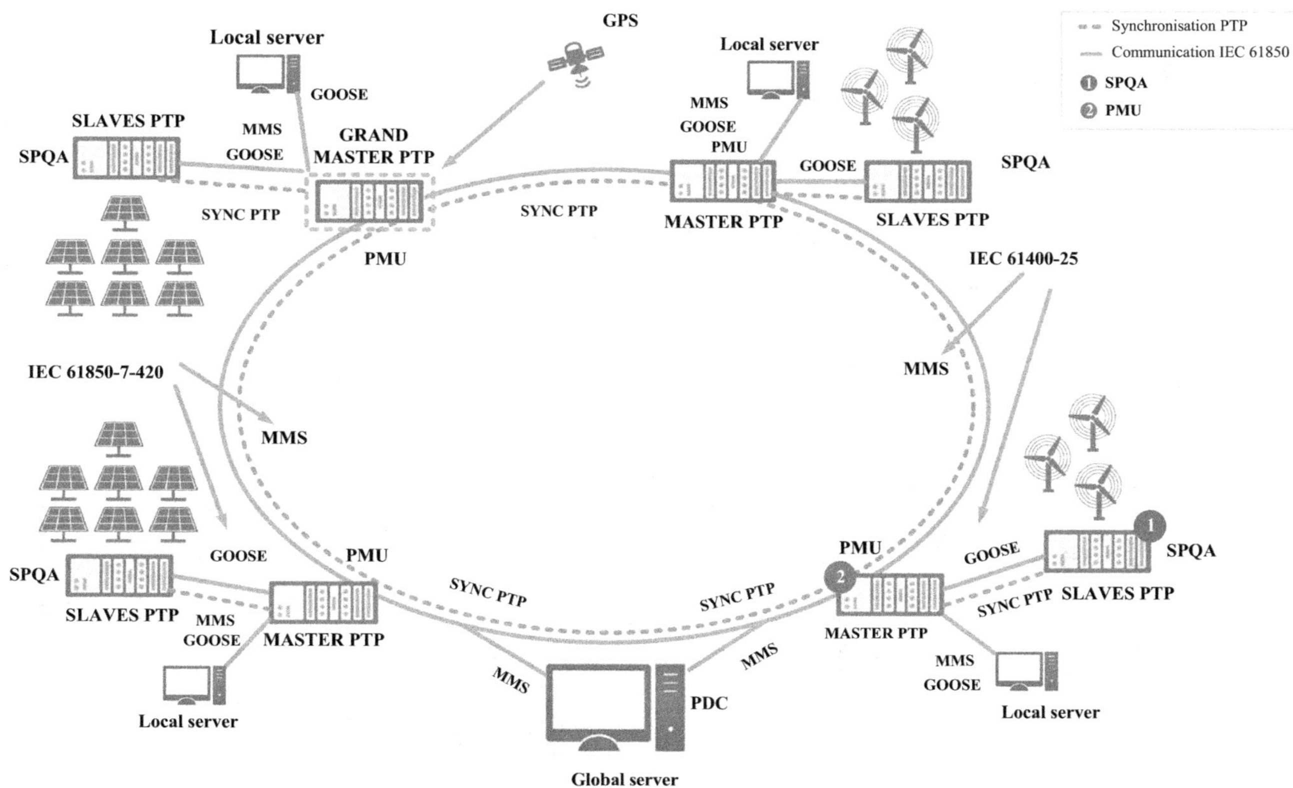

Figure 6.14 Wide-area network: Stability with synchrophasors, PTP synchronisation and IEC 61850 integration

For a distributed system, the greatest difficulty lies in a heterogeneous network composed of clocks with different types of oscillators and therefore with different degrees of frequency stability. To achieve this synchronism, the devices must have direct access to the synchronisation signals or synchronise their clocks individually to maintain a common time base for the whole system. When devices are relatively close, within a few meters of each other, sharing a clock signal is relatively easy and maintains a high degree of accuracy. For accurate use of the clock signal, a device must be calibrated to compensate for possible propagation delays that may occur from the clock source to the device.

Sharing a clock signal at a moderate distance of about 50 m may imply a significant cost for wiring and system configuration. In this situation, a distributed synchronisation system is required. An example of a distributed system is found in devices that are synchronised with GPS or a set of devices with an internal clock synchronised with an NTP server.

For distributed synchronisation, it is necessary to keep in mind the relation between the precision of the synchronism and the distance between the nodes of the system. Any system design must take into account the limitations imposed by these variables. As the transmission distance increases, it becomes more difficult to share the signal between systems to maintain synchronisation.

Figure 6.15 shows that the distance between nodes degrades the clock signal. Therefore, from a certain limit, it is necessary to use another method to synchronise the clocks. A GPS located at each point does not suffer degradation with distance. All points take a very precise common signal as reference. In the IEEE 1588 protocol, instead of directly sharing a clock signal, these devices periodically exchange information and adjust their local clock according to the clock that acts as a master.

The accuracy of the synchronism [24] in the electric sector depends on the type of task performed by the system. As explained above, electric energy must be consumed at the same time as it is generated, so the generation has to be

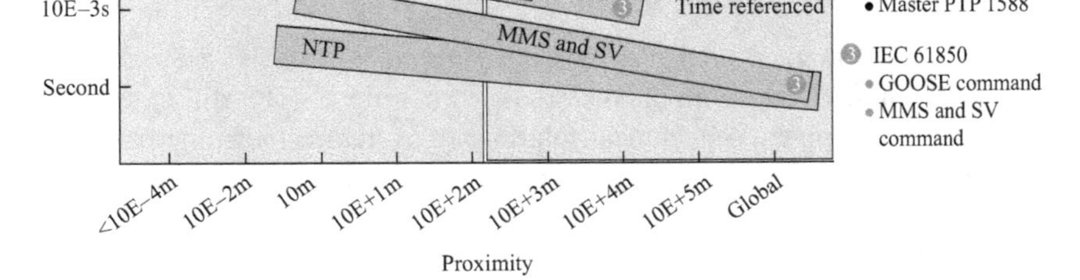

Figure 6.15 Synchronism accuracy versus distance among different technologies

Table 6.6 Synchronisation levels for electrical grids

System function	Measurement	Time requirement
Generation control	Generator phase	10 ms
Events recorder	Time tagging of record	1 ms
Stability controls	Phase angle, $\pm 1°$	55 μs
Network controls	Phase angle, $\pm 1°$	5.5 μs
Travelling wave fault locators	300 meter tower spacing	1 μs
Synchrophasor measurements	Phase angle, $\pm 0.018°$	1 μs

continuously balanced on the basis of the evolution of the load. As shown in Table 6.6, control of power generation requires synchronisation times of 10 ms, which is referred to as a frequency of 50 Hz. The memorisation of events or faults in the network requires a synchronisation of around 1 ms. Stability control equipment require equipment with 55.5-μs synchronisation (equivalent to $\pm 1°$ of phase angle for 50 Hz) to prevent downtime generation in systems, and the control of the network requires a synchronism of an order of magnitude of 5.5 μs (equivalent to $\pm 0.1°$ of phase angle for 50 Hz).

As the distance between the high-voltage towers is around 300 m, the required synchronism is 1 μs or the period equivalent to a wavelength of 300 m. New measurement techniques such as synchrophasor measurements require synchronisation times in reference to universal time coordinates of less than 1 μs corresponding to a phase-angle-based accuracy of 0.018° for a system of 50 Hz.

It is of little use to maintain such a high threshold to ensure the synchronisation of the phasor meters if the information transfer is done within an Ethernet network with conventional TCP/IP commands. Fortunately, the IEC 16850 standard has defined a new type of communication service independent of the TCP/IP protocol to achieve high performance with real-time communications. It is logical that critical events be transmitted locally with GOOSE messages. A comparison between MMS messages and GOOSE messages has been included in Figure 6.15 to highlight the differences in performance depending on the distance between the devices. An intelligent network must maintain a high degree of quality in both detection techniques and critical information transfer strategies.

6.5 Conclusions

Despite the technological advances that are currently available, the transition to future intelligent networks with the massive incorporation of DG still requires wide research and development of technical solutions to overcome the numerous existing barriers. Changes and adaptations in the present regulations are also necessary.

On one hand, future smart grids need to implement efficient monitoring and control technologies to improve their operational performance as well as optimal integration in the grid of their production that takes into account their specific fluctuating and unpredictable behaviour. In order for this type of facility to be dispatchable, it is vital for them to be perfectly characterised, and the information regarding energy flowing in and out of the DERs must be fully monitored. To this end, the monitoring systems of the DERs must collect, in real time, detailed information about the values of the different parameters and magnitudes that characterise the operation of these systems as well as all of the technical specifications corresponding to their different components. An example of a platform developed for supervising a PV plant is included in this chapter. An adequate characterisation of a DG system enables the following results to be obtained:

- To optimise production; diagnose; quantify and minimise the different loss mechanisms; increase the reliability of its components; and reduce the costs of operating and maintaining the plant by developing algorithms for self-diagnosis, identification and classification of deviations and failures in any element of the facility.
- To obtain information to optimise the installation and operation of future storage systems.
- The intelligent management of the interconnection between DG and the electrical system, which is one of the important keys to the implementation of the smart grid concept. For this, it is key to know the temporary requirements specified by the regulations regarding PQ and protections. For this, the characterisation of a DG system must also determine the quality of the electrical signal produced, by developing diagnostic methods to detect deviations, disturbances and events in the generated signal as well as the capacity to accommodate, in an appropriate way, the production in the electric grid.

In this new scenario, DERs must be able to actively manage the interconnection, allow bidirectional flows of energy and provide some of the intelligence of the global system. Proof of this is the behaviour demanded for the new inverters, which have to be able to generate the active and reactive power necessary to control the injection or absorption of reactive power in the point of interconnection according to the needs of electric service.

Therefore, the integration of DERs and the existence of a large number of active loads in the future grid, which will threaten the stability of the electricity network, will also require more sophisticated systems of protection and control. There are multiple systems based on voltage, current and frequency measurements, but there is often no standardisation for communications. The IEC 61850 standard is allowing the integration of a large number of generation and transport systems, especially when the equipment is located in very large areas. Compatibility between systems developed by different manufacturers is fundamental.

The requirements for detecting and controlling failures at the local level are much more demanding. At this level, the IEC 61850 standard provides for the rapid transmission of critical information. GOOSE messages are not dependent on the standard TCP/IP service, and the transfer rate is the maximum allowable by the routing equipment.

Another key is the timing of all of the systems involved. This chapter has justified the choice of the PTP standard for coordinating all of the equipment located in an installation as the most appropriate way of extending it to other installations in a wide geographical area. It is possible to reduce the amount of equipment equipped with GPS, but guaranteeing the synchronism with a quality equivalent to 1 μs.

An intelligent network must evolve dynamically according to demand. In this context, it is essential to maintain a control of the stability of the network in a greater number of control points. The current technology has lowered the costs of high-performance electronic systems. As discussed in this chapter, PMUs or synchrophasors are an essential piece, and their integration into the IEC 61850 standard allows a local and centralised management system to be defined that cooperates with other systems located in the network, such as SPQAs and intelligent inverters.

References

[1] Strasser T, Andrén F, Kathan J, *et al.* A Review of Architectures and Concepts for Intelligence in Future Electric Energy Systems. IEEE Trans Ind Electron 2015;62:2424–38. doi:10.1109/TIE.2014.2361486.

[2] Yu X, Cecati C, Dillon T, Simões M. The New Frontier of Smart Grids. IEEE Ind Electron Mag 2011;5:49–63. doi:10.1109/MIE.2011.942176.

[3] International Energy Agency (IEA). Technology Roadmap. Smart Grids 2011. https://www.iea.org/publications/freepublications/publication/smartgrids_roadmap.pdf (accessed 7 November 2016).

[4] Tannahill BK, Jamshidi M. System of Systems and Big Data Analytics – Bridging the Gap. Comput Electr Eng 2014;40:2–15. doi:10.1016/j.compeleceng.2013.11.016.

[5] Mikati M, Santos M, Armenta C. Modelado y Simulación de un Sistema Conjunto de Energía Solar y Eólica para Analizar su Dependencia de la Red Eléctrica. Rev Iberoam Automática E Informática Ind RIAI 2012;9:267–81. doi:10.1016/j.riai.2012.05.010.

[6] Moreno-Munoz A, De la Rosa JJG, Real-Calvo RJ, Pallares V. Embedding Measurement in Distribution Automation Systems. 2010 IEEE Int. Symp. Ind. Electron., IEEE; 2010, p. 3722–27. doi:10.1109/ISIE.2010.5637698.

[7] Real-Calvo RJ, Moreno-Munoz A, Pallares-Lopez V, Gonzalez-Redondo MJ, Moreno-Garcia IM. Design of an Intelligent Electronic Device to Control a Private Microgrid. 2012 IEEE Second Int. Conf. Consum. Electron. – Berlin, IEEE; 2012, p. 99–101. doi:10.1109/ICCE-Berlin.2012.6336492.

[8] Fischer JR, González SA, Carugati I, Judewicz MG, Carrica DO. Control Directo de Potencia Predictivo Robusto con Sincronismo Intrínseco. Rev Iberoam Automática E Informática Ind RIAI 2015;12:199–207.

[9] Elektrotechniker VD. Technical Minimum Requeriments for the Connection to and Parallel Operation with the Low-Voltage Distribution Networks, VDE Verlag; 2011.

[10] Guerrero-Martínez M., Romero-Cadaval E, Minambres-Marcos V, Milanés-Montero MI. Supercapacitor Energy Storage System for Improving the

Power Flow in Photovoltaic Plants. Inf Midem-J Microelectron Electron Compon Mater 2014;44(1):40–52.

[11] Bordons C, García-Torres F, Valverde L. Gestión Óptima de la Energía en Microrredes con Generación Renovable. Rev Iberoam Automática E Informática Ind RIAI 2015;12:117–32.

[12] Moreno-Muñoz A. Power Quality: Mitigation Technologies in a Distributed Environment. London, Springer Science & Business Media; 2007.

[13] Schneider KP, Chassin D, Chen Y, Fuller JC. Distribution Power Flow for Smart Grid Technologies. 2009 IEEE/PES Power Syst. Conf. Expo., IEEE; 2009, p. 1–7.

[14] Zhen H, Cheng W, Ruj S, Stojmenovic M, Nayak A. Modeling Cascading Failures in Smart Power Grid Using Interdependent Complex Networks and Percolation Theory. 2013 IEEE Eighth Conf. Ind. Electron. Appl., IEEE; 2013, p. 1023–28. doi:10.1109/ICIEA.2013.6566517.

[15] McGhee J, Goraj M. Smart High Voltage Substation Based on IEC 61850 Process Bus and IEEE 1588 Time Synchronization. 2010 First IEEE Int. Conf. Smart Grid Commun., IEEE; 2010, p. 489–94. doi:10.1109/SMARTGRID.2010.5622092.

[16] Martin KE. Synchrophasors in the IEEE C37.118 and IEC 61850. 2010 5th Int. Conf. Crit. Infrastruct., IEEE; 2010, p. 1–8. doi:10.1109/CRIS.2010.5617573.

[17] Ali I, Aftab MA, Hussain SMS. Performance Comparison of IEC 61850-90-5 and IEEE C37.118.2 based Wide Area PMU Communication Networks. J Mod Power Syst Clean Energy 2016;4:487–95. doi:10.1007/s40565-016-0210-y.

[18] Martin KE. Synchrophasor Measurements Under the IEEE Standard C37.118.1-2011 With Amendment C37.118.1a. IEEE Trans Power Deliv 2015;30:1514–22. doi:10.1109/TPWRD.2015.2403591.

[19] Nieves JC, Espinoza A, Penya YK, de Mues MO, Peña A. Intelligence Distribution for Data Processing in Smart Grids: A Semantic Approach. Eng Appl Artif Intell 2013;26:1841–53. doi:10.1016/j.engappai.2013.03.016.

[20] Phuangpornpitak N, Tia S. Opportunities and Challenges of Integrating Renewable Energy in Smart Grid System. Energy Procedia 2013;34:282–90. doi:10.1016/j.egypro.2013.06.756.

[21] Cleveland FM. IEC 61850-7-420 Communications Standard for Distributed Energy Resources (DER). 2008 IEEE Power Energy Soc. Gen. Meet. – Convers. Deliv. Electr. Energy 21st Century, IEEE; 2008, p. 1–4. doi:10.1109/PES.2008.4596553.

[22] Ancillotti E, Bruno R, Conti M. The Role of Communication Systems in Smart Grids: Architectures, Technical Solutions and Research Challenges. Comput Commun 2013;36:1665–97. doi:10.1016/j.comcom.2013.09.004.

[23] IEEE Standard for a Precision Clock Synchronization Protocol for Networked Measurement and Control Systems. IEEE 1588, IEEE Instrumentation and Measurement Society; 2008.

[24] Blaabjerg F, Teodorescu R, Liserre M, Timbus AV. Overview of Control and Grid Synchronization for Distributed Power Generation Systems. IEEE Trans Ind Electron 2006;53:1398–409. doi:10.1109/TIE.2006.881997.

[25] Chine W, Mellit A, Pavan AM, Kalogirou SA. Fault Detection Method for Grid-Connected Photovoltaic Plants. Renew Energy 2014;66:99–110. doi:10.1016/j.renene.2013.11.073.

[26] Homepage – European Photovoltaic Industry Association – EPIA, 2017. http://www.epia.org/home/ (accessed 22 October 2014).

[27] Silvestre S, da Silva MA, Chouder A, Guasch D, Karatepe E. New Procedure for Fault Detection in Grid Connected PV Systems Based on the Evaluation of Current and Voltage Indicators. Energy Convers Manage 2014;86:241–9. doi:10.1016/j.enconman.2014.05.008.

[28] Where the Energy Storage Expansion Will Happen First and Why, 2015. http://www.utilitydive.com/news/where-the-energy-storage-expansion-will-happen-first-and-why/346265/ (accessed January 19, 2015).

[29] Koohi-Kamali S, Rahim NA, Mokhlis H. Smart Power Management Algorithm in Microgrid Consisting of Photovoltaic, Diesel, and Battery Storage Plants Considering Variations in Sunlight, Temperature, and Load. Energy Convers Manage 2014;84:562–82. doi:10.1016/j.enconman.2014.04.072.

[30] Etherden N, Bollen MHJ. Dimensioning of Energy Storage for Increased Integration of Wind Power. IEEE Trans Sustain Energy 2013;4:546–53. doi:10.1109/TSTE.2012.2228244.

[31] Santiago I, Trillo-Montero D, Luna-Rodriguez JJ, Real-Calvo R, Moreno-Munoz A. Analysis of Variations in PV Production, Focussing on Storage and Dispatchability Decisions. 2015 IEEE Int. Conf. Ind. Technol., IEEE; 2015, p. 2884–9. doi:10.1109/ICIT.2015.7125523.

[32] Moreno-Garcia IM, Pallares-Lopez V, Gonzalez-Redondo M, Lopez-Lopez J, Varo-Martinez M, Santiago I. Implementation of a Real Time Monitoring System for a Grid-Connected PV Park. 2015 IEEE Int. Conf. Ind. Technol., IEEE; 2015, p. 2915–20. doi:10.1109/ICIT.2015.7125528.

[33] Trillo-Montero D, Santiago I, Luna-Rodriguez JJ, Real-Calvo R. Development of a Software Application to Evaluate the Performance and Energy Losses of Grid-Connected Photovoltaic Systems. Energy Convers Manage 2014;81:144–59. doi:10.1016/j.enconman.2014.02.026.

[34] Electromagnetic Compatibility (EMC) – Part 4-30: Testing and Measurement Techniques – Power Quality Measurement Methods. IEC 61000-4-30; International Electrotechnical Commission; 2008.

[35] Moreno-Munoz A, Pallares-Lopez V, Gonzalez de la Rosa JJ, Real-Calvo R, Gonzalez-Redondo M, Moreno-Garcia IM. Embedding Synchronized Measurement Technology for Smart Grid Development. IEEE Trans Ind Informatics 2013;9:52–61. doi:10.1109/TII.2012.2209659.

[36] Moreno-Garcia IM, Moreno-Munoz A, Domingo-Perez F, Pallares-Lopez V, Real-Calvo R-J, Santiago-Chiquero I. Implementation of a Smart Grid

Inverter through Embedded Systems. Electron Electr Eng 2013;19:3–6. doi:10.5755/j01.eee.19.3.1378.

[37] Moreno-Garcia IM, Moreno-Munoz A, Gil-de-Castro A, Real-Calvo R, Palacios-Garcia EJ, González-Redondo MJ. Grid-Tied Smart Inverter Safety Functionality: Fast Power Quality Event Detection. Inf Midem-J Microelectron Electron Compon Mater 2015;45(3):204–15.

[38] Pallares-Lopez V, Moreno-Muńoz A, Gonzalez de la Rosa J, *et al.* Synchrophasor for Smart Grid with IEEE 1588-2008 Synchronism. Przegląd Elektrotechniczny 2012;88:31–6.

[39] Pallares-Lopez V, Moreno-Munoz A, de la Rosa JJG, Real-Calvo R. A IEEE1588-based System for Synchronized PMUs and Protective Relaying Functions. SPEEDAM 2010, IEEE; 2010, p. 593–8. doi:10.1109/SPEEDAM.2010.5542201.

[40] Pallares-Lopez V, Moreno-Munoz A, Gonzalez-Redondo M, *et al.* Synchronism with Software-Based IEEE 1588–2008 for Smart Grid. 2011 IEEE Int. Symp. Ind. Electron., IEEE; 2011, p. 1237–42. doi:10.1109/ISIE.2011.5984335.

[41] Gonzalez-Redondo MJ, Moreno-Munoz A, Pallares-Lopez V, Real-Calvo RJ. Influence of Data-Related Factors on the Use of IEC 61850 for Power Utility Automation. Electr Power Syst Res 2016;133:269–80. doi:10.1016/j.epsr.2015.12.028.

[42] Gonzalez-Redondo MJ, Moreno-Munoz A, Pallares-Lopez V, Real-Calvo RJ, Lopez MAO, Moreno-Garcia IM. IEC 61850 GOOSE Transfer Time Measurement in Development Stage. 2013 IEEE Int. Symp. Ind. Electron., IEEE; 2013, p. 1–6. doi:10.1109/ISIE.2013.6563762.

[43] Pallares-Lopez V, Moreno-Munoz A, Gonzalez-Redondo M, Real-Calvo R, Garcia IM, de la Rosa JJG. Synchrophasor Integration in IEC 61850 Standard for SmartGrid and Synchronism with PTP-base System. 2011 6th IEEE Conf. Ind. Electron. Appl., IEEE; 2011, p. 1507–12. doi:10.1109/ICIEA.2011.5975829.

[44] System Performance Monitoring-Guidelines for Measurements, Data Exchanges and Analysis. EN 61724; European Committee for Standards; 2008.

[45] Moreno-Garcia IM, Palacios-Garcia E, Pallares-Lopez V, *et al.* Real-Time Monitoring System for a Utility-Scale Photovoltaic Power Plant. Sensors 2016;16:770. doi:10.3390/s16060770.

[46] Photovoltaic System Performance – Part 3: Energy Evaluation Method. IEC TS 61724-3 2016.

[47] Communication Networks and Systems for Power Utility Automation – Part 7-420: Basic Communication Structure – Distributed Energy Resources Logical Nodes. IEC 61850-7-420; International Electrotechnical Commission; 2009.

[48] Serrano J, Alvarez P, Cattin M, *et al.* The White Rabbit Project. Proc. ICALEPCS TUC004, Kobe, Japan, Kobe; 2009.

[49] Moreno-Garcia IM, Palacios-Garcia EJ, Santiago I, Pallares-Lopez V, Moreno-Munoz A. Performance Monitoring of a Solar Photovoltaic Power Plant Using an Advanced Real-Time System. 2016 IEEE 16th Int. Conf. Environ. Electr. Eng., IEEE; 2016, p. 1–6. doi:10.1109/EEEIC.2016.7555473.

[50] Gong S, Zhang Z, Trinkle M, Dimitrovski AD, Husheng Li. GPS Spoofing Based Time Stamp Attack on Real Time Wide Area Monitoring in Smart Grid. 2012 IEEE Third Int. Conf. Smart Grid Commun., IEEE; 2012, p. 300–5. doi:10.1109/SmartGridComm.2012.6486000.

[51] Caruana C, Sattar A, Al-Durra A, Muyeen SM. Real-Time Testing of Energy Storage Systems in Renewable Energy Applications. Sustain Energy Technol Assessments 2015;12:1–9. doi:10.1016/j.seta.2015.08.001.

[52] Deng R, Yang Z, Hou F, Chow M-Y, Chen J. Distributed Real-Time Demand Response in Multiseller–Multibuyer Smart Distribution Grid. IEEE Trans Power Syst 2015;30:2364–74. doi:10.1109/TPWRS.2014.2359457.

[53] Dong Han D, Zheng Yan Z, Yiqun Song Y, Libing Yang L, Yuanrui Hong Y. A Robust Optimization Approach to Evaluate the Impact of Smart Grid Technologies on Generation Plans. 2014 Int. Conf. Power Syst. Technol., IEEE; 2014, p. 1706–11. doi:10.1109/POWERCON.2014.6993510.

[54] Benoit C, Mercier A, Besanger Y, Wurtz F. Deterministic Optimal Power Flow for Smart Grid Short-Term Predictive Energy Management. 2013 IEEE Grenoble Conf., IEEE; 2013, p. 1–7. doi:10.1109/PTC.2013.6652502.

[55] Ransom DL. Upgrading Relay Protection? – Be Prepared. 2013 66th Annu. Conf. Prot. Relay Eng., IEEE; 2013, p. 126–42. doi:10.1109/CPRE.2013.6822032.

[56] Zavoda F, Abbey C, Brissette Y, Lemire R. Universal IED for distribution Smart Grids. 22nd International Conference and Exhibition on Electricity Distribution (CIRED 2013), Stockholm, 2013, pp. 1–4. doi:10.1049/cp.2013.0976.

[57] IEEE Standard for Interconnecting Distributed Resources with Electric Power Systems; 2003, p. 1–28. doi:10.1109/IEEESTD.2003.94285.

[58] IEEE Recommended Practice for Utility Interface of Photovoltaic (PV) Systems; 2000, p. i. doi:10.1109/IEEESTD.2000.91304.

[59] Photovoltaic (PV) Systems – Characteristics of the Utility Interface. IEC 61727;2004.

[60] Test Procedure of Islanding Prevention Measures for Utility-Interconnected Photovoltaic Inverters. IEC 62116;2014.

[61] Automatic Disconnection Device between a Generator and the Public Low-Voltage Grid. VDE V 0126-1-1;2006.

[62] Dash PK, Padhee M, Barik SK. Estimation of Power Quality Indices in Distributed Generation Systems During Power Islanding Conditions. Int J Electr Power Energy Syst 2012;36:18–30. doi:10.1016/j.ijepes.2011.10.019.

[63] Mahat P, Chen Z, Bak-Jensen B. Review of Islanding Detection Methods for Distributed Generation. 2008 Third Int. Conf. Electr. Util. Deregul. Restruct. Power Technol., IEEE; 2008, p. 2743–8. doi:10.1109/DRPT.2008.4523877.

[64] Colak I, Kabalci E, Fulli G, Lazarou S. A Survey on the Contributions of Power Electronics to Smart Grid Systems. Renewable Sustainable Energy Rev 2015;47:562–79. doi:10.1016/j.rser.2015.03.031.

[65] Enslin J. Integration of Photovoltaic Solar Power – The Quest Towards Dispatchability. IEEE Instrum Meas Mag 2014;17:21–6. doi:10.1109/MIM.2014.6810041.

[66] Su W, Eichi H, Zeng W, Chow M-Y. A Survey on the Electrification of Transportation in a Smart Grid Environment. IEEE Trans Ind Informatics 2012;8:1–10. doi:10.1109/TII.2011.2172454.

[67] Western Utilities Call for Smart Solar Inverters. Greentech Media, 2013 https://www.greentechmedia.com/articles/read/western-utilities-call-for-smart-solar-inverters.

[68] Katiraei F, Sun C, Enayati B. No Inverter Left Behind: Protection, Controls, and Testing for High Penetrations of PV Inverters on Distribution Systems. IEEE Power Energy Mag 2015;13:43–9. doi:10.1109/MPE.2014.2380374.

[69] Resolución de 4 de octubre de 2006, de la Secretaría General de Energía, por la que se aprueba el procedimiento de operación 12.3 Requisitos de respuesta frente a huecos de tensión de las instalaciones eólicas 2006:37017–9.

[70] Garcia Sanchez T, Gomez Lazaro E, Molina Garcia A Integración De Recursos Renovables Y. Requerimientos De Conexión En El Sistema Eléctrico Español: Análisis De Datos En Instalaciones Fotovoltaicas FOTOVOLTAICAS. Dyna Ing E Ind 2014;89:649–55. doi:10.6036/7114.

[71] Real-Calvo R, Moreno-Munoz A, Gonzalez-De-La-Rosa J, Pallares-Lopez V, Gonzalez-Redondo M, Moreno-Garcia IM. An Embedded System in Smart Inverters for Power Quality and Safety Functionality. Energies 2016;9:219. doi:10.3390/en9030219.

[72] Romero-Cadaval E, Minambres-Marcos VM, Moreno-Munoz A, Real-Calvo RJ, Gonzalez de la Rosa JJ, Sierra-Fernandez JM. Active Functions Implementation in Smart Inverters for Distributed Energy Resources. 2013 Int. Conf. Compat. Power Electron., IEEE; 2013, p. 52–7. doi:10.1109/CPE.2013.6601128.

[73] Colak I, Fulli G, Sagiroglu S, Yesilbudak M, Covrig C-F. Smart Grid Projects in Europe: Current Status, Maturity and Future Scenarios. Appl Energy 2015;152:58–70. doi:10.1016/j.apenergy.2015.04.098.

[74] Legarreta AE, Figueroa JH, Bortolin JA. An IEC 61000-4-30 Class a – Power Quality Monitor: Development and Performance Analysis. 11th Int. Conf. Electr. Power Qual. Util., IEEE; 2011, p. 1–6. doi:10.1109/EPQU.2011.6128813.

[75] Radil T, Ramos PM. Power Quality Detection and Classification Method for IEC 61000-4-30 Class A Instruments. 2010 IEEE Instrum. Meas. Technol. Conf. Proc., IEEE; 2010, p. 691–6. doi:10.1109/IMTC.2010.5488085.

[76] Baetens R, De Coninck R, Van Roy J, *et al*. Assessing Electrical Bottlenecks at Feeder Level for Residential Net Zero-Energy Buildings by Integrated System Simulation. Appl Energy 2012;96:74–83. doi:10.1016/j.apenergy.2011.12.098.

[77] Real Calvo RJ, Moreno Munoz A, Pallares Lopez V, Gonzalez Redondo MJ. Intelligent Electronic Device for Distributed Energy Resources. IEEE Lat Am Trans 2016;14:3270–7. doi:10.1109/TLA.2016.7587630.

[78] Mehrasa M, Pouresmaeil E, Mehrjerdi H, Jørgensen BN, Catalão JPS. Control Technique for Enhancing the Stable Operation of Distributed Generation Units within a Microgrid. Energy Convers Manage 2015;97:362–73. doi:10.1016/j.enconman.2015.03.078.

[79] Sučić S, Havelka JG, Dragičević T. A Device-Level Service-Oriented Middleware Platform for Self-manageable DC Microgrid Applications Utilizing Semantic-Enabled Distributed Energy Resources. Int J Electr Power Energy Syst 2014;54:576–88. doi:10.1016/j.ijepes.2013.08.013.

[80] Niitsoo J, Taklaja P, Palu I, Klüss J. Power Quality Issues Concerning Photovoltaic Generation and Electrical Vehicle Loads in Distribution Grids. Smart Grid Renewable Energy 2015;06:164–77. doi:10.4236/sgre.2015.66015.

[81] Alharbi W, Bhattacharya K. Accommodating High Levels of Renewable Generation in Remote Microgrids under Uncertainty. 2014 IEEE Electr. Power Energy Conf., IEEE; 2014, p. 60–4. doi:10.1109/EPEC.2014.11.

[82] Zhang L, Gari N, Hmurcik LV. Energy Management in a Microgrid with Distributed Energy Resources. Energy Convers Manage 2014;78:297–305. doi:10.1016/j.enconman.2013.10.065.

[83] Siano P. Demand Response and Smart Grids – A Survey. Renewable Sustainable Energy Rev 2014;30:461–78. doi:10.1016/j.rser.2013.10.022.

[84] Cucinotta T, Mancina A, Anastasi GF, *et al.* A Real-Time Service-Oriented Architecture for Industrial Automation. IEEE Trans Ind Informatics 2009;5:267–77. doi:10.1109/TII.2009.2027013.

[85] Carta A, Locci N, Muscas C. A PMU for the Measurement of Synchronized Harmonic Phasors in Three-Phase Distribution Networks. 2008 IEEE Instrum. Meas. Technol. Conf., IEEE; 2008, p. 426–31. doi:10.1109/IMTC.2008.4547073.

[86] Carta A, Locci N, Muscas C. GPS-Based System for the Measurement of Synchronized Harmonic Phasors. IEEE Trans Instrum Meas 2009;58: 586–93. doi:10.1109/TIM.2008.2005075.

[87] Ridwan MI, Miswan NS, Shokri MSM, Noran MN, Lajim RM, Awang HN. Interoperability in Smart Grid using IEC 61850 Standard: A Power Utility Prospect. 2014 IEEE Innov. Smart Grid Technol. – Asia (ISGT ASIA), IEEE; 2014, p. 261–6. doi:10.1109/ISGT-Asia.2014.6873800.

[88] IEEE Standard for Synchrophasors for Power Systems; 2006, p. 0_1–57. doi:10.1109/IEEESTD.2006.99376.

[89] IEC 61850-90-5. Use of IEC 61850 to Transmit Synchrophasor Information According to IEEE C37.118. Int Electrothecnical Com; 2012.

[90] Pallares-Lopez V, Moreno-Munoz A, Moreno Garcia IM, de la Rosa JJG. A Experimental IEEE1588-BASED System for Synchronized Phasor Measurement in Electric Substation. 2010 5th IEEE Conf. Ind. Electron. Appl., IEEE; 2010, p. 942–7. doi:10.1109/ICIEA.2010.5515698.

Chapter 7

DC distribution systems and microgrids

Tomislav Dragicevic[1], *Amjad Anvari-Moghaddam*[1], *Juan C. Vasquez*[1] *and Josep M. Guerrero*[1]

7.1 Introduction

The idea of merging variable sources with energy storage and controllable loads into flexible systems called microgrids (MGs) has been presented more than a decade ago [1]. MGs can operate autonomously or be grid-connected, and depending on the type of voltage in the point of common coupling, AC and DC MGs can be distinguished. Although a remarkable progress has been made in improving the performance of AC MGs during the past decade [2–5], DC MGs have been recognized as more attractive for numerous uses due to their higher efficiency, more natural interface to many types of renewable energy sources (RESs) and energy storage systems (ESSs), and better compliance with consumer electronics [6]. Besides, when components are coupled around a DC bus, there are no issues with reactive power flows, power quality and frequency regulation, resulting in a notably less complex control system [7–10]. Therefore, DC MGs have been gained a high momentum over the past couple of years both in academia and industry. A broad class of traditional DC distribution applications, such as traction, telecom, vehicular and distributed power systems can be classified under DC MG framework and on-going development, and expansion of the field is largely influenced by concepts used over there. This chapter aims to present a technology overview of DC MGs both from the aspects of hardware design and control.

7.2 DC microgrid system overview

One of the first DC MG experimental prototypes was proposed by Ito *et al.* He claimed simpler control while superior efficiency and reliability with regard to its AC counterpart [11]. Later on, several works followed up on it and demonstrated usage of flexible DC MGs for telecom systems [12], distributed renewable and storage plants [13–16], data centres [17] and residential households [18–22]. In the next years, accompanied by the rapid emergence of consumer electronics, integration of

[1]Aalborg University, 9220 Aalborg East, Denmark

ever more RESs, and expansion of EV market with associated charging infrastructure, DC MGs have firmly established as a research field itself. Driven by practical requirements of specific DC MG applications, a number of hardware topologies have been proposed. Selection of particular topology for given application was done according to required control flexibility, robustness and reliability. As these performance indicators are often opposed to each other, different trade-offs need to be taken into account before deciding the most suitable configuration for a given application. In general, the topologies of DC MG can be classified into three categories, that is, single-bus topology, multi-bus topology and reconfigurable topology, respectively. Finally, as DC MGs are often coupled with the AC MGs and grids, hybrid AC–DC architectures are very common as well.

7.2.1 Single-bus topologies

Single-bus topology is the most commonly used one for DC MGs. Figure 7.1 shows an exemplary schematic diagram of the type which is frequently deployed in practical industrial applications. It is based on the singular DC bus to which ESS, typically an electrochemical battery stack, is directly connected [23–25]. The number of series battery cells is determined depending on the required voltage. Electrical power supply systems for telecom applications have been historically using this configuration operating at −48 V. Many types of converters that can be seamlessly connected to the bus operating around this nominal voltage are available in the market. However, despite inherent dynamic stability of the system, an uncontrollable voltage in the common DC bus, which depends mostly on the state of charge (SoC) and current of the battery, limits its application only to dense, singular bus systems [26]. Apart from that, it suffers from the practical problem of unregulated battery charging

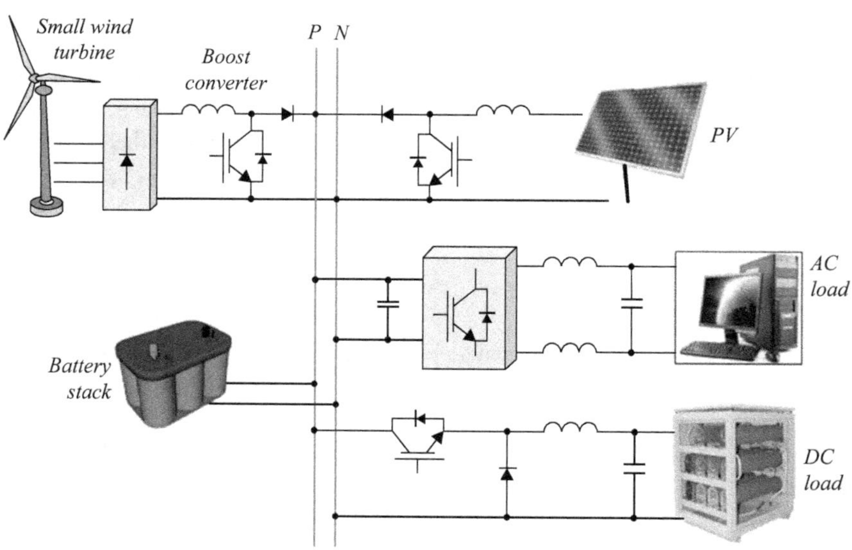

Figure 7.1 Single-bus DC microgrid with direct battery connection

as it needs to be performed co-ordinately by a number of paralleled converters that have inherent imperfections in bus voltage measurement. This causes a circulating current problem that leads to uneven loading of those converters [27–30], and also to accelerated wear and tear of stationary batteries [26].

On the other side, connection of the ESS through dedicated converter interface to a regulated low-voltage DC bus (see Figure 7.2) allows application of more flexible control and possibility of connecting multiple buses for enhancing the reliability of the system or to supply loads in a wider area. Moreover, the reliability of this system can be increased by using multiple battery stacks. This topology, operated at low voltage, has by now been the most widely studied in the academic literature (e.g. [31–34]). However, its attractive features are partially counter-balanced with several associated technical problems. First, there now appears a need for careful design of circuit and control parameters due to a fact that equivalent capacitance of DC bus gets much smaller than in the case of direct battery connection. Second, there exists only one DC bus, compelling the consumers to be powered from it. As a practical remedy to the aforementioned problem, several alternative concepts have been examined. Kakigano *et al.* proposed a bipolar single regulated bus structure (see Figure 7.3) that operates at ±170 V [35]. The main benefit of this configuration is a fact that load-side DC–DC converters have an option to choose the source voltages between 340, +170 and −170 V. Therefore, the reliability of the system is increased as in the case of failure of one line, it is still possible to supply power by the other two lines and an auxiliary converter.

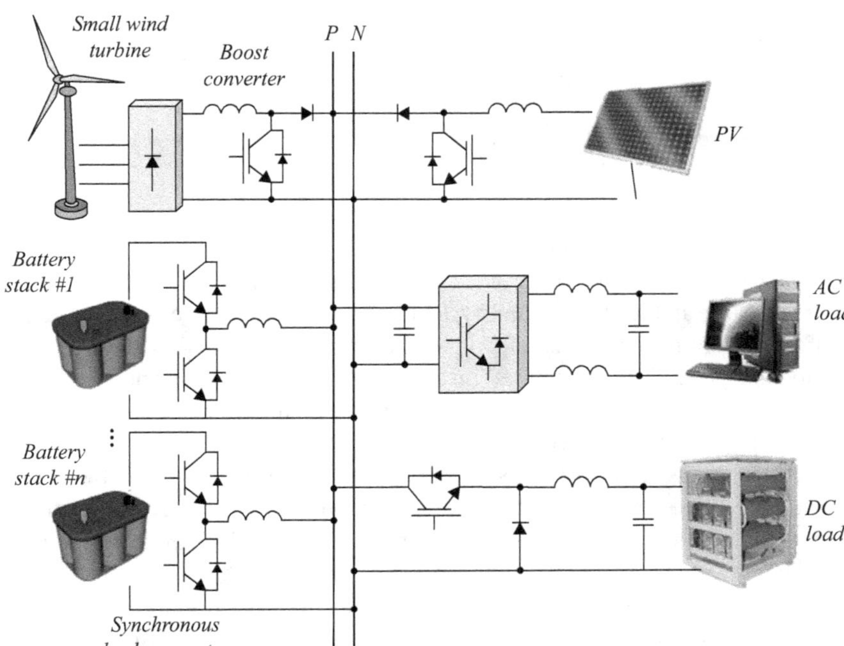

Figure 7.2 Single-bus DC microgrid with controllable DC bus

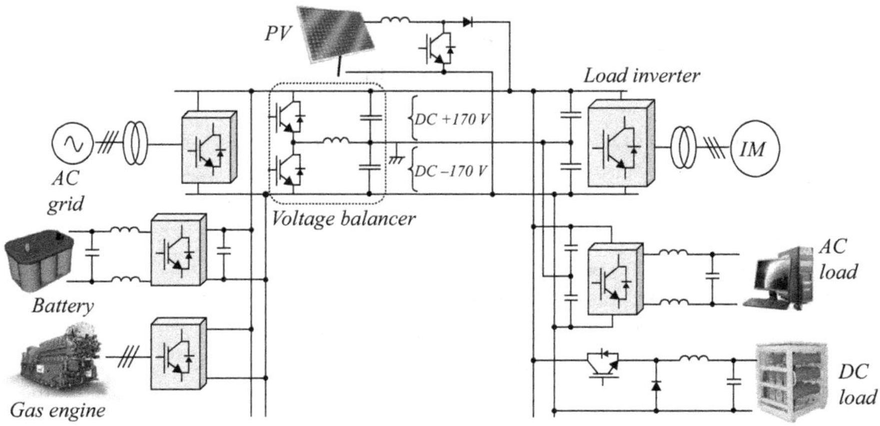

Figure 7.3 Bipolar-bus DC microgrid

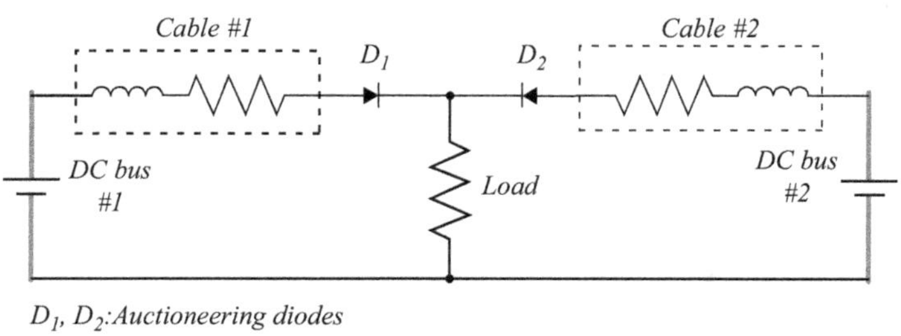

Figure 7.4 Dual-bus selection based on auctioneering diodes

7.2.2 Multi-bus topologies

Besides the single-bus topologies, DC MGs can be extended to multi-bus configurations aiming at higher availability and reliability. An interesting topological modification for enhancing the reliability of the system in that sense was proposed by Balog and Krein [36]. It is based on the automated hot-swap principle between more buses using the method of auctioneering diodes. In that work, the case of two redundant buses was analysed. The location of intervention is on the load side, where critical loads automatically select the bus to be supplied from on the basis of higher voltage. The structure of this system is shown in Figure 7.4. Similar strategy for selecting the most appropriate supply bus is presented in [37], but it uses a game-theoretic approach. In that case, loads have more flexibility to define the way of selecting desired bus, and several objective functions are tested for that matter. Besides, simultaneous supply from multiple buses is possible in order to enhance the total efficiency of the system. Multiple DC MG cluster configuration (see Figure 7.5) is an alternative redundant solution [6,38]. In this way, every MG is

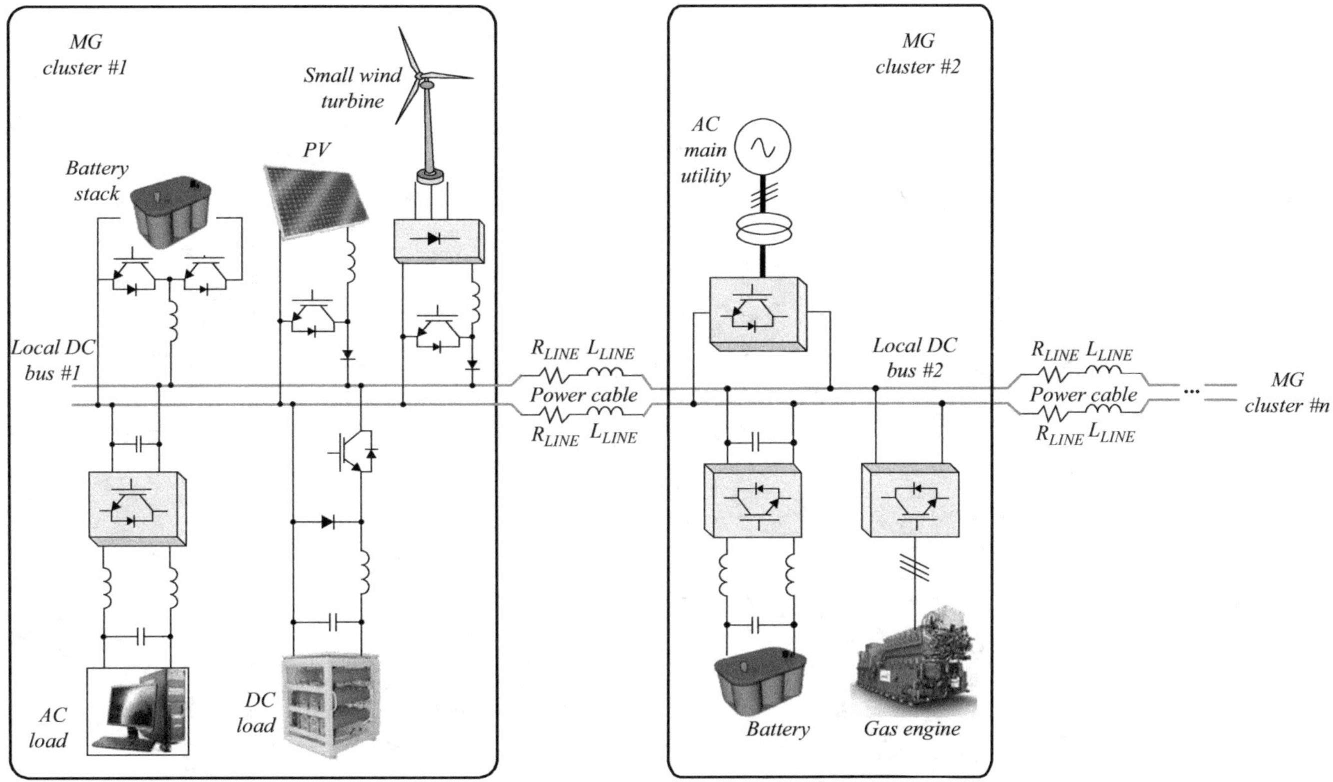

Figure 7.5 Multiple microgrid cluster configuration

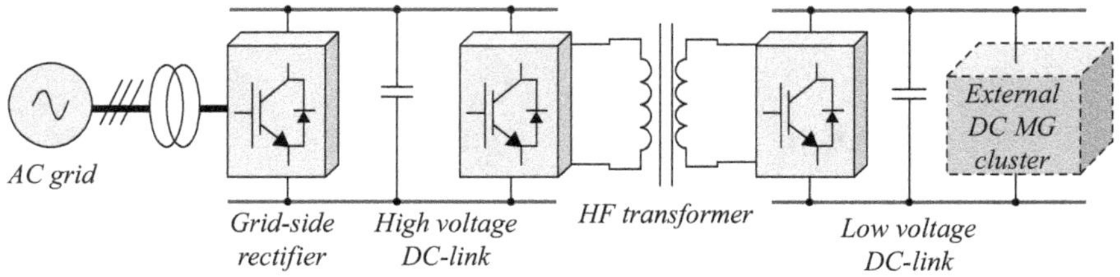

Figure 7.6 Solid-state transformer enabled DC microgrid

able to absorb or inject power from its neighbouring MGs in the case of shortage or surplus of power, respectively. In addition, depending on the configuration in which the MGs are connected, some corrupted buses can be automatically isolated from the system in the case of failure. Power exchanges between multiple DC buses are regulated by imposing appropriate local voltage deviations. However, by dint of dedicated communication technologies, total average voltage can be regulated to a nominal value.

Low-voltage DC distribution systems (possibly comprising multiple-MG clusters) can be interfaced to medium-voltage AC utility mains through solid-state transformer (SST), as shown in Figure 7.6. With connection of a number of SSTs to medium-voltage AC network, it is envisioned that the energy management in lower voltage levels will be completely in SST domain, largely simplifying the task of system operator in the AC grid above SST [39]. However, although it is widely deemed that SSTs will bring a new revolution to future distribution systems, serving as the true enablers of DC distribution architecture, they are still in the early development stage [40]. Recognizing its potential, many academic and industrial projects around the world are currently underway, focusing on pushing this technology to the next technical readiness level. Speed of its further evolution will to a large extent determine the significance of the role of DC distribution in future power systems.

7.2.3 Reconfigurable topologies

Besides the aforementioned architectures of DC MGs, reconfigurable topologies were also proposed for the purpose of increased flexibility during the faults or periodic equipment maintenance periods. Figure 7.7 shows a reconfigurable DC ring bus architecture [41]. Each node and the link between neighbouring nodes are controlled by intelligent electronic devices (IEDs). The main merit of this configuration is high availability. Since the load connected to the common DC bus can be fed bi-directionally, alternative path is provided at the ring bus in the case of fault. When encountering a fault in this DC MG type, the fault is isolated by IEDs first, and then the load power turns to be supplied by the sources in the alternative path.

Another DC MG architecture that features high reliability is based on zonal configuration, as shown in Figure 7.8 [42]. It is seen that this type of DC MG comprises multiple DC distribution units, with each one of them being connected in series to form a zonal architecture. Note that each DC distribution unit in the zonal

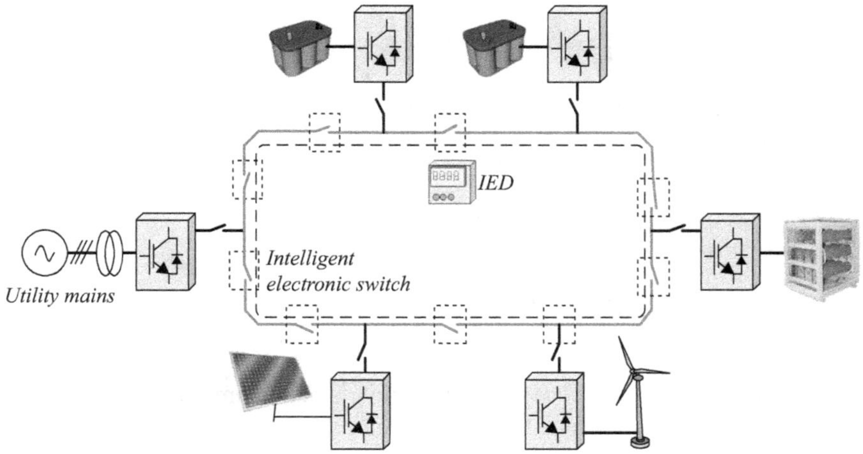

Figure 7.7 Ring-bus DC microgrid architecture

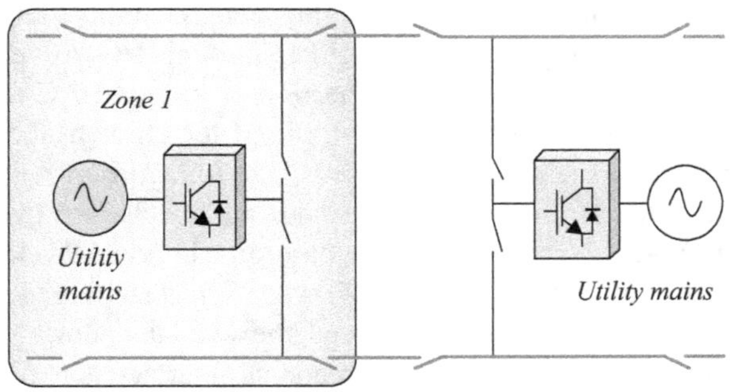

Figure 7.8 Zonal DC microgrid architecture

configuration has two DC buses. These two buses form the redundant configuration and the required load power can be flexibly obtained by either one of them. In particular, assuming that the load power is supplied by the upper bus, in the case of its failure the switches at the upper side are turned OFF and the switches at the lower side are turned ON to change the source bus and ensure the normal operation of the load. Meanwhile, since each DC distribution unit in the zonal architecture is powered by the utility grid separately, fault can be isolated within each unit without influence on the operation of other parts of the zonal system. This topology is similar to the redundant bus system proposed in [36], but it has better flexibility due to higher number of switches.

Multi-terminal DC (MTDC) system is proposed in [43–45], as shown in Figure 7.9. It should be noted that the MTDC architecture was originally proposed for high-voltage DC (HVDC) systems, for example, off-shore wind farms. In this configuration, the DC distribution network is fed by multiple input terminals. The system usually features mesh configuration. Compared to conventional radial

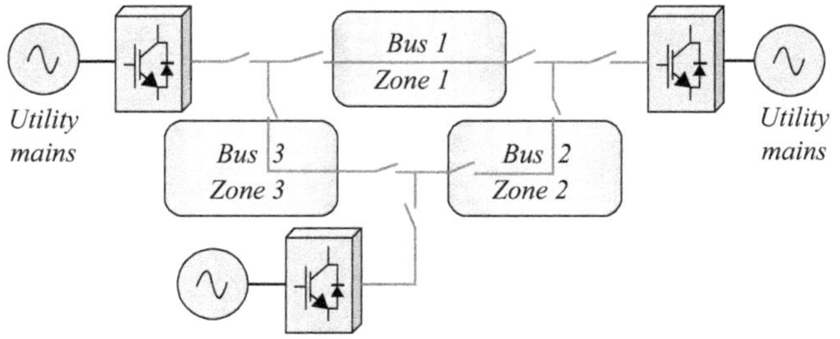

Figure 7.9 Multi-terminal DC system

configuration, power flow in mesh configuration is more complicated. However, it can be also more flexible since the system connection provides multiple paths for power transmission.

7.2.4 Hybrid AC/DC MGs

In order to ensure the high power availability and reliability, DC MGs can be interconnected with existing AC infrastructures through DC–AC converter interfaces. In that way, either an AC grid connection or a hybrid AC/DC MG can be established. In any case, DC link is typically used for efficient local coupling of DC-type components. The interface towards the AC grid is reflected in most architectures presented in previous sections and can be achieved by unidirectional or bidirectional DC–AC converters. In the case of a hybrid MG, DC MG can be regulated to absorb power from the AC MG when it has surplus of power. On the other hand, when load requirements exceed the available power, DC MG can deliver it back to the AC side (only if bidirectional interface is used). Focusing on the power architectures of the interconnecting energy conversion systems between DC MGs and AC grids, some work has been done in the existing literature. In general, the interconnecting system can be classified into three categories, that is, single, parallel and aggregated converter system. The single converter to link the DC MG and AC grid is usually used in small-scale systems.

The capability of DC MG of interconnecting with AC utility grid or AC MG depends primarily on the availability of high enough DC-link voltage level. As a final remark, it should be noted that although the power architectures were presented in this section within the MG framework, they have a quite general structure. In that sense, the vast majority of them can be used in a very broad voltage and power range, given the proper selection of semiconductor devices and/or their arrangement within the multilevel converter structures. For that matter, a resemblance with typical architectures used in HVDC systems can be readily observed (see, i.e. [46]). In the respective reference, singular bus, bipolar bus multiple bus, multi-terminal and ring bus structures have been shown to be used in HVDC systems up to levels of 350 kV. The exception is the topology presented in Figure 7.1, which has a direct connection of the battery stack that is not used for HVDC systems because it is not viable to increase the number of cells indefinitely due to problems associated with balancing of individual cells and reliability. To the best knowledge of the authors, the highest reported voltage of a series battery stack by now is 6 kV [47]. The thorough discussion about the applications of each topology specifically for four emerging DC MG applications is provided in the following section. It presents the typical performance requirements and applicability of corresponding power architectures, together with the summary of suitable control strategies in that they can achieve the imposed objectives. Moreover, the typical voltage and power limitation limits for each application are given.

7.3 Operation and control of DC microgrids

In order to guarantee stable and efficient operation of a DC MG, effective control strategies should be developed. The general structure of a DC MG control system is shown in Figure 7.10. In general, MG consists of a number of parallel converters that should work in harmony. Local control functions of these converters typically cover the following: (1) current, voltage and droop control for each unit; (2) source-dependent functions, for example, maximum power point tracking (MPPT) for photovoltaic (PV) modules and wind turbines, or a SoC estimation for ESSs; and (3) decentralized coordination functions, such as local adaptive calculation of virtual resistances (VRs), distributed DC bus signalling (DBS) or power line signalling (PLS).

At a global MG level, a dedicated communication-based coordinated control can be implemented to achieve advanced energy management functions. It can be realized either in a centralized or a distributed fashion, via central controller (CC) or sparse communication network, respectively [48,49]. In the case of distributed control, variables of interest are exchanged only between local controllers (LCs). Consensus algorithm can then be used to calculate either the average of all the variable values in distributed LCs or the exact value of any variable present in a specific LC. Some of the functionalities that can be accomplished by using DCLs include secondary/tertiary control, real-time optimization, unit commitment and internal operating mode changing (see Figure 7.10 for more details) [6].

From the communication perspective, overall control of DC MGs can be divided into the following three categories. (1) Decentralized control: DCLs do not exist and power lines are used as the only channel of communication. (2) Centralized control: data from distributed units are collected in a centralized aggregator, processed and feedback commands are sent back to them via DCLs. (3) Distributed control: DCLs exist, but are implemented between units and coordinated control strategies are processed locally. The basic configuration of each of these control structures is depicted in Figure 7.11. A more detailed overview of the significant features of local and coordinated control strategies is provided in the following sections.

7.3.1 Local control functionalities

As previously mentioned, the control framework of a DC MG consists of local and coordinated control levels. In this section, a local control level is discussed in detail. Basic functions which include current, voltage and droop control are

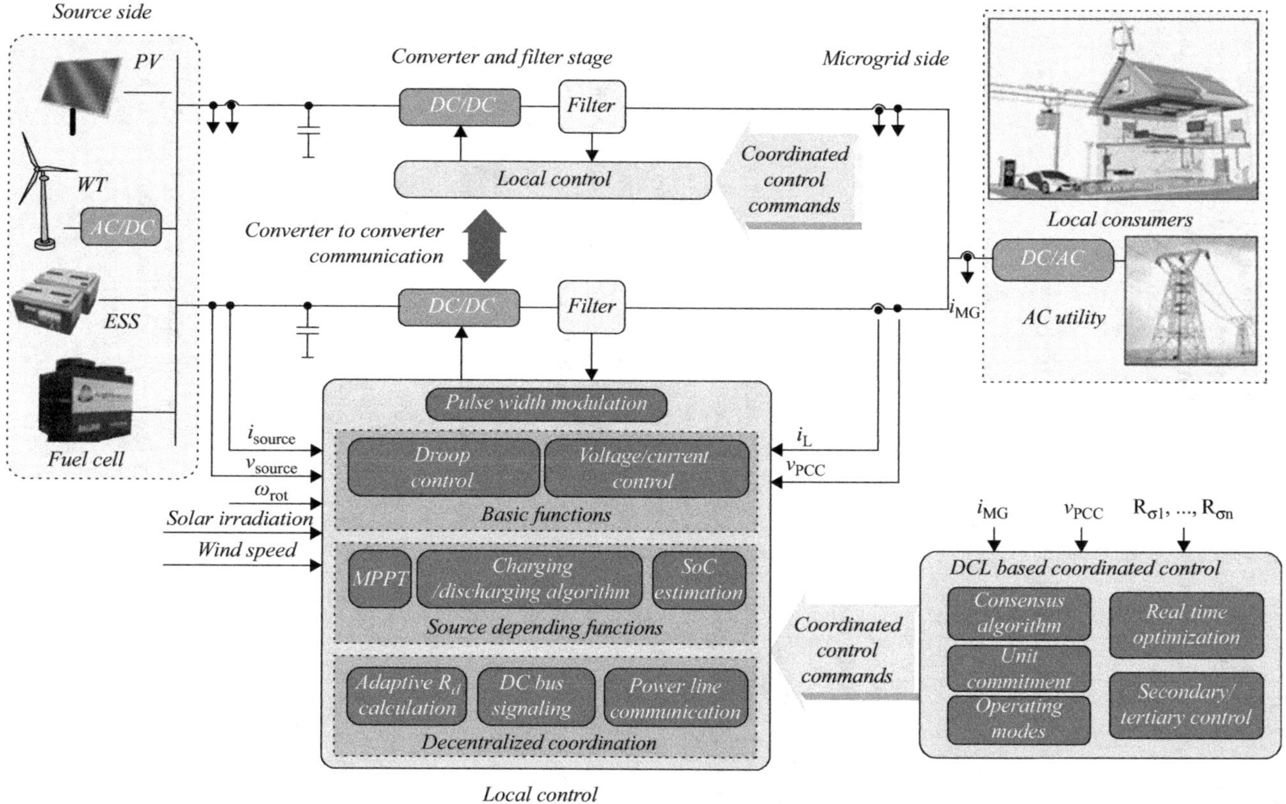

Figure 7.10 General control structures in DC microgrids

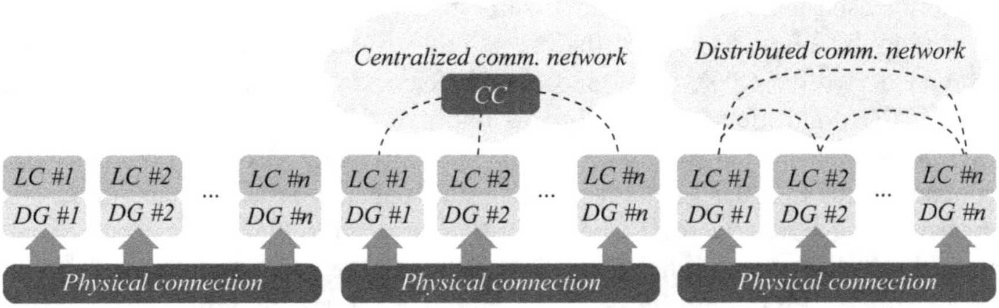

Figure 7.11 Differences between decentralized, centralized and distributed control architectures

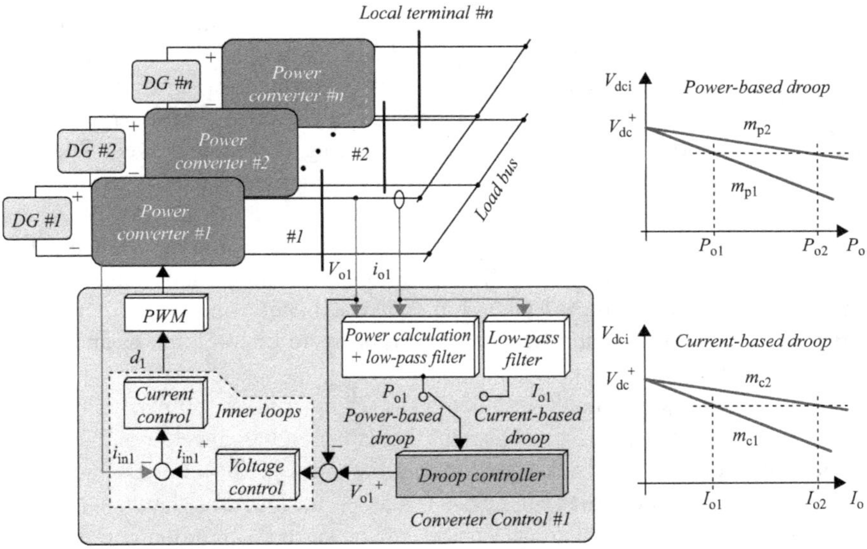

Figure 7.12 Inner control loops for DC microgrids

reviewed. A review of MPPT and charging algorithms has been omitted here. More details on charging algorithms for batteries can be found in [50], whereas an in-depth analysis of MPPT algorithms has been presented in a number of references, for example, in [51]. As a backbone of a DC MG, the interface converters play an important role in efficient and reliable operation of the overall system. In order to ensure not only proper local operation, but also to enable coordinated interconnection between different modules in a DC MG, flexible local current and voltage control should be employed and accurate power sharing among parallel connected converters should be achieved.

The basic local control diagram is shown in Figure 7.12, including local current and voltage controllers and a droop control loop. For local DC current and voltage control systems in DC MGs, proportional-integral (PI) controllers are commonly used since they introduce zero steady-state error, can be easily tuned and are highly robust. However, the use of other types of controllers, such as proportional derivative (PD), fuzzy and boundary controllers has also been reported [52–54]. PD controllers can be used to improve the phase margin (PM) of the system, but they do not eliminate steady-state error and also need to have high-frequency poles in order to attenuate the high-frequency noise. Hence, rather than appearing in a pure PD form, the derivative term in a PD controller is usually replaced by a high-pass digital filter. By combining the beneficial effects of PI and PD controllers, PID controllers can be employed. Fuzzy control is designed to emulate a human being's conclusion–deduction process based on the stimulus he/she gets from the environment and his/her own embedded knowledge. In the engineering world, it can be defined as a knowledge-based control method that can simultaneously take advantage of both static and dynamic properties of the system, and it has therefore good adaptability towards in environments with varying conditions. For the purpose of local voltage and current regulation, fuzzy controllers can either be used as principal regulators that process the error signal or in a series with feedback loops. To ensure fast convergence and extreme robustness, nonlinear control strategies based on state-dependent switching can be employed. They present simple implementation, but their detailed performance analysis can be quite complex. It should be noted that alternative control methods for DC MGs have recently drawn a lot of attention in the academic circles. However, their practical application should be elaborately justified by performing modelling, analysis, simulation, implementation, as well as a full cost–benefit analysis. For instance, increased production cost and lead time often prove to be too large of an obstacle for their deployment.

Droop control is commonly installed on top of inner loops, primarily for current sharing purposes. Figure 7.12 demonstrates that either output power or output current can be selected as the feedback signal in droop control [55,56]. For DC MGs with power-type load, output power can be used as droop feedback, as shown in (7.1). On the other hand, when current signal is used, as shown in (7.2), droop coefficient m_c can be regarded as a virtual internal resistance. In that case, the implementation and design of the parallel converter system in a DC MG can be simplified to some extent as the control law is linear [55]. The principle of current-based droop control was also extensively used in distributed power systems (DPSs) for paralleling multiphase converters that supply computer CPUs. Here, droop control is commonly known as adaptive voltage positioning (AVP) [57–59]. The calculations of references for voltage controller in the two aforementioned cases are as follows:

$$v_{DCi}^* = v_{DC}^* - m_p \cdot P_{oi} \tag{7.1}$$

$$v_{DCi}^* = v_{DC}^* - m_c \cdot i_{oi} \tag{7.2}$$

where v_{DCi}^* is the output of the droop controller, that is the reference value of DC output voltage of converter #i , v_{DC}^* is the rated value of DC voltage; m_p and m_c are the droop coefficients in power- and current-based droop controllers, whereas P_{oi} and i_{oi} are the output power and current of converter #i, respectively.

7.3.2 Coordinated control

Although the local interface converter control is an essential part of a DC MG which provides regulation of physical quantities, coordinated control is needed to achieve extended objectives. As already mentioned, depending on the means of communication between the interface converters, it can be realized either by using decentralized, centralized or distributed control the local interface converter control is an essential part of a DC MG, coordinated control should be implemented in order to achieve an intelligent control system with extended objectives. As already mentioned, depending on the means of communication between the interface converters, it can be realized either by using centralized, decentralized or distributed control.

7.3.2.1 Centralized coordination

Centralized coordination control can be implemented in DC MGs by employing a CC and a dedicated communication network to connect it with sources and loads, as shown in the *DCL-based coordination control* window of Figure 7.10 and in Figure 7.11. For small-scale DC MGs, each unit can be directly controlled by the CC that employs a high bandwidth communication using a master/slave approach [23,60]. However, for larger scale DC MGs, hierarchical control is often a preferred choice since it introduces a certain degree of independence between different control levels. It is more reliable, as it can be designed to continue the operation even in the case of failure of centralized control. Hierarchical control is achieved by simultaneously using local converter control and DCL-based coordinated control, which are separated by at least an order of magnitude in control bandwidth [55]. Coordinated functions can include secondary/tertiary regulation of DC voltage, power flow control and different grid-interactive control objectives such as unit commitment, changing operating modes, global optimization aimed at maximizing efficiency, minimizing operating cost and others.

It should be noted that centralized control provides the best foundation for employment of advanced control functionalities since all relevant data can be collected and processed in a single controller. However, the most obvious disadvantage of this strategy is that it has a single point of failure. In particular, if the CC or any key communication link fails, the commands from/to the controller will not be transmitted and corresponding control objectives will likely not be achieved. For mission critical applications, redundant communication systems can be installed in order to reduce the possibility of failure, but this needs to be justified by a cost-benefit analysis. Another option to increase the reliability of the system is to combine decentralized and centralized control methods into a hierarchical control structure [38,61–65]. In that case, basic functions of DC MGs can be retained even if the centralized controller fails.

7.3.2.2 Decentralized coordination

Decentralized coordination strategies are achieved exclusively by LCs, as shown in Figures 7.10 and 7.11. The most common types of decentralized control are DBS, adaptive adjustment of droop coefficients and PLS. Although their advantage is simplicity of control and independence from dedicated communication technology, they inherently have performance limitations due to lack of information from other units. Moreover, as these methods are invariably based on the interpretation of the voltage in the common DC bus, the accuracy of voltage sensors impacts their effectiveness and reliability.

Originally proposed in [10] and the follow-up in [15,66], DBS is the most prominent decentralized coordination method for DC MGs. By using the DBS approach, coordinated operation of different units in DC MGs is realized by imposing and identifying variations in the common DC-bus voltage. The transitions between different modes are triggered by different pre-set DC-bus voltage values.

The method of adaptive droop coefficient calculation has been mainly used for power balancing of distributed ESSs, as shown above. The main limitation of the method is potential instability induced by improperly designed droop curves. To that end, there always exists a trade-off between the permissible voltage deviation and stability properties of the system, that is, higher voltage deviation is associated with the higher PM. It should also be noted that some of the SoC balancing methods mentioned above [67,68] were originally proposed for AC MGs. However, since the same principle can apply to DC MGs (in AC MGs, the frequency is normally used as a system level coordinating signal), they are also included here.

PLS is another decentralized method that can be deployed for coordinated control. In particular, sinusoidal signals of specific frequency are injected through amplifiers into the DC bus, allowing each device to send and receive information on its status, performance, history or internal operational mode. Although PLS relies on dedicated communication, here it is categorized as decentralized since the power network is the only communication medium. It should be noted that in power systems literature, this particular way of communication is sometimes also referred to as power line communication (PLC) [69]. REbus, an open standard for DC electricity distribution in homes, commercial buildings, campuses and other settings, uses PLS as a primary communication carrier [70]. Alternative methods that exchange information between devices without using dedicated amplifiers were presented in [71,72]. Signals are generated by PWM of DC–DC converters in those works. They are injected in open loop in [72], whereas dedicated proportional-resonant (PR) controllers are used during injection periods in [71] to avoid the steady-state error in the DC bus.

PLS is generally more complex to implement compared to other decentralized methods. Moreover, it is commonly used only for changing operating modes or shutting corrupted components of the system, and it is not suitable for power sharing. However, as opposed to permanent voltage deviation in the common DC bus which is inherent for DBS and adaptive droop methods, sinusoidal signals are only periodically injected into the system. Therefore, the quality of the voltage waveform can be considered to some extent improved compared to other methods.

7.3.2.3 Distributed coordination

Distributed control indicates the control principle where central control unit does not exist and LCs communicate only among themselves through dedicated DCLs, as shown in Figure 7.11. The main advantage of this approach is that the system can maintain full functionality, even if the failure of some communication links occurs, provided that communication network remains connected.[1] Therefore, distributed control is immune to single point of failure. The functionalities that can be achieved by this approach resemble those of centralized control and are also represented in the *DCL-based coordination control* window of Figure 7.10.

However, in order to enable these functionalities, the information exchanged through DCLs first needs to be appropriately processed. In particular, information directly exchanged between LCs can contain only locally available variables. In other words, if the two units are not connected by a DCL, they do not have direct access to each other's data and their observation of the system is quite limited. In order to circumvent this problem and to make the level of awareness of an LC similar to that of a CC, a consensus algorithm can be used. In its basic form, a consensus algorithm is a simple protocol installed within every LC which continuously adds up all algebraic differences of some variable(s) of interest present in a given LC and those present in LCs adjacent to it. If we look at LC #i, this definition can also be expressed by the following equation:

$$\dot{x}_i(t) = \sum_{j \in N_i} \left[x_i(t) - x_j(t)\right] + b_i(t) \tag{7.3}$$

where $x_i(t)$ and $x_j(t)$ are the values of variables of interest in LC #i and LC #j, respectively. Here, j is iterated through the whole set of neighbours of LC #i, which is represented by N_i. Finally, $b_i(t)$ is an optional input bias of LC, which can be used to declare it as a virtual leader. It can be seen from (7.3) that $x_i(t)$ is interactively adapted with respect to the values of its neighbouring units. Likewise, variables in any other controller adapt with respect to the values of their own neighbours. Consequently, it can be analytically proved that, if the communication network is connected, all variable values will converge to a common average after a certain amount of time [73,74]. Another option is to use a non-zero input bias in one of the LCs. In that case, variables of all other LCs will converge to his respective bias [75]. In either case, the ability of consensus to share information in such a manner has wider applicability than simple data averaging. For instance, if every LC has information on the number of other active LCs, an exact value of any specific variable can be calculated directly from the average.

The collective dynamics of communication system realized via consensus protocol can be represented by the following equation:

$$\dot{x}(t) = -\Theta \cdot x(t) \tag{7.4}$$

where $\Theta = [\theta_{ij}]$ is the graph Laplacian of the network whose elements are defined as follows:

$$\theta_{ij} = \begin{cases} -1 & j \in N_i \\ |N_i| & j = i \end{cases} \tag{7.5}$$

[1]For the exact definition and a more in-depth discussion of the connectivity of communication networks, please refer to [74].

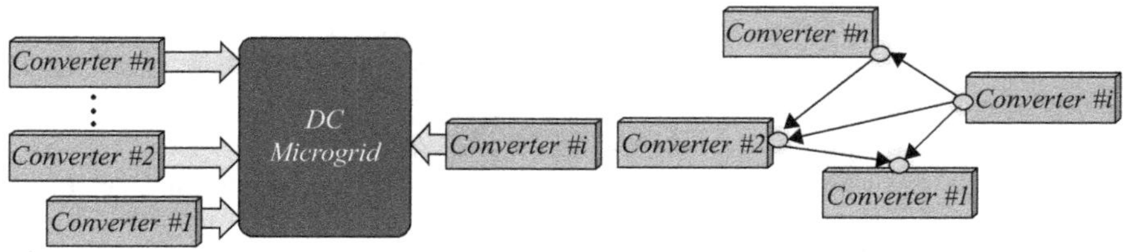

Figure 7.13 Configurations of exemplary physical and communication networks

where $|N_i|$ denotes the number of neighbours of node i. The topology of communication network is explicitly reflected by graph Laplacian, and it is also possible to design weights of the respective matrix to control the convergence speed [76]. Figure 7.13 shows the configuration of physical and an exemplary sparse communication network.

In summary, it can be concluded that distributed control can achieve information awareness comparable to that of centralized control. Therefore, objectives such as output current sharing, voltage restoration, global efficiency enhancement, SoC balancing and others can be easily realized. In that sense, distributed control offers much wider functionalities than decentralized control but remains protected from the single point of failure. Its main limitation is complexity of analytical performance analysis, that is, assessment of convergence speed and stability margins, especially in non-ideal environments characterized by communication time delays and measurement errors.

7.4 DC microgrid system protection

As opposed to many advantages that DC systems bring in relation to AC, design of their protection represents a considerable challenge. The most important aggravating circumstance in that view is problem with extinguishment of DC arc which appears upon the tripping of protective devices. Moreover, there is still a general lack of understanding and experience with operational issues of these systems [42]. Therefore, widely accepted protection standards and guidelines are yet to be defined.

The aim of this section is to update the current status in DC MG protection area. For that matter, four main aspects are addressed: (1) types of faults, (2) grounding options, (3) protective devices, and (4) design of protection systems.

7.4.1 Types of faults

Figure 7.14 shows two basic types of faults that can exist in DC power systems, that is line to ground and line-to line-faults [77]. It should be noted that although less serious, line-to-ground faults are the ones most frequently appearing in industrial distribution systems [78]. Line-to-ground fault can have either low or high fault

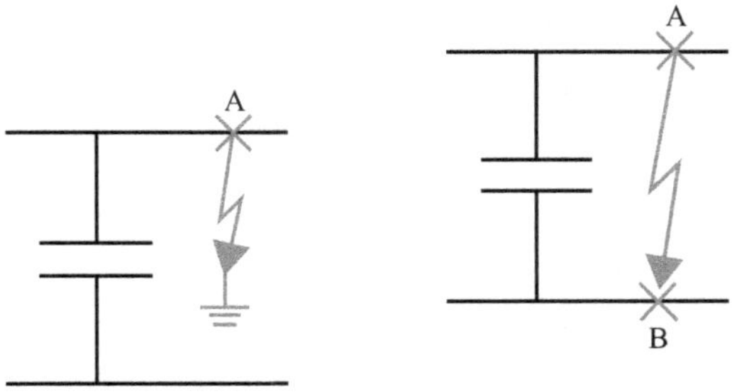

Figure 7.14 Two basic types of faults

impedance, whereas line-to-line fault typically has a low impedance [79]. Moreover, a fault can occur at various points in the system, and each one of them will have different implications. For instance, unlike feeder faults that can be rapidly isolated from the main bus, faults at the main bus itself present a greater danger since all the feeders connected to it get affected [79].

When designing the protection system, it is important to identify all possible fault locations and to calculate the fault current for every given location [80]. All active sources within the system participate in the formation of the total fault current and their actual contributions depend on the effective impedances between the respective source and the fault locations. For instance, in the case of bus fault, the contribution of a battery which is directly connected to the common DC bus via cable can be expressed as follows:

$$i_{\mathrm{bat}}(t) = \frac{u_{\mathrm{bat}}}{R_{\mathrm{bat}} + R_{\mathrm{cable}}}\left(1 - e^{-t/\tau_{\mathrm{eff}}}\right) \tag{7.6}$$

where u_{bat} is the battery open-circuit voltage at the time of the fault, R_{bat} and R_{cable} are battery and connecting cable internal resistances, respectively, whereas $\tau_{\mathrm{eff}} = (L_{\mathrm{bat}} + L_{\mathrm{cable}})/(R_{\mathrm{bat}} + R_{\mathrm{cable}})$ is the time constant of the effective battery/cable system. On the other hand, if a battery (or any other type of source) is connected to the bus through power electronics converter interface, current resulting from quick discharge of its filter capacitor plays a pivotal role in the fault current:

$$i_{\mathrm{c}}(t) = \frac{u_{\mathrm{DC}}}{R_{\mathrm{c}}} e^{-t/\tau_{\mathrm{c}}} \tag{7.7}$$

where u_{DC} is the common DC bus voltage at the time of the fault, R_{c} is the capacitor series resistance and $\tau_{\mathrm{c}} = C_{\mathrm{DC}}R_{\mathrm{c}}$ is the time constant of the effective converter/common DC-bus RC system. If there exists a cable of considerable length between a converter and common DC bus, corresponding parameters of the cable should be included in calculation as well, resulting in a higher order dynamic system [78]. Similarly, faults that appear on feeders can be treated just as special cases of bus faults and their current contribution can be obtained by looking into effective impedances [78,79,81].

Total fault current can be calculated as a summation of current contributions from the battery which is directly connected to the common bus and all other source converter currents [77]:

$$i_{\text{fault}}(t) = i_{\text{bat}}(t) + \sum_{j=1}^{N} i_{c,j}(t) \tag{7.8}$$

with N being the number of source converters connected to the common bus. It should be noted that $i_{\text{bat}}(t)$ in equation above is optional and exist only in systems which include direct interface of the battery to the common DC bus (e.g. system shown in Figure 7.1). On the other hand, the impact of line to ground faults on system performance depends heavily on the grounding method. More details about the grounding methods are given in the next subsection.

7.4.2 Grounding

The most important objectives of the system grounding are to facilitate fault detection, minimize stray currents (leakage current from conductor to the ground) and to ensure the personnel safety by minimizing the common mode voltage (potential difference between device conductor and ground) [82].

Corrosion is the main problem associated with stray current. The phenomenon of stray-current induced corrosion was first observed at the end of nineteenth century, in DC-powered electrified rail transit systems [83]. The problem has been initially attributed to the chemical structure of the soil surrounding the rails, but it was soon realized that the soil cannot be the sole contributor to it since corrosion has been observed in all metallic structures around the rail base. Therefore, it has been concluded that the leakage or stray current is the main culprit for significant amplification of this undesirable effect. Stray currents cause corrosion at the locations of current leakage from conductor into the soil. Since the current flow in conductor is primarily electronic, and ionic in the soil (i.e. ground), an electron to ion transfer is inevitable as current changes its transmitting medium. In particular, this electron-producing or oxidation reaction can be represented by the following equation [84]:

$$\text{Fe} \rightarrow \text{Fe}^{2+} + 2\text{e}^{-} \tag{7.9}$$

The reaction shown above is known as corrosion if it takes place over a prolonged period of time. In a broad sense, corrosion of metallic objects will occur always at places where current transfers from them to the surrounding environment. By undertaking the measures to reduce the stray current, the corrosion rate will be reduced as well and severe structural damages to conductors avoided. Apart for traction applications, modelling, assessment and stray current control methods have been recently also proposed for large photovoltaic plants [85,86].

Another important metric of grounding system performance is the level of common mode voltage. In traction applications, this voltage is commonly referred to as the touch potential [82]. Common mode voltage presents a personnel safety concern if it gets exceeds the value of 60 V, which is generally considered as a safety threshold [82].

It is important to note that the common mode voltage and the level of stray current are related to each other by the grounding resistance. More precisely, for a very high grounding resistance, there will essentially be no stray current, but the common mode voltage will take its maximum value. On the contrary, if the system is solidly grounded, there will be virtually no common mode voltage, but stray current will be the highest. Therefore, the objectives to minimize both of them at the same time are conflicting with each other. However, with a proper grounding strategy, this trade-off can be explicitly taken into account.

Different grounding alternatives have been proposed for DC power systems. Some references suggest ungrounded operation, especially for low voltage applications [42,87,88]. The reasons for that are fact that the common mode voltage will typically not be on such a high level to present a danger for personnel safety, stray current will be essentially inexistent, and system will continue to operate even in the case of a single phase-to-ground fault. Still, it is instrumental to discover and correct a possible second ground fault or it will create a line-to-line fault [89].

In grounded mode, either a solid, low or high resistance grounding can be applied [79]. Solid grounding is rarely used in modern systems because of the pronounced stray-current-induced corrosion [82]. There are a number of options regarding selection of the point in the system to be grounded; it can be either a positive pole, midpoint or negative pole of the common DC link [78]. For that matter, Jacobson and Walker [90] suggest to use midpoint grounding in order to minimize the ground potential and enhance safety. Also, the circulating current can be greatly reduced in this way in relation to AC-side neutral grounded system. IEC 60364 standard [91] defines three grounding strategies which give specifications on how the source bus and conductive parts of the system are connected to earth. The respective methods are denoted by double letter abbreviations, that is TN, IT and TT. Here, first letter reveals the type of connection of source bus to the ground, whereas the second one the conductive part type of ground connection. T, N and I indicate direct connection to earth, direct connection to neutral and isolation, respectively.

In the case of TN, power line and exposed conducting parts are commonly earthed through respective mid-points, whereas in IT the power line is not earthed while conducting parts are commonly earthed either on positive or negative pole. On the other hand, there are multiple grounding points in TT systems. In general, it can be said that TN systems require low grounding impedance and that it is easy to detect a line fault when this grounding strategy is applied. However, the touch voltage may get unacceptably high depending on the voltage level at the time when fault occurs. IT systems do not need low grounding impedance and have very low fault current if there is a single fault. However, another ground fault is likely to induce much higher current [77]. Moreover, it is complex to detect the line fault. It should be noted that positive pole is grounded by IT method in telecom systems in order to minimize the corrosion effect [79]. Installation of grounding system is straightforward in case of TT grounding, and the faults do not migrate [77]. Nevertheless, the high-voltage stress problem still exists and circulating current paths are likely to occur.

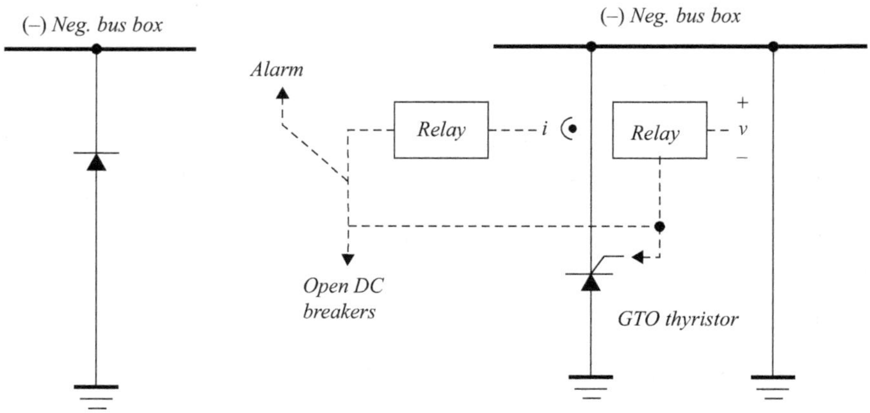

Figure 7.15 Diode and thyristor grounding options

Unlike methods presented above, Paul [82] suggests several reconfigurable grounding options with a possibility to operate the system normally in ungrounded mode in order to minimize stray current induced corrosion, but to automatically ground the system upon sensing dangerously high common mode voltage. After the clearance of abnormal operating condition is achieved, the system is automatically switched back to ungrounded mode. Block diagrams of two examples of reconfigurable grounding systems, that is, diode and thyristor grounding, are depicted in Figure 7.15.

Diode grounding comprises a direct metallic connection of the negative portion of DC bus to the ground, but via a diode circuit. In that way, current is allowed to flow from the ground towards the negative bus when the voltage across diode exceeds some threshold value. Once it happens, it gets dissipated. Since corrosion is not completely eliminated in these kinds of systems, they require a periodical maintenance [92].

Contrary to the diode scheme, bidirectional thyristor scheme offers active control over the grounding instances. This is achieved by triggering the thyristor gate only when the negative-to-ground voltage exceeds a pre-set value. Moreover, a current sensor is installed in series with the grounding conductor. If the decay of current is observed, it means that the system can be returned to ungrounded mode. On the contrary, if the current would maintain its amplitude then a positive to ground fault is the most likely reason. In that case, DC feeder breakers are tripped in that case. The status of the system can be communicated with environment through the alarm signal. The basic advantage of the thyristor grounding strategy is that the system maintains operation in ungrounded mode in every other instance, keeping the stray current minimized [82].

7.4.3 Protective devices

Protection devices that are presently commercially available for DC systems include fuses and circuit breakers (CBs) [79]. However, they inherently introduce large time constants and time delays before activation, respectively. In addition, interruption of the current in both cases is accompanied by the appearance of the arc. Although arc gets extinguished naturally in AC systems within the half cycle after tripping by first crossing of the current through zero, it presents a challenge in DC systems since the current has a steady value. Arc occurrence presents a dangerous condition not only from safety point of view, but also causes contact erosion in CBs and consequently a short lifetime and high maintenance costs [42,93].

Fuses operate on a principle of melting down the fuse link in a heat-absorbing material. Their ratings are specified in terms of RMS values of voltages and currents, and are hence equally applicable to AC and DC systems. However, it needs to be ensured that the time constant of current increase during the fault is below certain limits since slow increase of temperature may prevent the heat-absorbing material to extinguish the arc [94].

Standard moulded-case CBs (MCCB) that consist of a contactor, a quenching chamber and a tripping device can be used as well. They can have either a thermal-magnetic or electronic tripping device. In either case, sufficient voltage blocking capability can be achieved by connecting contactors in series [79]. One potential problem with MCCBs in systems dominated by power-electronic converters, as is the typical case in DC MGs, is that short circuit currents are supplied mainly by filter capacitors. Their quick discharge following the fault causes currents of very high magnitude, but which last for a very short periods of time. Therefore, the force generated by these currents may be insufficient for opening the contacts of MCCB [42].

Due to inherent tripping time delays and other problems associated with fuses and CBs, protections utilizing only these devices have inherent performance limitations. In order to cope with that, usage of power electronics-based protection devices has been proposed. They are commonly referred to as the solid state CBs (SSCBs) [95,96]. Several semiconductor devices such as gate turn-off thyristors (GTOs), insulated-gate bipolar transistors (IGBTs) and insulated-gate commutated thyristors (IGCTs) have been used in this context, with each one of them having some specific advantages and drawbacks. For that matter, Meyer *et al.* [95] summarize and compare a number of different switch topologies under technical and economic aspects. It is demonstrated that GTOs and IGCTs have much lower on-losses than IGBTs. Also, it is argued that IGCT in rectifier configuration has by far the highest reliability.

In order to be able to interrupt the current, there needs to exist an element to which SSCB dissipates energy during interruption. Typical elements installed for that matter are metal–oxide varistors (MOVs), capacitors and switched resistors or their combinations [96–98]. This brings a different perspective of current interruption when compared to conventional breakers where the electric power during tripping the device is manifested in an arc that needs to be extinguished using an arc chute [97,99].

Exploiting the capability of fast switching, this dedicated power dissipation element can be readily used for current limiting as well. In that sense, SSCBs can actively regulate the fault current to a selected level, rather than leaving it dependent on characteristics of supplying sources. Limited fault current provides less thermal and magnetic stress on the surrounding components resulting in a lower installation

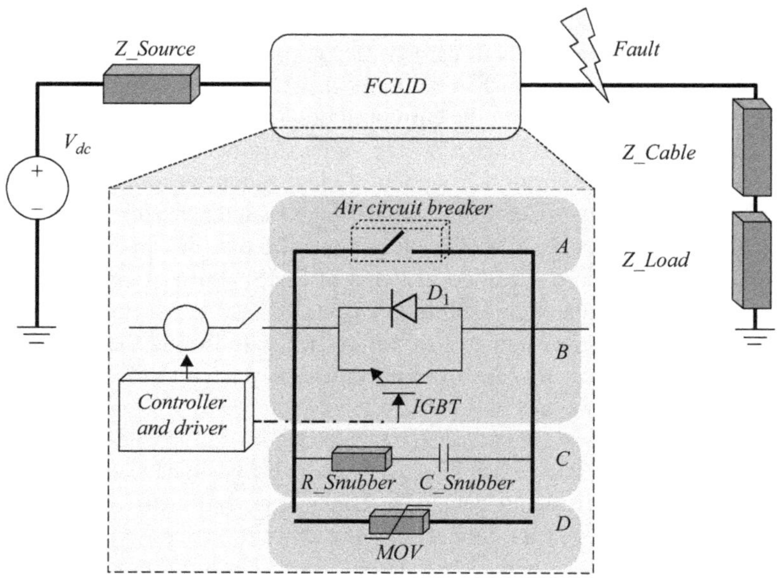

Figure 7.16 Solid-state circuit breaker

cost, as well as safer operation of the system in general [93,100,101]. Moreover, current limiting circuits installed at strategic places, for example, radial line supplying a high-power density load centre, allow usage of commercially available protection devices with low power rating in the entire downstream network. It should be noted that current limitation functionality can also be achieved by superconducting elements [102]. However, they need to be actively cooled and the current limiting capacity with present technology is inferior to SSCB solutions [101].

A typical example of IGBT-based SSCB with current limiting capability is depicted in Figure 7.16 [96]. When the semiconductor device (IGBT) switches off, the current flowing towards the fault location decays since the power gets dissipated in parallel combination of a snubber circuit and MOV. On the other hand, when the IGBT turns on, the current raises up again. The hysteresis-based controllers with preselected upper and lower thresholds for the fault current are the most commonly deployed to achieve current limiting.

7.4.4 Design of protection systems

The objective of protection system is to detect, locate and isolate any fault so that the DC MG can continue operating its uncorrupted part. In order to achieve this goal, knowledge about possible faults should be incorporated to select the most appropriate protection devices and grounding methods into a complete protection system. More specifically, desirable characteristics of such a system are reliability, selectivity, speed, performance, economics and simplicity [42]. In the remainder of this subsection, several prominent approaches that try to meet these features are summarized.

An insightful example of selecting appropriate protection devices for different components of a DC MG was presented in [79]. According to specific fault-withstanding capability of each associated unit, it was concluded that converter protection must be very fast in order to limit the fault current through IGBT diodes whereas battery protection can allow slower reaction time as it can withstand considerably higher fault current without damage. For that matter, ultra-fast hybrid CB was proposed for converter protection and regular CBs for battery protection. On the other hand, fuses and MCCBs were used for feeder protection. In particular, it was claimed that it is better to use MCCBs closer to the loads since they can simultaneously trip both poles, whereas fuses are more appropriate for installation closer to the bus due to their magnetic sensing that gives good selectivity. DC-link voltage information together with fault current was used for accurate fault detection.

A handshake method was proposed in [103] to split the DC system into three possible zones and isolate the possible fault on any of the buses. The drawback of this method is that it needs to completely shut down the system following the fault, which may be unnecessary in most conditions. Alternative method that avoids the shutdown was suggested in [77], where a loop-type DC bus was used together with protection that is able to detect the fault and separate only the faulted part, so that the rest of the system keeps operating. The method was realized by using communication system where one master unit receives current measurements from two slave units and gives commands to control the bus switches depending on the difference between the two measured currents. A protection scheme that relies on solid-state converter with fast-acting current-limiting capability was proposed in [104]. The respective converter uses two Emitter Turn-Off devices in anti-parallel configuration to realize both switching and protection. Furthermore, a relay-based coordination and backup protection schemes were addressed. Coordination was realized in a fully decentralized fashion, by using appropriate devices such as fuses and diodes.

7.5 Application of DC microgrids to future smart grids

Taking into account the advantages of DC MGs, their usage has been proposed for a number of different applications in future SG, with each one of them having its own specific requirements. In this section, a critical view on performance requirements of four modern DC MG applications is presented. They include high-efficiency households, renewable energy parks, hybrid ESSs and EV fast charging stations, as shown in Figure 7.17. Although it should be acknowledged that the considered list is by no means complete, we have limited our discussion only to these novel applications of DC MGs.

Considering the specific characteristics of various power architectures that were reviewed in Section 7.2 of this chapter together with control strategies that were reviewed in the first part of these two articles, a critical judgment in selecting the most suitable ones for every given application is provided below.

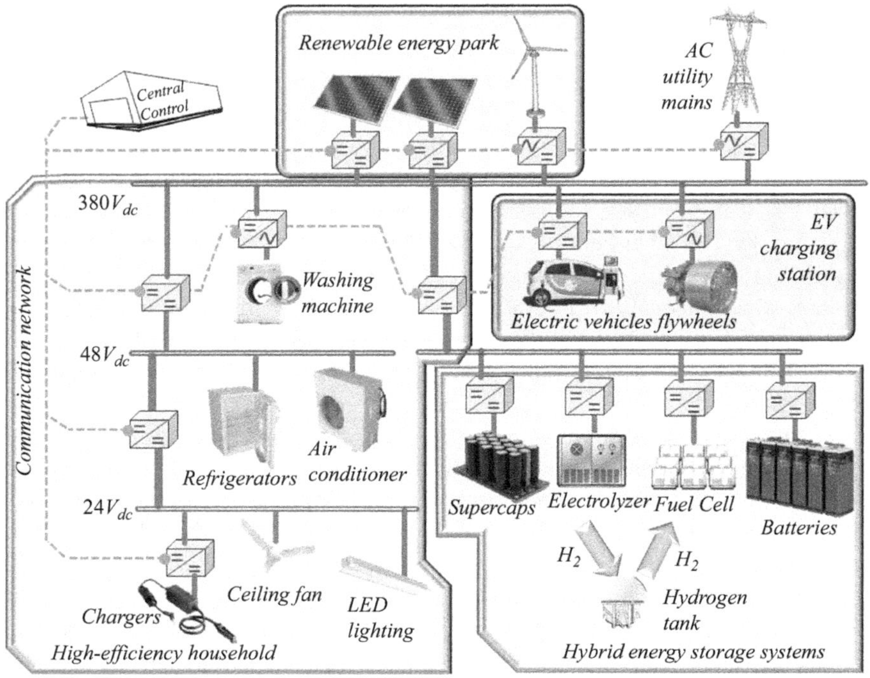

Figure 7.17 DC microgrid type systems in various smart grid applications

7.5.1 High-efficiency households

As the majority of modern household energy sources and loads, for example, PV panels, batteries, LEDs and consumer electronics operate naturally at DC, it is more efficient to connect them around the DC bus to form a DC MG [105]. There are several demo projects of DC high-efficiency houses all over the world shown in the existing publications and project websites, such as USA [106,107], Japan [35], Korea [108] and Denmark [109]. In contrast to the traditional AC system, DC system has higher efficiency. The energy-saving opportunities by changing infrastructure from AC to DC can be up to 15% [110].

Except the benefits from efficiency point of view, the control simplicity in the case of DC makes it a much better option for integration of energy management systems. For this reason, DC MGs are excellent candidates for households that could greatly facilitate the achievement of objectives imposed by a number of present SG roadmaps, which state that increase in flexibility[2] of household consumers is a key for sustainable operation of the future grids with high penetration of renewables [111,112]. For that matter, real-time price signal is predicted to be the

[2]Flexibility of consumers mentioned here is related to controllable power extraction from the grid and is different from the hardware flexibility addressed in Section 7.3.1.

main driving factor that will motivate local households to rethink their distribution of energy.

7.5.2 Renewable energy parks

A number of RESs can be clustered around the common DC bus to form an energy park. In that way, the system efficiency can be increased and issues such as reactive power circulation and harmonics avoided [105]. These kinds of DC systems are sometimes also referred to as the collector grids and are used for both photovoltaic and wind power applications [113,114]. Compared to households, they present a different perspective since they inject large amounts of power into the conventional AC power system and have more direct implications on its operation. Therefore, they need to be able to provide a number of ancillary services defined by well-established grid codes [115,116]. As compliance with grid codes often implies reduction of active power injection to the grid, coordination between multiple renewable generators, which should achieve this task simultaneously, is mandatory. Several applications of DC collector grids are presented in the following.

7.5.3 Hybrid ESS

ESS technologies are characterized by several important performance indicators, such as energy efficiency, energy and power density, cycle life and rate of self-discharge [117]. In that sense, for situations when a single ESS is not able to meet prescribed requirements of specific application, it is advisable to combine two or more technologies that complement each other. Since DC distribution offers a natural interface for practically all types of ESS, these kinds of hybrid ESS are frequently deployed around the common DC bus.

7.5.4 EV fast charging stations

Electric vehicles (EVs) have gained increasing attention due to potential of reducing the greenhouse gas emissions and conventional fuel consumption. However, it has been shown that their large-scale integration could have a significant adverse effect on the electric grid [118] and DC MGs have been conceived as a promising solution to support the operation of the power system by aggregating multiple EVs [119]. Fleets of PHEV chargers represent a specific group of consumers with an outstanding potential from the increase of flexibility standpoint.

7.6 Conclusions

A qualitative overview of different hardware topologies and control systems for DC MGs has been presented in this chapter. Some challenges and design considerations of DC protections systems have also been discussed. Finally, applications of DC MGs in emerging smart grid applications have been summarized. Due to its attractive characteristics in terms of compliance with modern generation, storage and electronic load technologies, high reliability and current carrying capacity, as well as simple control, DC systems are already an indispensable part of power systems. Moreover, the existing challenges such as protection issues will be effectively resolved in the near future due to fast progress of semiconductor technology which is a key enabler of cheap and reliable future DC solid-state protection systems. Therefore, it is the view of the author that more and more DC systems will appear in different industries and gradually lead to new ways of rethinking of the future power distribution philosophies, especially with the emergence of SSTs. Research in DC systems, especially in the power electronics-based technologies will be highly attractive in the future.

References

[1] R. H. Lasseter, "MicroGrids," in 2002 *IEEE Power Engineering Society Winter Meeting. Conference Proceedings (Cat. No. 02CH37309)*, vol. 1, pp. 305–308.

[2] N. Pogaku, M. Prodanovic, and T. C. Green, "Modeling, Analysis and Testing of Autonomous Operation of an Inverter-Based Microgrid," *IEEE Trans. Power Electron.*, vol. 22, no. 2, pp. 613–625, 2007.

[3] R. Majumder, B. Chaudhuri, A. Ghosh, G. Ledwich, and F. Zare, "Improvement of Stability and Load Sharing in an Autonomous Microgrid Using Supplementary Droop Control Loop," *IEEE Trans. Power Syst.*, vol. 25, no. 2, pp. 796–808, 2010.

[4] K. De Brabandere, B. Bolsens, J. den Keybus, A. Woyte, J. Driesen, and R. Belmans, "A Voltage and Frequency Droop Control Method for Parallel Inverters," *IEEE Trans. Power Electron.*, vol. 22, no. 4, pp. 1107–1115, 2007.

[5] J. Rocabert, A. Luna, F. Blaabjerg, and P. Rodríguez, "Control of Power Converters in AC Microgrids," *IEEE Trans. Power Electron.*, vol. 27, no. 11, pp. 4734–4749, 2012.

[6] T. Dragicevic, J. C. Vasquez, J. M. Guerrero, and D. Skrlec, "Advanced LVDC Electrical Power Architectures and Microgrids: A Step Toward a New Generation of Power Distribution Networks," *Electrif. Mag. IEEE*, vol. 2, no. 1, pp. 54–65, 2014.

[7] H. Kakigano, Y. Miura, and T. Ise, "Distribution Voltage Control for DC Microgrids Using Fuzzy Control and Gain-Scheduling Technique," *IEEE Trans. Power Electron.*, vol. 28, no. 5, pp. 2246–2258, 2013.

[8] K. Sun, L. Zhang, Y. Xing, and J. M. Guerrero, "A Distributed Control Strategy Based on DC Bus Signaling for Modular Photovoltaic Generation Systems with Battery Energy Storage," *IEEE Trans. Power Electron.*, vol. 26, no. 10, pp. 3032–3045, 2011.

[9] P. Karlsson and J. Svensson, "DC Bus Voltage Control for a Distributed Power System," *IEEE Trans. Power Electron.*, vol. 18, no. 6, pp. 1405–1412, 2003.

[10] J. Schonberger, R. Duke, and S. D. Round, "DC-Bus Signaling: A Distributed Control Strategy for a Hybrid Renewable Nanogrid," *IEEE Trans. Ind. Electron.*, vol. 53, no. 5, pp. 1453–1460, 2006.

[11] Y. Ito, Y. Zhongqing, and H. Akagi, "DC Microgrid Based Distribution Power Generation System," in *Power Electron. and Motion Control Conference, 2004. IPEMC 2004. The Fourth International*, 2004, vol. 3, pp. 1740–1745.

[12] A. Kwasinski and P. T. Krein, "A Microgrid-Based Telecom Power System Using Modular Multiple-Input DC–DC Converters," in *Telecommunications Conference, 2005. INTELEC'05. Twenty-Seventh International*, 2005, pp. 515–520.

[13] H. Kakigano, Y. Miura, T. Ise, and R. Uchida, "DC Micro-Grid for Super High Quality Distribution—System Configuration and Control of Distributed Generations and Energy Storage Devices," in *Power Electron. Specialists Conference, 2006. PESC'06. 37th IEEE*, 2006, pp. 1–7.

[14] A. Anvari-Moghaddam, J. Carlos Vasquez, J. C. Vasquez, and J. M. Guerrero, "Load Shifting Control and Management of Domestic Microgeneration Systems for Improved Energy Efficiency and Comfort," *41st Annual Conference of the IEEE Industrial Electronics Society*, Yokohama, Japan, 2015, pp. 96–101.

[15] G. Mokhtari, A. Anvari-Moghaddam, and G. Nourbakhsh, "Distributed Control and Management of Renewable Electric Energy Resources for Future Grid Requirements," in D. E. L. Mihet (Ed.): *Energy Management of Distributed Generation Systems*, Rijeka, Croatia, InTech Open Access Publishers, 2016.

[16] A. Anvari-Moghaddam, T. Dragicevic, J. C. Vasquez, and J. M. Guerrero, "Optimal Utilization of Microgrids Supplemented with Battery Energy Storage Systems in Grid Support Applications," in *2015 IEEE First International Conference on Direct Current Microgrids, ICDCM 2015*, 2015.

[17] D. Salomonsson, L. Soder, and A. Sannino, "An Adaptive Control System for a DC Microgrid for Data Centers," *IEEE Trans. Ind. Appl.*, vol. 44, no. 6, pp. 1910–1917, 2008.

[18] P. Biczel, "Power Electronic Converters in DC Microgrid," in *Compatibility in Power Electron., 2007. CPE'07*, 2007, pp. 1–6.

[19] A. Anvari-Moghaddam, G. Mokhtari, and J. M. Guerrero, "Coordinated Demand Response and Distributed Generation Management in Residential Smart Microgrids," in D. E. L. Mihet (Ed.): *Energy Management of Distributed Generation Systems*, Rijeka, Croatia: InTech Open Access Publishers, 2016.

[20] A. Anvari-Moghaddam, H. Monsef, and A. Rahimi-Kian, "Cost-Effective and Comfort-Aware Residential Energy Management under Different Pricing Schemes and Weather Conditions," *Energy Build.*, vol. 86, pp. 782–793, 2015.

[21] A. Anvari-Moghaddam, H. Monsef, and A. Rahimi-Kian, "Optimal Smart Home Energy Management Considering Energy Saving and a Comfortable Lifestyle," *IEEE Trans. Smart Grid*, vol. 6, no. 1, pp. 324–332, 2015.

[22] A. Anvari-Moghaddam, H. Monsef, A. Rahimi-Kian, J. M. Guerrero, and J. C. Vasquez, "Optimized Energy Management of a Single-House Residential Micro-Grid with Automated Demand Response," *IEEE Eindhoven PowerTech, PowerTech 2015*, 2015.

[23] F. Valenciaga and P. F. F. Puleston, "Supervisor Control for a Stand-Alone Hybrid Generation System Using Wind and Photovoltaic Energy," *IEEE Trans. Energy Convers.*, vol. 20, no. 2, pp. 398–405, 2005.

[24] A. M. Knight and G. E. Peters, "Simple Wind Energy Controller for an Expanded Operating Range," *IEEE Trans. Energy Convers.*, vol. 20, no. 2, pp. 459–466, 2005.

[25] S. J. Chiang, H.-J. Shieh, and M.-C. Chen, "Modeling and Control of PV Charger System with SEPIC Converter," *IEEE Trans. Ind. Electron.*, vol. 56, no. 11, pp. 4344–4353, 2009.

[26] T. Dragicevic, *Hierarchical Control of a Direct Current Microgrid with Energy Storage Systems in a Distributed Topology*. PhD Thesis, Electrical Engineering and Computing, University of Zagreb, 2013.

[27] J. S. Glaser and A. F. Witulski, "Output Plane Analysis of Load-Sharing in Multiple-Module Converter Systems," *IEEE Trans. Power Electron.*, vol. 9, no. 1, pp. 43–50, 1994.

[28] J.-W. Kim, H.-S. Choi, and B. H. Cho, "A Novel Droop Method for Converter Parallel Operation," *IEEE Trans. Power Electron.*, vol. 17, no. 1, pp. 25–32, 2002.

[29] R. Ayyanar, R. Giri, and N. Mohan, "Active Input-Voltage and Load-Current Sharing in Input-Series and Output-Parallel Connected Modular DC–DC Converters Using Dynamic Input-Voltage Reference Scheme," *IEEE Trans. Power Electron.*, vol. 19, no. 6, pp. 1462–1473, 2004.

[30] D. Sha, Z. Guo, and X. Liao, "Cross-Feedback Output-Current-Sharing Control for Input-Series-Output-Parallel Modular DC–DC Converters," *IEEE Trans. Power Electron.*, vol. 25, no. 11, pp. 2762–2771, 2010.

[31] G. AlLee and W. Tschudi, "Edison Redux: 380 VDC Brings Reliability and Efficiency to Sustainable Data Centers," *Power Energy Mag. IEEE*, vol. 10, no. 6, pp. 50–59, 2012.

[32] D. Boroyevich, I. Cvetkovic, D. Dong, R. Burgos, F. Wang, and F. Lee, "Future Electronic Power Distribution Systems: A Contemplative View," in *International Conference on Optimization of Electrical and Electronic Equipment*, 2010, pp. 1369–1380.

[33] V. Nasirian, S. Moayedi, A. Davoudi, and F. L. Lewis, "Distributed Cooperative Control of DC Microgrids," *IEEE Trans. Power Electronics*, vol. 30, no. 4, pp. 2288–2303, 2015.

[34] S. Anand, B. G. Fernandes, and J. Guerrero, "Distributed Control to Ensure Proportional Load Sharing and Improve Voltage Regulation in Low-Voltage DC Microgrids," *IEEE Trans. Power Electron.*, vol. 28, no. 4, pp. 1900–1913, 2013.

[35] H. Kakigano, Y. Miura, and T. Ise, "Low-Voltage Bipolar-Type DC Microgrid for Super High Quality Distribution," *IEEE Trans. Power Electron.*, vol. 25, no. 12, pp. 3066–3075, 2010.

[36] R. S. Balog and P. T. Krein, "Bus Selection in Multibus DC Microgrids," *IEEE Trans. Power Electron.*, vol. 26, no. 3, pp. 860–867, 2011.

[37] N. C. Ekneligoda and W. W. Weaver, "Game-Theoretic Cold-Start Transient Optimization in DC Microgrids," *IEEE Trans. Ind. Electron.*, vol. 61, no. 12, pp. 6681–6690, 2014.

[38] Q. Shafiee, T. Dragicevic, J. C. Vasquez, and J. M. Guerrero, "Hierarchical Control for Multiple DC-Microgrids Clusters," *IEEE Trans. Energy Conversion*, vol. 29, no. 4, pp. 922–933, 2014.

[39] R. W. De Doncker, "Power Electronic Technologies for Flexible DC Distribution Grids," in *Power Electronics Conference (IPEC-Hiroshima 2014 – ECCE-ASIA), 2014 International*, 2014, pp. 736–743.

[40] X. She, A. Q. Huang, and R. Burgos, "Review of Solid-State Transformer Technologies and Their Application in Power Distribution Systems," *Emerg. Sel. Top. Power Electron., IEEE J.*, vol. 1, no. 3, pp. 186–198, 2013.

[41] J.-D. Park, J. Candelaria, L. Ma, and K. Dunn, "DC Ring-Bus Microgrid Fault Protection and Identification of Fault Location," *Power Deliv. IEEE Trnas.*, vol. 28, no. 4, pp. 2574–2584, 2013.

[42] R. M. Cuzner and G. Venkataramanan, "The Status of DC Micro-Grid Protection," in *Industry Applications Society Annual Meeting, 2008. IAS'08. IEEE*, 2008, pp. 1–8.

[43] N. R. Chaudhuri, R. Majumder, and B. Chaudhuri, "System Frequency Support through Multi-Terminal DC (MTDC) Grids," *IEEE Trans. Power Syst.*, vol. 28, no. 1, pp. 347–356, 2013.

[44] R. T. Pinto, P. Bauer, S. F. Rodrigues, E. J. Wiggelinkhuizen, J. Pierik, and B. Ferreira, "A Novel Distributed Direct-Voltage Control Strategy for Grid Integration of Offshore Wind Energy Systems through MTDC Network," *IEEE Trans. Ind. Electron.*, vol. 60, no. 6, pp. 2429–2441, 2013.

[45] C. Gavriluta, J. Ignacio Candela, C. Citro, J. Rocabert, A. Luna, and P. Rodriguez, "Decentralized Primary Control of MTDC Networks with Energy Storage and Distributed Generation," *IEEE Trans. Ind. Appl.*, vol. 50, no. 6, pp. 4122–4131, 2014.

[46] N. Flourentzou, V. G. Agelidis, and G. D. Demetriades, "VSC-Based HVDC Power Transmission Systems: An Overview," *IEEE Trans. Power Electron.*, vol. 24, no. 3, pp. 592–602, 2009.

[47] M. Bragard, N. Soltau, S. Thomas, and R. W. De Doncker, "The Balance of Renewable Sources and User Demands in Grids: Power Electronics for Modular Battery Energy Storage Systems," *IEEE Trans. Power Electron.*, vol. 25, no. 12, pp. 3049–3056, 2010.

[48] A. Anvari-Moghaddam, Q. Shafiee, J. C. Vasquez, and J. M. Guerrero, "Optimal Adaptive Droop Control for Effective Load Sharing in AC Microgrids," *42nd Annual Conference of the IEEE Industrial Electronics Society (IECON'16)*, pp. 2–7, 2016.

[49] A. Anvari-Moghaddam, J. M. Guerrero, A. Rahimi-Kian, and M. S. Mirian, "Optimal Real-Time Dispatch for Integrated Energy Systems: An Ontology-Based Multi-Agent Approach," in *2016 IEEE Seventh International Symposium on Power Electronics for Distributed Generation Systems, PEDG 2016*, 2016.

[50] D. Linden and T. B. Reddy, *Handbook of Batteries*, Third Edition. New York: McGraw-Hill, 2002.

[51] E. Koutroulis, K. Kalaitzakis, and N. C. Voulgaris, "Development of a Microcontroller-Based, Photovoltaic Maximum Power Point Tracking Control System," *IEEE Trans. Power Electron.*, vol. 16, no. 1, pp. 46–54, 2001.

[52] A. Kwasinski and C. N. Onwuchekwa, "Dynamic Behavior and Stabilization of DC Microgrids with Instantaneous Constant-Power Loads," *IEEE Trans. Power Electron.*, vol. 26, no. 3, pp. 822–834, 2011.

[53] N. L. Diaz, T. Dragicevic, J. C. Vasquez, and J. M. Guerrero, "Intelligent Distributed Generation and Storage Units for DC Microgrids—A New Concept on Cooperative Control without Communications Beyond Droop Control," *IEEE Trans. Smart Grid*, vol. 5, no. 5, pp. 2476–2485, 2014.

[54] C. N. Onwuchekwa and A. Kwasinski, "Analysis of Boundary Control for Buck Converters with Instantaneous Constant-Power Loads," *IEEE Trans. Power Electron.*, vol. 25, no. 8, pp. 2018–2032, 2010.

[55] J. M. Guerrero, J. C. Vasquez, J. Matas, L. G. de Vicuna, and M. Castilla, "Hierarchical Control of Droop-Controlled AC and DC Microgrids—A General Approach Toward Standardization," *IEEE Trans. Ind. Electron.*, vol. 58, no. 1, pp. 158–172, 2011.

[56] X. Lu, K. Sun, J. M. M. Guerrero, J. C. C. Vasquez, and L. Huang, "Double-Quadrant State-of-Charge-Based Droop Control Method for Distributed Energy Storage Systems in Autonomous DC Microgrids," *IEEE Trans. Smart Grid*, vol. 6, no. 1, pp. 147–157, 2015.

[57] M. Lee, D. Chen, K. Huang, C.-W. Liu, and B. Tai, "Modeling and Design for a Novel Adaptive Voltage Positioning (AVP) Scheme for Multiphase VRMs," *IEEE Trans. Power Electron.*, vol. 23, no. 4, pp. 1733–1742, 2008.

[58] C.-J. Chen, D. Chen, C.-S. Huang, M. Lee, and E. K.-L. Tseng, "Modeling and Design Considerations of a Novel High-Gain Peak Current Control Scheme to Achieve Adaptive Voltage Positioning (AVP) for DC Power Converters," *IEEE Trans. Power Electron.*, vol. 24, no. 12, pp. 2942–2950, 2009.

[59] H.-H. Huang, C.-Y. Hsieh, J.-Y. Liao, and K.-H. Chen, "Adaptive Droop Resistance Technique for Adaptive Voltage Positioning in Boost DC–DC Converters," *IEEE Trans. Power Electron.*, vol. 26, no. 7, pp. 1920–1932, 2011.

[60] F. Valenciaga and P. F. Puleston, "High-Order Sliding Control for a Wind Energy Convers. System Based on a Permanent Magnet Synchronous Generator," *IEEE Trans. Energy Convers.*, vol. 23, no. 3, pp. 860–867, 2008.

[61] T. Dragicevic, J. M. Guerrero, J. C. Vasquez, and D. Skrlec, "Supervisory Control of an Adaptive-Droop Regulated DC Microgrid with Battery Management Capability," *IEEE Trans. Power Electron.*, vol. 29, no. 2, pp. 695–706, 2014.

[62] L. Che and M. Shahidehpour, "DC Microgrids: Economic Operation and Enhancement of Resilience by Hierarchical Control," *IEEE Trans. Smart Grid*, vol. 5, no. 5, pp. 2517–2526, 2014.

[63] X. Lu, J. M. M. Guerrero, K. Sun, J. C. C. Vasquez, R. Teodorescu, and L. Huang, "Hierarchical Control of Parallel AC–DC Converter Interfaces for Hybrid Microgrids," *IEEE Trans. Smart Grid*, vol. 5, no. 2, pp. 683–692, 2014.

[64] B. Wang, M. Sechilariu, and F. Locment, "Intelligent DC Microgrid with Smart Grid Communications: Control Strategy Consideration and Design," *IEEE Trans. Smart Grid*, vol. 3, no. 4, pp. 2148–2156, 2012.

[65] C. Jin, P. Wang, J. Xiao, Y. Tang, and F. H. H. Choo, "Implementation of Hierarchical Control in DC Microgrids," *IEEE Trans. Ind. Electron.*, vol. 61, no. 8, pp. 4032–4042, 2014.

[66] Y. Gu, X. Xiang, W. Li, and X. He, "Mode-Adaptive Decentralized Control for Renewable DC Microgrid with Enhanced Reliability and Flexibility," *IEEE Trans. Power Electron.*, vol. 29, no. 9, pp. 5072–5080, 2014.

[67] A. Gkountaras, S. Dieckerhoff, and T. Sezi, "Performance Analysis of Hybrid Microgrids Applying SoC-Adaptive Droop Control," in *Power Electronics and Applications (EPE'14-ECCE Europe), 2014 16th European Conference on*, 2014, pp. 1–10.

[68] E. Liegmann and R. Majumder, "An Efficient Method of Multiple Storage Control in Microgrids," *IEEE Trans Power Syst*, vol. 30, no. 6, pp. 3437–3444, Nov. 2015.

[69] H. C. Ferreira, L. Lampe, J. Newbury, and T. G. Swart, *Power Line Communications: Theory and Appl. for Narrowband and Broadband Communications over Power Lines*. Hoboken, NJ, USA: John Wiley & Sons, 2010.

[70] "REbus DC Microgrid," 2013. Available from: http://rebuspower.com/.

[71] T. Dragicevic, J. M. Guerrero, and J. C. Vasquez, "A Distributed Control Strategy for Coordination of an Autonomous LVDC Microgrid Based on Power-Line Signaling," *IEEE Trans. Ind. Electron.*, vol. 61, no. 7, pp. 3313–3326, 2014.

[72] W. Stefanutti, S. Saggini, P. Mattavelli, and M. Ghioni, "Power Line Communication in Digitally Controlled DC–DC Converters Using Switching Frequency Modulation," *IEEE Trans. Ind. Electron.*, vol. 55, no. 4, pp. 1509–1518, 2008.

[73] R. Olfati-Saber, J. A. Fax, and R. M. Murray, "Consensus and Cooperation in Networked Multi-Agent Systems," *Proc. IEEE*, vol. 95, no. 1, pp. 215–233, 2007.

[74] R. Olfati-Saber and R. M. Murray, "Consensus Problems in Networks of Agents with Switching Topology and Time-Delays," *IEEE Trans. Autom. Control.*, vol. 49, no. 9, pp. 1520–1533, 2004.

[75] Z. Meng, W. Ren, Y. Cao, and Z. You, "Leaderless and Leader-Following Consensus with Communication and Input Delays under a Directed Network Topology," *IEEE Trans. Syst. Man, Cybern. Part B Cybern.*, vol. 41, no. 1, pp. 75–88, 2011.

[76] L. X. L. Xiao and S. Boyd, "Fast Linear Iterations for Distributed Averaging,"*42nd IEEE Int. Conf. Decis. Control (IEEE Cat. No.03CH37475)*, vol. 5, 2003.

[77] J.-D. Park and J. Candelaria, "Fault Detection and Isolation in Low-Voltage DC-Bus Microgrid System," *Power Deliv. IEEE Trnas.*, vol. 28, no. 2, pp. 779–787, 2013.

[78] J. Yang, J. E. Fletcher, and J. O'Reilly, "Short-Circuit and Ground Fault Analyses and Location in VSC-Based DC Network Cables," *IEEE Trans. Ind. Electron.*, vol. 59, no. 10, pp. 3827–3837, 2012.

[79] D. Salomonsson, L. Soder, and A. Sannino, "Protection of Low-Voltage DC Microgrids," *IEEE Trans. Power Deliv.*, vol. 24, no. 3, pp. 1045–1053, 2009.

[80] B. Ram, *Power System Protection and Switchgear*. Pennsylvania Plaza, New York City, USA: McGraw-Hill Education (India) Pvt Limited, 2011.

[81] A. Berizzi, A. Silvestri, D. Zaninelli, and S. Massucco, "Short-Circuit Current Calculations for DC Systems," *IEEE Trans. Ind. Appl.*, vol. 32, no. 5, pp. 990–997, 1996.

[82] D. Paul, "DC traction power system grounding," *IEEE Trans. Ind. Appl.*, vol. 38, no. 3, pp. 818–824, 2002.

[83] S. A. Memon and P. Fromme, "Stray Current Corrosion and Mitigation: A Synopsis of the Technical Methods Used in DC Transit Systems," *IEEE Electrif. Mag.*, vol. 2, no. 3, pp. 22–31, 2014.

[84] I. Cotton, C. Charalambous, P. Aylott, and P. Ernst, "Stray Current Control in DC Mass Transit Systems," *IEEE Trans. Veh. Technol.*, vol. 54, no. 2, pp. 722–730, 2005.

[85] J.-M. Shen, H.-L. Jou, and J.-C. Wu, "Novel Transformerless Grid-Connected Power Converter with Negative Grounding for Photovoltaic Generation System," *IEEE Trans. Power Electron.*, vol. 27, no. 4, pp. 1818–1829, 2012.

[86] C. Charalambous, N. Kokkinos, and A. Demetriou, "Impact of Photovoltaic Oriented DC Stray Current Corrosion on Large Scale Solar Farms' Grounding and Third-Party Infrastructure: Modelling and Assessment," *IEEE Trans. Ind. Appl.*, vol. 51, no. 6, pp. 5421–5430, Nov.-Dec. 2015.

[87] M. E. Baran and N. R. Mahajan, "DC Distribution for Industrial Systems: Opportunities and Challenges," *IEEE Trans. Ind. Appl.*, vol. 39, no. 6, pp. 1596–1601, 2003.

[88] J. Park, "Ground Fault Detection and Location for Ungrounded DC Traction Power Systems," *IEEE Trans. Veh. Technol.*, vol. 64, no. 12, pp. 5667–5676 Dec. 2015.

[89] M. Valdes, T. Papallo, and B. Premerlani, "Finding Fault—Locating a Ground Fault in Low-Voltage, High-Resistance Grounded Systems via the Single-Processor Concept for Circuit Protection," *IEEE Ind. Appl. Mag.*, vol. 13, no. 5, pp. 24–30, 2007.

[90] B. Jacobson and J. Walker, "Grounding Considerations for DC and Mixed DC and AC Power Systems," *Nav. Eng. J*, vol. 119, no. 2, pp. 49–62, 2007.

[91] "IEC 60364-1 Low-Voltage Electrical Installations—Part 1: Fundamental Principles, Assessment of General Characteristics, Definitions." IEC 60364-1, 2005.

[92] C.-H. Lee and C.-J. Lu, "Assessment of Grounding Schemes on Rail Potential and Stray Currents in a DC Transit System," *IEEE Trans. Power Deliv.*, vol. 21, no. 4, pp. 1941–1947, 2006.

[93] J.-M. Meyer and A. Rufer, "A DC Hybrid Circuit Breaker with Ultra-Fast Contact Opening and Integrated Gate-Commutated Thyristors (IGCTs)," *IEEE Trans. Power Deliv.*, vol. 21, no. 2, pp. 646–651, 2006.

[94] "IEEE Guide for the Protection of Stationary Battery Systems," *IEEE Std* 1375–1998, 1998. DOI: 10.1109/IEEESTD.1998.87900.

[95] C. Meyer, S. Schroder, and R. W. De Doncker, "Solid-State Circuit Breakers and Current Limiters for Medium-Voltage Systems Having Distributed Power Systems," *IEEE Trans. Power Electron.*, vol. 19, no. 5, pp. 1333–1340, 2004.

[96] D. Lawes, L. Ran, and Z. Xu, "Design of a Solid-State DC Circuit Breaker for Light Rail Transit Power Supply Network," in *Energy Conversion Congress and Exposition (ECCE), 2014 IEEE*, 2014, pp. 350–357.

[97] S. Krstic, E. L. Wellner, A. R. Bendre, and B. Semenov, "Circuit Breaker Technologies for Advanced Ship Power Systems," in *Electric Ship Technologies Symposium, 2007. ESTS'07. IEEE*, 2007, pp. 201–208.

[98] Z. J. Shen, G. Sabui, Z. Miao, and Z. Shuai, "Wide-Bandgap Solid-State Circuit Breakers for DC Power Systems: Device and Circuit Considerations," *IEEE Trans. Electron Devices*, vol. 62, no. 2, pp. 294–300, 2015.

[99] J. Jebramcik and F. Berger, "Observations on Switching Characteristics of Arc Chutes in DC Contactors," in *ICEC 2014; The 27th International Conference on Electrical Contacts; Proceedings of*, 2014, pp. 1–6.

[100] C. Jin and R. Dougal, "Current Limiting Technique Based Protection Strategy for an Industrial DC Distribution System," in *Industrial Electronics, 2006 IEEE International Symposium on*, 2006, vol. 2, pp. 820–825.

[101] F. Luo, J. Chen, X. Lin, Y. Kang, and S. Duan, "A Novel Solid State Fault Current Limiter for DC Power Distribution Network," in *Applied Power Electronics Conference and Exposition, 2008. APEC 2008. Twenty-Third Annual IEEE*, 2008, pp. 1284–1289.

[102] K. Yasuda, A. Ichinose, A. Kimura, *et al.*, "Research Development of Superconducting Fault Current Limiter in Japan," *IEEE Trans. Appl. Supercond.*, vol. 15, no. 2, pp. 1978–1981, 2005.

[103] L. Tang and B.-T. Ooi, "Locating and Isolating DC Faults in Multi-Terminal DC Systems," *IEEE Trans. Power Deliv.*, vol. 22, no. 3, pp. 1877–1884, 2007.

[104] M. E. Baran and N. R. Mahajan, "Overcurrent Protection on Voltage-Source-Converter-Based Multiterminal DC Distribution Systems," *IEEE Trans. Power Deliv.*, vol. 22, no. 1, pp. 406–412, 2007.

[105] J. Johnston, J. Counsell, G. Banks, and M. J. Stewart, "Beyond Power over Ethernet: The Development of Digital Energy Networks for Buildings," London: CIBSE Technical Symposium, Imperial University, 18th and 19th April 2012.

[106] D. Dong, I. Cvetkovic, D. Boroyevich, W. Zhang, R. Wang, and P. Mattavelli, "Grid-Interface Bidirectional Converter for Residential DC Distribution Systems—Part One: High-Density Two-Stage Topology," *IEEE Trans. Power Electron.*, vol. 28, no. 4, pp. 1655–1666, 2013.

[107] D. Dong, F. Luo, X. Zhang, D. Boroyevich, and P. Mattavelli, "Grid-Interface Bidirectional Converter for Residential DC Distribution Systems—Part 2: AC and DC Interface Design with Passive Components Minimization," *IEEE Trans. Power Electron.*, vol. 28, no. 4, pp. 1667–1679, 2013.

[108] M.-H. Ryu, H.-S. Kim, J.-H. Kim, J.-W. Baek, and J.-H. Jung, "Test Bed Implementation of 380 V DC Distribution System Using Isolated Bidirectional Power Converters," in *Energy Conversion Congress and Exposition (ECCE), 2013 IEEE*, 2013, pp. 2948–2954.

[109] "Intelligent DC Microgrid Living Lab." Available from:http://www.et.aau.dk/research-programmes/microgrids/activities/intelligent-dc-microgrid-living-lab/

[110] D. J. Becker and B. J. Sonnenberg, "DC Microgrids in Buildings and Data Centers," in *Telecommunications Energy Conference (INTELEC), 2011 IEEE 33rd International*, 2011, pp. 1–7.

[111] "Smart Grid Strategy: The Intelligent Energy System of the Future, Danish Ministry of Climate, Energy Building," May 2013, [online]. Available from: http://old.efkm.dk/sites/kebmin.dk/files/climate-energy-and-building-policy/denmark/energy-supply-and-efficiency/smart-grid-the-intelligent-0/smart_grid_strategy_uk_summary_web.pdf [Accessed 9 November 2015].

[112] "The Smart Grid: An Introduction", the U.S. Department of Energy, 2009, [online] Available from: https://energy.gov/sites/prod/files/oeprod/DocumentsandMedia/DOE_SG_Book_Single_Pages%281%29.pdf [Accessed 8 July 2014].

[113] H. A. B. Siddique, S. M. Ali, and R. W. De Doncker, "DC Collector Grid Configurations for Large Photovoltaic Parks," in *Power Electronics and Applications (EPE), 2013 15th European Conference on*, 2013, pp. 1–10.

[114] S. Chuangpishit, A. Tabesh, Z. Moradi-Sharbabk, and M. Saeedifard, "Topology Design for Collector Systems of Offshore Wind Farms with Pure DC Power Systems," *IEEE Trans. Ind. Electron.*, vol. 61, no. 1, pp. 320–328, 2014.

[115] B.-I. Craciun, T. Kerekes, D. Sera, and R. Teodorescu, "Overview of Recent Grid Codes for PV Power Integration," in *Optimization of Electrical and Electronic Equipment (OPTIM), 2012 13th International Conference on*, 2012, pp. 959–965.

[116] M. Tsili and S. Papathanassiou, "A Review of Grid Code Technical Requirements for Wind Farms," *Renew. Power Gener. IET*, vol. 3, no. 3, pp. 308–332, 2009.

[117] S. Vazquez, S. M. Lukic, E. Galvan, L. G. Franquelo, and J. M. Carrasco, "Energy Storage Systems for Transport and Grid Applications," *IEEE Trans. Ind. Electron.*, vol. 57, no. 12, pp. 3881–3895, 2010.

[118] K. Clement-Nyns, E. Haesen, and J. Driesen, "The Impact of Charging Plug-In Hybrid Electric Vehicles on a Residential Distribution Grid," *IEEE Trans. Power Syst.*, vol. 25, no. 1, pp. 371–380, 2010.

[119] A. Ravichandran, P. Malysz, S. Sirouspour, and A. Emadi, "The Critical Role of Microgrids in Transition to a Smarter Grid: A Technical Review," in *Transportation Electrification Conference and Expo (ITEC), 2013 IEEE*, 2013, pp. 1–7.

Chapter 8

Distributed energy resources integration and demand response: the role of stochastic demand modelling

Emilio J. Palacios-García[1], Antonio Moreno-Muñoz[1], Isabel Santiago-Chiquero[1], José María Flores-Arias[1] and Francisco J. Bellido-Outeiriño[1]

8.1 Introduction

One of the main problems with it comes to the integration of distributed energy resources is the estimation and prediction of the energetic demand that those energy sources must be able to supply [1]. This is especially difficult nowadays due to the upward trend observed in all the projections for the energy consumption for major energy end-use sectors (residential, commercial, industrial, and transportation) [2].

From all these end-use sectors, probably the one that presents the greatest interest in the modelling field is the residential one. Whereas the consumption patterns and daily load profiles of the commercial, industrial and transportation sectors might be relatively easy to predict depending on the activity, the wide amount of appliances, household types and consumers' behaviours in the residential sector generate stochastic consumption profiles, being almost impossible to find two households with the same habits [3].

Moreover, in order to accurately estimate the integration of new energy resources, detailed knowledge of the consumption profiles is needed. Nowadays estimations and predictions for the global consumption are available, but the temporal resolution is not enough since the great variety of consumption patterns between the different sectors, as well as within them, makes the task of balancing production and demand not easy, and much less when non-dispatchable energy sources are included in the system [4,5]. In addition, this higher resolution is justified from the point of view of demand response and energy management as the time granularity needed for the development of the strategies has to be between 5 and 30 min and submetering programmes are not usually carried out, so only aggregated measurement are available [6].

[1]Universidad de Córdoba, Córdoba, Spain

In this context, stochastic modelling techniques have been presented as the most suitable ones as they are able to create the diversity needed to take into account the behaviour of the residents while keeping the general observed statistical trends [7]. These methods are based on the individual modelling of each appliance and the consumer behaviour patterns so not can only high-resolution profiles be generated, but also the end-use of the energy can be specifically determined. That is not even possible nowadays with the new advanced metering infrastructure, unless non-intrusive load monitoring programmes are applied, which also makes these models a great tool for the assessment of energy policies and the impact of including new appliances at home [8].

In this chapter, the usage and application of stochastic demand models will be addressed, with a focus on the prediction and estimation of the energy consumption, the assessment of different demand response policies and the integration of distributed energy resources at small scale. For this aim, first, a short overview of the main modelling techniques employed in the residential sector will be given to the reader in order to contextualise this type of models. Subsequently, the methodology used in these models will be exposed, showing the different blocks that usually built up the estimation process. Finally, the usability for assessing energy policies and integrating distributed energy resources will be discussed.

8.2 Overview of modelling techniques for energy demand prediction

The great diversity of factors that influence the residential energy consumption leads to a wide range of modelling techniques and methodologies. One of the most accepted criteria to classify them is the hierarchical relationship between the consumption that is being modelled and the input data for the model. This classification was first proposed by Swan and Ugursal [9], and it distinguishes basically two types of models named as top-down and bottom-up models respectively, also defining different categories within each of these two types depending on the methodologies employed.

A similar approach was followed by Grandjean *et al.* [10] in a more recent work, nevertheless, the proposed classification of the different load curve models was based on the manner how the diversity among dwellings is obtained. The idea of this basic classification criterion is illustrated in Figure 8.1, where the most usual input data and their relationship with the household consumption estimation are represented together with the two suggested approaches, which will be discussed in the following sections.

8.2.1 Top-down models

The top-down models are those where the consumption estimation is based on variables or input data that belong to the whole sector. Therefore, the residential consumption is regarded as an energy sink without taking into account the individual characteristics of the demands that account for this consumption. Thus, the main purpose of these modelling strategies is planning and estimating the possible

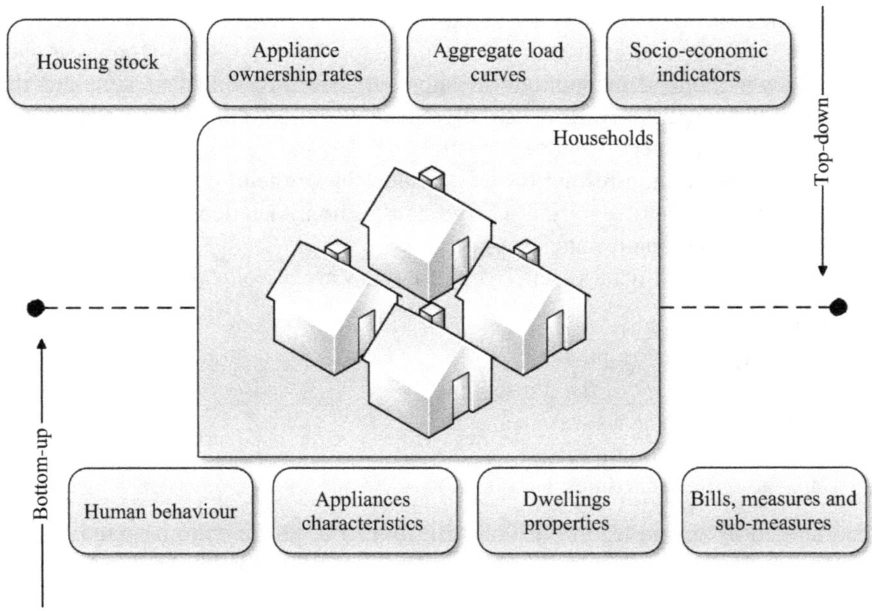

Figure 8.1 Conceptual approach. Top-down vs bottom-up models

trends and evolution of the sectors by using relatively simple and widely available aggregated data. However, its ability for predicting new emerging end-uses or disruptive technologies is limited.

Among the principal inputs variables or data for these models that can be highlighted we found as follows:

- Socio-economic indicators such as the gross domestic product
- Unemployment rates
- Electricity prices
- Weather conditions
- Housing stock
- Appliances usage rates

Regarding the possible subdivisions in the top-down category, these models were classified by Swan and Ugursal [9] as either Econometric or Technological models, although many times the characteristics are mixed. Econometric models generally make use of prices of electricity or appliances, whereas technological models base their predictions on characteristics such as appliances usage rates or housing stock.

8.2.2 Bottom-up models

With an opposite philosophy, the bottom-up models use input data which are in a lower hierarchical level than the energetic consumption itself. Therefore, it is the aggregation of detailed data what builds up the load demand curve, although some global indicators might also be included for the prediction, such as weather conditions of housing stock in the same line of the top-down models. In contrast with the top-down models, the bottom-up approach has the ability to simulate the occupants' behaviour, as well as the individual consumption of different appliances which makes them very suitable for studying demand response strategies, new energy policies or high-resolution simulations. Nevertheless, the complexity of these models is higher, requiring a large data sample, detailed input information and more computationally intensive tasks.

Some usual input data for this type of models are as follows:

- Human behaviour
- Electricity or energy bills
- Thermal properties of the households
- Appliances characteristics
- Equipment penetration rates
- Daily activities schedules

According to Swan and Ugursal [9] classification criteria, the bottom-up models are divided into two large categories which are statistics-based models and engineering-based models. The first ones make use of regression analysis techniques, conditional demand analysis or neural network for modelling the energy usage, whereas the engineering-based models are focused on appliances ratings, archetypes definitions or detailed sampling of the population. Nevertheless, alternative classifications can be found such as the one proposed by Grandjean *et al.* [10].

8.2.3 Comparison

After this initial categorization, it is clear that the decision of using any of the introduced methodologies will eventually be determined by the end goal. Therefore, when reliable long-term forecasting is needed, without taking into account possible technological disruptions and without knowledge of the explicit end-uses that compose the global consumption, the top-down models are the most suitable. Nevertheless, when an understanding of the energy end-uses is required, as well as the possibility of assessing the impact of new technologies the bottom-up methods are the only ones with this capability.

Subsequently, since the goal of this chapter is to address the usage of modelling techniques for the integration of renewable resources and assessing different demand response strategies and energy policies, the chapter will be focused on the bottom-up modelling techniques. More specifically, the engineering-based bottom-up models that utilise distribution data concerning appliances and usage rates will be the ones exposed, since they will allow simulating the individual appliances, as well as the influence of each dwelling behaviour. Moreover, these models were also classified by Grandjean *et al.* [10] as time of used based bottom-up models, a name that results to complete the definition, since they do not only use distribution data, but they are also based on monitoring campaigns on a sample of households such as the time use surveys (TUSs), where the time devoted to certain activities is recorded.

8.3 Time of use based bottom-up models

The conceptual architecture of the system is presented in Figure 8.2. The system has a set of input parameters such as the number of residents in the household, the type of day, the period of the year or the location. With these inputs, it will generate the daily consumption for the given household distinguishing between the different end-uses.

Each of the grey blocks represents a particular model aiming to estimate the occupancy of the household or the consumption due to appliances, lighting systems, and cooling or heating equipment, having defined a set of appliances, lamps or HVAC (heating, ventilation and air conditioning) systems (red blocks) that are responsible for consuming the energy and that can usually be randomly selected on the basis of distribution statistics or self-defined sets. In addition, some specifics statistics data about the activities performed at home or the occupancy probability (Transition Matrices), as well as meteorological information, is provided (yellow blocks).

This basic structure will be further discussed in the following sections, where the occupancy model and the three mentioned consumption models will be individually addressed, explaining the basic methodology of a real implemented model, the input information required for the model to run, and finally the actual simulation procedure.

8.3.1 Occupancy and consumers' behaviour

The energy consumption in the residential sector is strongly influenced by the occupancy patterns as most of the equipment at home requires the direct intervention of the residents, with only a few exceptions such as refrigerators, freezers,

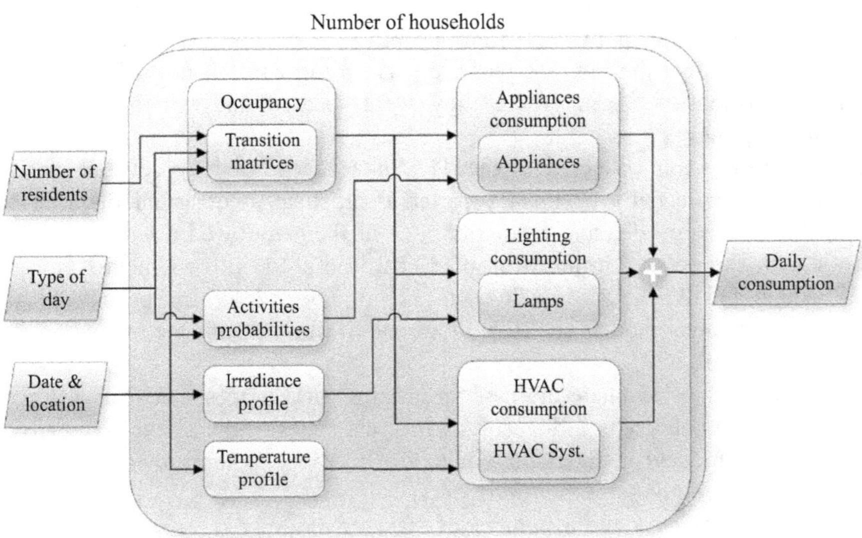

Figure 8.2 Block diagram of the simulation system

or appliances in standby mode. Therefore, following the previously exposed bottom-up philosophy, it is reasonable that the cornerstone for many of these models is the estimation or simulation of the consumers' behaviour.

However, the occupancy modelling has the main inconvenience of being affected by a wide range of external factors such as the working hours, the day of the week, or the number of residents in each household. Thus, it is almost impossible to find two dwellings with the same occupancy patterns or one dwelling whose occupancy profile is exactly the same for two consecutive days. Nevertheless, when the aggregate occupancy profiles are observed, some general trends can be deducted.

The models for estimating the consumers' behaviour aims to simulate individual households in a deterministic way, but keeping the results statistically accurate. The fundamental of these models, the main input parameters and an example of simulation process will be presented in the next sections.

8.3.1.1 Model basics

In the commercial and industrial sectors, presence control is usually common by means of access cards or other methods. Moreover, even when no control methods exist, the working hours are usually well defined and consequently, the occupancy patterns can be known with a high degree of confidence. Nevertheless, if the occupancy profiles of the residential sector are to be studied, a monitoring campaign is required, using a sample of households as a representative of the entire population. This is commonly performed through surveys, where the residents write down their daily activities and habits.

In the context of Europe, the main effort to collect this information is the harmonised European TUSs. These surveys have been carried out in most of the European countries in different years such as in the United Kingdom (2000), Sweden (2010), and Spain (2010) among others. The surveys include the different activities that the interviewees are performing with 10-min resolution, as well as information about where they were located and whether someone accompanied them.

The information collected in the TUS have been used by several authors aiming to generate daily occupancy patterns statistically consistent that can be employed for energy demand prediction. All of them followed a similar methodology as it can be read in the work of Richardson *et al.* in the United Kingdom [11], Widén and Wäckelgård in Sweden [12], or Lopez *et al.* in Spain [13]. The proposed occupancy models are all built around Markov Chains theory and Monte Carlo simulations.

Markov Chains simulate the random progress of a process between different states whose probability is influenced by the past states of the system. Therefore, this statistical process includes the concept of memory as the previous states or history define in some term the future ones.

Therefore, given a finite number of states, $S = \{s_1, s_2, \ldots s_n\}$, of a defined process that starts with one of those states and keeps moving only between the space of states S, it is called a Markov process if the probability of moving to

Table 8.1 Transition probability matrix for a three-resident households

Current state (07.30 h)	Next state (07.40 h)			
	0	1	2	3
0	0.7288	0.1864	0.0848	0
1	0.0938	0.7500	0.1563	0
2	0.0588	0.2941	0.5294	0.1177
3	0.1429	0.2857	0	0.5714

another state j in the time $t+1$ depend on the current state i at time t. This is defined by the probabilities p_{ij} which are included in the so-called matrix of transition probability or transition matrix. If this matrix is constant for any time step t, the process is said to be homogeneous, whereas if the probabilities of the transition matrix depend on the time $p_{ij}(t)$, the process is non-homogeneous and, therefore, for a process with R states and N time steps, there will be N matrices with a size $R \times R$.

In the case of active occupancy simulation, all the studies use the non-homogeneous Markov process for defining the transition probabilities. Specifically, due to the 10-min resolution of the TUS, 144 transition matrices are defined, one for each time step and ranging from 1 to 6 residents. Moreover, additional parameters such as the region, the quarter of the year or the day of week can be and have been used for estimating the transition matrices in different circumstances.

In order to illustrate the transition matrices, an example of one of them can be observed in Table 8.1, where the matrix of transition probabilities for a three-resident household located in Spain during a weekday and in the period from 7.30 to 7.40 h is represented. As it can be seen, there are four possible states ranging from 0 active occupants to 3 active occupants. Depending on the current state, the distribution probability of one of the rows will be considered. Therefore, if the current occupancy is 0, is very likely that it remains 0 in the next step too ($p_{00} = 72.88\%$), whereas if the occupancy in the current state is 2, there are around a half of chances of keeping the same occupancy value ($p_{22} = 52.94\%$), but it is also likely to be increased ($p_{23} = 11.77\%$) or decreased ($p_{21} = 29.41\%$). It should also be pointed out that the sum of each row must be equal to 1 since each row actually represents an individual discrete probability distribution.

8.3.1.2 Input parameters

The input parameters for the occupancy models will mainly depend on the input information available from the TUS. Nevertheless, some of the common information that can be found in all of them is as follows:

- **Number of residents**. Each dwelling in the survey is characterised with the number of residents that share the property.
- **Location**. The location of the households is also accounted in the survey, generally in terms of region, province or municipality.

- **Type of day**. The day of the week is written down by the interviewees in the dairies. Thus, weekdays and weekends can be distinguished in the modelling process.
- **Period of the year**. Some of these surveys distinguish between each quarter of the year, making possible to compare the behaviour in different seasons.
- **Other information**. The survey also collects information about the age, the work sector, the employment status etc. Those data can also be included in the model.

The main concerns when selecting the influence input variables for the transition matrices are two. On the one hand, the increment in the complexity of the model should be justified by an increment in the accuracy or sensibility of the results. On the other hand, a fine characteristic selection can derivate into small data sets of the sample that might not be representative of the population being studied.

8.3.1.3 Simulation algorithm

Using the Markov Chain methodology and the input parameters that have been explained before, the simulation algorithm is implemented as is illustrated in Figure 8.3. This algorithm starts by loading the transition matrices to be used during the process, which are defined depending on the given input information previously commented.

Nevertheless, as was exposed in the Markov Chain methodology, the initial state should be defined before the process starts. For this aim, the occupancy

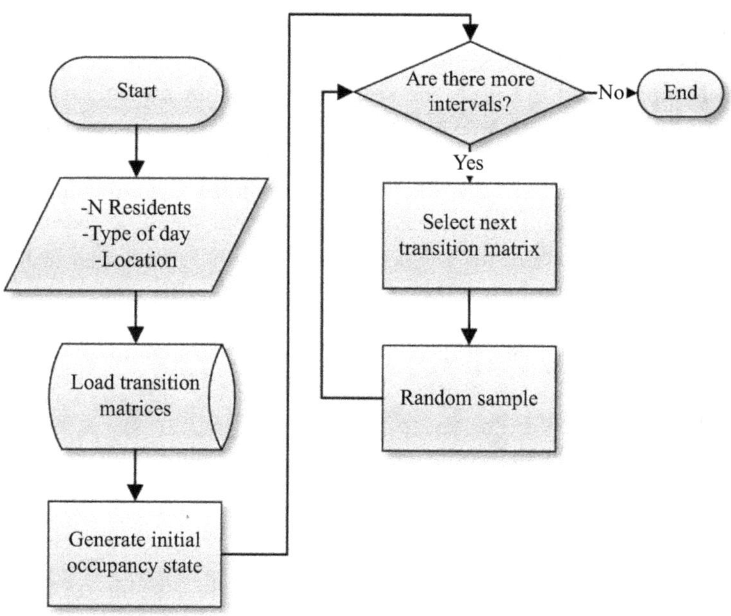

Figure 8.3 Flowchart of residents' behaviour simulation algorithm

probability distribution for the given starting point is used in order to generate the initial state with a random Monte Carlo sampling.

Once the initial state is defined, the process iterates through all the simulation intervals, first, selecting the appropriate transition matrix for this time step, and subsequently generating a random sampling over the row of the transition matrix corresponding with the current number of active occupants in order to determine the next state of the system. If this process is repeated for the 144 time steps of a day with 10-min resolution, the occupancy profile of a house with a given number of residents and some characteristics can be randomly obtained in a way that the results are completely deterministic, but they represent a significant sample of the population, and, therefore, they can be used in the following models for estimating the different end consumptions.

8.3.2 Lighting system consumption

The energy demand due to lighting systems comes to represent around a fifth part of the total electrical consumption in the residential sector [3,14]. In Spain, the figure is around 470 kWh a year per household, whereas in other countries like the United Kingdom, it might be even higher, 715 kWh a year reported. Due to the importance and the specific influence factors that determine this consumption, many different lighting demand models have been developed by many authors such as Stokes *et al.* [15] and Richardson *et al.* [16] for the United Kingdom, Widén *et al.* [17] for Sweden, and Gago *et al.* [18] and Palacios-Garcia *et al.* [8] for Spain. The methodology for developing a model focused on lighting consumption is subsequently exposed.

8.3.2.1 Model basics

The active occupancy of the household is a necessary condition for the usage of lighting systems, which require the direct intervention of the residents. Nevertheless, it is not enough to determine the period of consumption due to the direct relationship between the daylight availability and the needs of artificial light. Therefore, not only is the occupancy profile necessary for the estimation of lighting consumption, but also the external irradiance along the day or studied period should be known.

This external irradiance profile varies along the day from dawn to sunset and is also influenced by the meteorological conditions such as clouds or precipitations, resulting in different levels of daylight availability during the day. For a given threshold, this irradiance will be so low that the artificial lighting systems must be used. This has been supported by other authors such as Reinhart or Hund in the offices' sector [19,20], who have set the value for this threshold as normally distributed with mean 60 W/m^2 and standard deviation 10 W/m^2.

Thus, the occupancy and the irradiance can be considered the main factors that influence the lighting consumption and, consequently, the principal input data for the model. Nevertheless, if they are taken into account with no further restrictions the consumption is likely to be overestimated due to two main problems: the relative usage of different lighting spots and the sharing of them between the residents.

The relative usage is related to the fact that some lighting spots are more likely to be turned on than others. For instance, practically, the lamps installed in the living rooms or kitchens are more often used than the ones situated in corridors, bathrooms or bedrooms. Subsequently, each lighting systems should be adequately weighted according to the relative usage within the household.

The other factor, the shared usage, refers to the contribution of each active occupant to the total energy consumption. If the proportion of active occupants is considered to be proportionally related to the energy demand, this would mean that the consumption, when two residents are active, is double than when only a single person is in the household. However, this relation is not linear since some of the lighting systems are shared between the occupants, and it tends to saturate when the number of active occupants is increased.

In order to quantify this, a factor named as effective occupancy is usually inserted into the model, not only for the lighting consumption but in other consumption groups, as it will be seen in the following sections. This factor is calculated on the basis of the energy consumption figure of households with a different number of residents, obtaining the relative increment in the consumption with the increment in the number of people living in the same house.

Finally, if all the parameters determine that the lighting system should be turned on, then it is necessary to establish the period of time that it will remind in this state. Those times can only be determined by means of an experimental study such as the one performed by Stokes *et al.* [15]. All this methodology has been successfully used by other authors in previous works such as [8,16,17] showing accurate results.

8.3.2.2 Input data

Regarding the above-mentioned influence parameters, a set of input parameters for the model can be defined. Some of them have been already used in previous works; others can be used for improving the current results obtained by these models.

- **Active occupancy**. The number of occupants with 10-min resolutions is used for simulating the behaviour based on the previously discussed model.
- **External irradiance**. The global irradiance profile is used for estimating the probability of using the artificial light systems. This irradiance can be extracted from historical databases, or other prediction or forecasting models. Nevertheless, they must have at least the same temporal resolution as the occupancy data, although in some previous works it has been proved that fast irradiance variations due to clouds have a small impact on the residents' behaviour [16].
- **Lighting systems**. Following the bottom-up methodology, the consumption will be generated by the simulation of each lighting spot in the household. Therefore, information about the number of lighting units, their power, as well as additional information such as the technology or the possible grouping has to be defined. Different approaches can be followed for that. In some cases, depending on the type of household or family a set of lighting units can be selected. It is also possible to individually define the lighting units in each

household. Finally, another approach can be the development of a statistical model that is able to randomly establish the distribution of lighting technologies and power within the dwellings.

- **Additional information**. As in the case of the occupancy sector, additional information can be included to improve the estimation. Due to the high correlation between the daylight availability and the orientation of the windows in the household, as well as the type of household (detached, semi-detached, block etc.), it might be possible to gain accuracy in the model if parameters such as the geometry of the house or the archetype of the household are taken into account.

8.3.2.3 Simulation algorithm

The simulation process is illustrated in Figure 8.4. As can be observed, the first step performed by the system is the estimation of the active occupancy. This process can be done in advance if a daily occupancy profile is to be generated, but it can also be obtained dynamically for each simulation step. In addition, information about the external irradiance and the characteristic of the lighting spots should also be given using any of the methodologies exposed in the previous section. Once those data are available for the simulation process, the iteration through all the time steps and all the given lighting spot starts.

First, the number of active occupants in the dwelling is obtained since the existence of active occupancy is a necessary condition for the lighting spots to be turned on. Otherwise, all the lighting systems in the household will be considered to be off due to the absence of occupants. That might not be consistent with the reality in some cases when the residents forgot to turn off the lighting system. Hence, the model includes a small percentage of uncertainty to account for this, as was previously proposed by other authors [16].

In the case at least one active occupant is at home the algorithmic process continues normally, checking the operation of each installed bulb in the household. If the lighting point is already working and the occupancy is not null, the operation cycle of the bulb is decreased. In contrast, if the light is off, the comparison process of the external irradiance and the internal lighting threshold is performed.

Thus, if the global external irradiance is lower than the selected threshold, the lighting point is likely to be turned on. This threshold is defined as a normally distributed value, so for each household a random value is sampled from the normal distribution in order to define the lighting threshold, emulating with this the random behaviour of different people.

If the comparison in the previous step results to be true, the lighting point will be turned on, providing the relative usage of this lighting spot is high enough and the influence of the effective active occupancy determines that. In addition, a scalar value is usually included in the decision process in order to keep the energy figure comparable with the historical average values of consumption in the studied region.

Finally, if after all the decision process the lamp has to be turned on, the operation cycle has to be assigned. This is usually performed on the bass of a

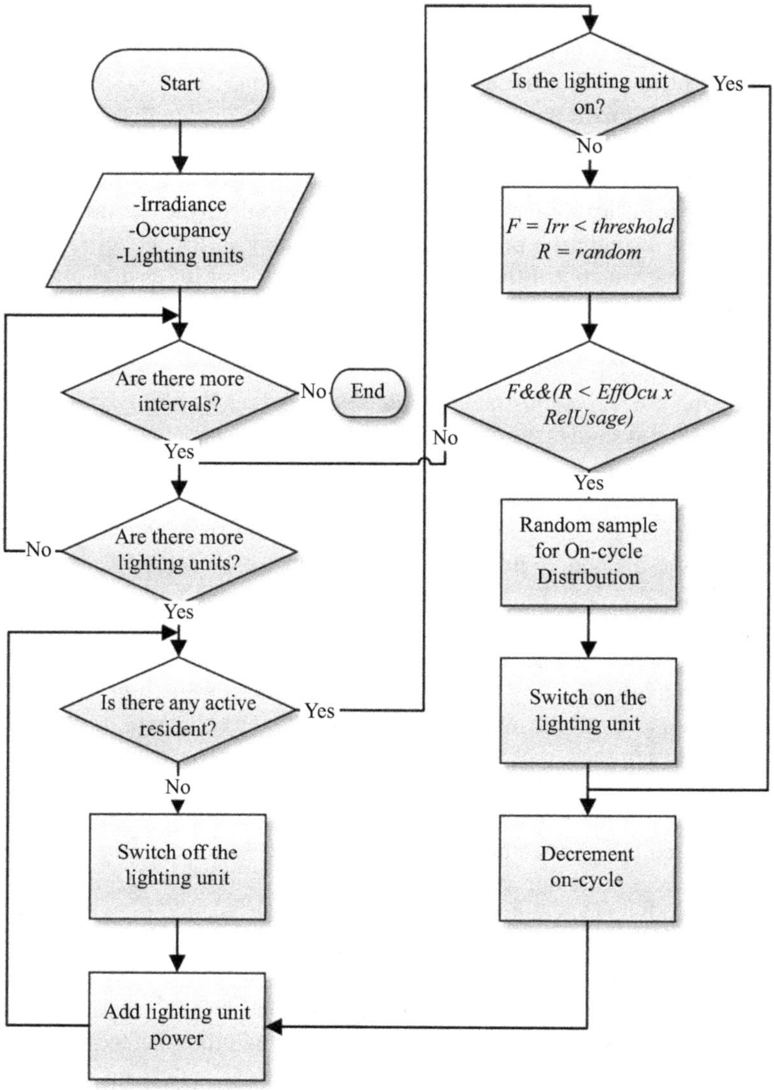

Figure 8.4 Flowchart of the lighting consumption simulation algorithm for each household

probability distribution of on cycles. The whole process, as it can be seen in the flowchart of Figure 8.4 is then repeated for each lighting spot in the household and for each time interval in order to simulate the lighting consumption profile of a household. Thus, if a given number of households (N) is to be simulated, the above-described process has to be repeated for each house.

8.3.3 General appliances consumption

The following group of sub-consumptions that are generally considered in the bottom-up models is the one related to the general appliances, meaning those that are related neither to the lighting systems nor to the heating or cooling appliances. As far as electrical energy is concerned, these consumptions usually present the highest demand as they cover the appliances devoted to food refrigeration and freezing, cooking systems, laundry appliances, entertainment devices and others additional equipment.

In the European households, this consumption has an average value of 2,600 kWh/year as it was shown in the results of the REMODECE project [21]. Nevertheless, the figures vary significantly from one to another country with maximums of around 3,500 kWh/year for Denmark and Norway, and minimums of around 1,500 kWh/year for the Czech Republic or Romania [3].

This model keeps using the active occupancy as the main input parameter, which determines the probability of using or not some equipment. Nevertheless, in order to associate each appliance with a certain behaviour, the concept of activity is now introduced as it will be exposed in the following section.

8.3.3.1 Model basics

The main concept that is introduced in the equipment consumption model is the idea of activities in the household. Most of the appliances are linked with a determined activity performed in the household, with the exception of some equipment such as refrigerators, freezers or appliances in standby mode, that do not depend on the activity. All these activities together with the active occupancy determine the interval where the different appliances are more likely to be turned on. The activities are linked to the TUS data, where the interviewees wrote down for each 10-min interval the primary and the secondary activity that they were carrying out.

Some of the activities that are recorded in the diaries are ironing, house cleaning, cooking, laundry, personal care, watching TV or PC usage. As it can be depicted, many of those activities are directly linked with some appliances. For instance, an iron is needed in order to perform the activity ‘ironing’. On the other hand, some relationships with the energy consumption are not so obvious, but if some laundry activity is performed, either a watching machine or a tumble dryer is likely to be employed. Therefore, most of the appliances usually installed in a household can be related to the activities found in TUS.

The methodology used for obtaining the activity probabilities is illustrated with the example given in Table 8.2. First, for each time interval, the number of households with the selected number of active occupants has to be determined. As can be observed in Table 8.2, the number of household with one active occupant at 13.00 h are 2,367, whereas for two active occupants the number is 956. After that, the number of households where at least one of the active residents is carrying out the activity has to be accounted. This means that from the 2,367 households where one occupant is active, only in 822 these occupants are performing the activity Cooking. Likewise, from the 956 households with two active residents at the same

Table 8.2 Probability for the activity cooking for a weekday at 13.00 h. Calculation example

Number of active occupants	Number of household with the given number of active occupants	Number of household where at least on residents perform the activity	Activity probability
1	2,367	822	822/2,367 = 0.35
2	956	429	429/956 = 0.45

time, only in 429 of them at least one occupant is carrying out the activity. By means of this calculation, the activity probability for each time step and depending on the number of active occupants can be calculated.

It should be pointed out that this method, as can be observed in the resulting probabilities in Table 8.2, intrinsically takes into account the concept of appliance sharing since the condition is that at least one occupant is performing the activity. So as the results depict, the probability for one active occupant to carry out the activity is 35%, whereas for two active occupants this figure is not the double but only 45%.

8.3.3.2 Input data

The input parameters for this model include again the occupancy profile, but taking also into account the activities and the appliances that are responsible for the energy consumption.

- **Occupancy profile**. As in the previous model, the number of active occupants in the dwelling will be needed in order to estimate the intensity of the power consumption.
- **Activities probability**. For each defined activity from the TUS, the probability distribution is established, distinguishing between the numbers of active occupants, so the probability of performing a selected activity can be determined for any interval of time during the day and depend on additional parameters such as the day of the week, the region or the period of the year.
- **Appliances characteristics**. Following the bottom-up methodology, each appliance consumption is basically defined as an individual power rate (reactive power might also be included) and an on-cycle. In addition, a standby power, as well as an off-delay, can be assigned. For other appliances, such as the dishwasher or the washing machine, the actual working cycle is defined as a variable power along the on period.

8.3.3.3 Simulation algorithm

In Figure 8.5, the simulation process of this specific end-use is illustrated. As in the case of the lighting systems consumption, the estimation iterates over each time interval and for each appliance, deciding if it should be turned on, kept in on state, kept in off state, or switched off, depending on the decision factors. Those decision factors are the active occupancy and the dwelling activities.

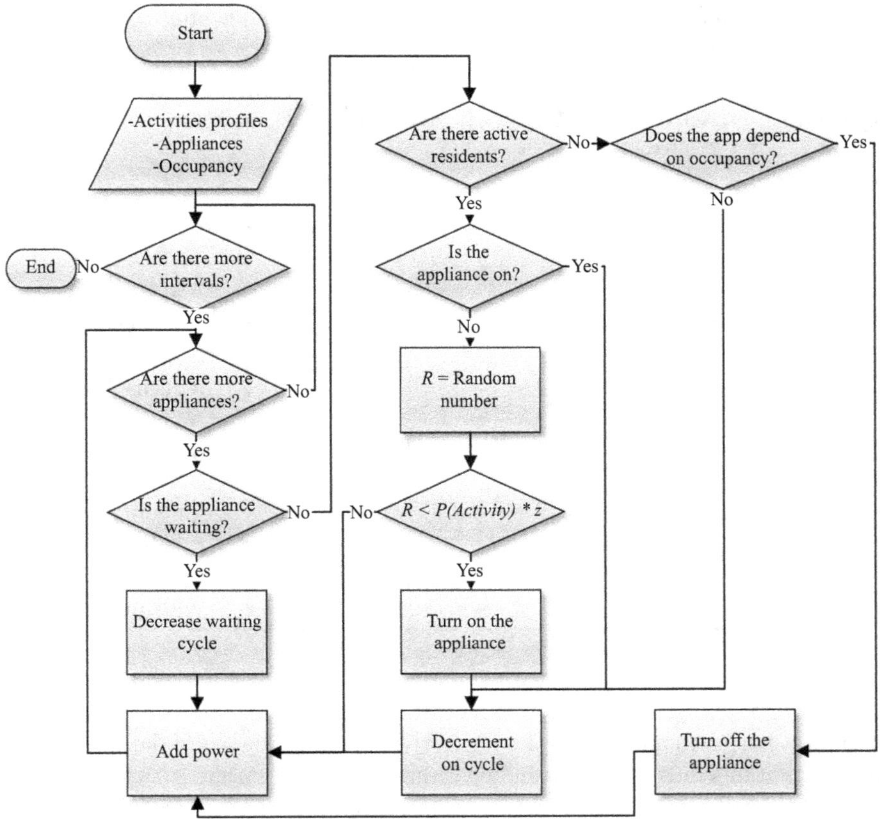

Figure 8.5 Flowchart of the general appliances consumption simulation algorithm for each household

Thus, the state of every appliance is first checked. If the appliance has a delay cycle such as the refrigerator, the freezer and other thermostat controller equipment, the first test is to check if it is in the waiting state and, if this is the case, the waiting cycle should be decreased. If the appliance has no waiting time, then the process continues.

In the second stage, the active occupancy is checked, so if the number of active residents is zero either the appliance is turned off if it depends on occupancy and it is active, or the appliance keeps on working with the normal operation if it does not depend on occupancy. Opposite, when there is active occupancy, the state of the appliances is tested.

Subsequently, if the appliance was already turned on, the system will only decrease the on cycle and account for the energy consumed. However, if the appliance is off the stochastic process for determining if it should be switched on starts. In this process, the probability of the activity associated with the appliance $P(Activity)$, together with a calibration scalar z, calculated to keep the annual

Table 8.3 Simulation case for the general appliance consumption simulation algorithm

Appliance state	Activity related	Delay cycle	Occupancy	Decision
OFF	–	$\neq 0$	–	• Decrease delay cycle • Add standby power
OFF	NO	=0	–	• Turn on appliance • Add on power
ON	NO	=0	=0	• Remain ON • Add on power
OFF	YES	=0	=0	• Remain OFF • Add standby power
OFF	YES	=0	$\neq 0$	• Stochastic test with activity probability
ON	YES	=0	=0	• Turn off appliance • Add Standby power
ON	YES	=0	$\neq 0$	• Remain ON • Add on power

consumption consistent with the observed historical figures, are compared with a randomly generated number R. The result of this test will decide if the appliance is turned on at this interval or if it should remain in the off state.

In order to clarify the process, all the possibilities are summarised in Table 8.3 depending on the appliance state, the operation cycles the type of equipment and the occupancy, showing the decision that will be taken by the simulation procedure.

Thus, with this process, the individual operation of each appliance is simulated with a given time interval. This simulation has the main advantages of being based on the relationship between human behaviour and actual devices, which makes possible to define the consumption units with a big degree of details, as well as introduce new technologies, something fundamental in the prediction of new consumption trends due to emerging and disruptive appliances.

8.3.4 Heating and cooling consumption

The last element or simulation block that the proposed model takes into account is the energy demand due to heating and cooling appliances. Those consumptions strongly depend on the climate zone and the season. Therefore, the observed figures vary significantly even inside the same country as it has been exposed in the European ENTRANZE project [22]. For the heating consumption, the values can range from around 40 to 100 kWh/m^2/year for Mediterranean countries up to 160–240 kWh/m^2/year in countries such as Germany, Czech Republic or Finland. In the case of cooling consumption, the trend is opposite with values around 40–70 kWh/m^2/year for Mediterranean cities and almost non-existent demand in countries such as Germany (9 kWh/m^2/year), Czech Republic (5 kWh/m^2/year) or Finland (1 kWh/m^2/year).

In addition, whereas lamps and domestic appliances are mainly supplied with electricity, in the case of the heating systems a wide range of technologies and different energy sources such as natural gas, propane, fuel oil and hot water (district heating) can also be employed. This fact makes the modelling process more complex than for other energy end-uses. The methodology, input data and simulation process, for a proposed modelling technique, will be addressed in the following sections, pointing out its strength and weaknesses.

8.3.4.1 Model basics

The heating and cooling consumptions are not only linked with the human behaviour but also with the insulation of the households and the geometric characteristics. Subsequently, if an accurate model is to be developed, a large number of variables such as the overall heat transfer coefficients of the building construction elements (*U*-values), infiltration and other specifications will be needed, also making the modelling process non-applicable to other houses other than the studied. On the other hand, if not information at all is supplied to the model about the characteristics of the household, it might be highly estimative with no useful results.

Depending on the information that is known about the building or house to be simulated, the consumption models for heating and cooling are usually classified as white, black or grey-box models. In the white-box or physic-driven models a wide knowledge of the building structure, as well as the parameters and characteristic of the construction, is needed as they make use of the system physics for modelling the dynamic. On the other hand, the black-box or data-driven models only try to find a relationship between the inputs and the output data usually by means of statistical techniques and with no knowledge about the building itself. Finally, somewhere in between, the grey-box models consider some physical parameters of the building in the estimation, but they also make use of statistical approaches.

The white-box models provide good generalisation capabilities, whereas the black-box models are very accurate. The grey-box models benefit from the characteristic of both of them [23]. In addition to these models, many calculation tools that accurately simulate the building dynamic can be found, which include models of the most used HVAC systems. Some of these programmes are TRNSYS, EnergyPlus, DOE-2 or ESP-r to name a few.

The goal of the proposed modelling methodology is to simulate the diversity of the household in order to generate detailed individual heating and cooling consumption profiles. Therefore, according to the above-mentioned criteria, the model can be classified as a grey-box model, as a few characteristic of the household will me mixed with statistical data in order to generate diversity among the different residents and dwellings, but trying to keep the results statistically consistent.

Thus, the main reference input parameter for the modelling process will be the external temperature. It has been proved that the average external temperature for each day of the year can be used to determine the intensity of the energetic consumption, by comparing the value with a reference comfort temperature. The main problem of this comfort value is that it has to be empirically determined for each region. The flexibility of this figure has been shown in previous works such as the 21 °C (69.8 °F) obtained by Sailor in Florida [24], the 15 °C (59 °F) estimated by Sarak and Satman for heating usage in Turkey [25], the 22 °C (71.6 °F) proposed by Giannakopoulos and Psiloglu for Greece and the 18 °C (64.4 °F) determined by Valor *et al.* for Spain [26]. In addition, a 3 °C hysteresis is observed around these values, where the consumption is insensitive to temperature.

These comfort values are usually used as a reference for the calculation of the heating degree-days (HDD) and the cooling degree-days (CDD) depending on the usage of either heating or cooling appliances respectively. Therefore, using the reference comfort value with the hysteresis term and the deviation of the average temperature from this one, the HDD and CDD can be calculated for a given day and this value subsequently related to the energy intensity needed [27].

However, the degree-days only allow us to account for the energy intensity, but the period of the day where those consumptions take place cannot be determined unless the daily temperature profile is used. This instantaneous temperature together with the insulation level designated for the dwelling will allow us to estimate the heat gains and losses in order to establish the switch on probability or the thermostat cycling operation of the appliances.

If a more detailed knowledge about the specific region or set of household to be simulated is obtained, more constraints can be added to these models such as the division into zones of the dwelling with different temperatures. Nevertheless, the more detailed the model, the more computationally intensive the tasks will be, and probably the diversity generated by the simulation will decrease, what might be a problem when the human behaviour wants to be simulated.

8.3.4.2 Input data

- **Occupancy profile**. As in the other models, the occupancy determines the main periods where the consumption might take place. In addition, when the thermal gains are considered for the cooling consumption, the active occupancy in the dwelling has a direct impact due to the metabolic heat gain from persons.
- **Daily temperature profile**. The external temperature mainly influences the energy needs to be required to maintain the comfort conditions inside the household. The average, minimum and maximum temperatures of the day can help the model to estimate the demand of this day and simulate the seasonality. Nevertheless, due to the high resolution of the model, the daily profile of temperatures is required so the probabilities of turning on the heating and cooling devices can be determined.
- **Heating and cooling appliances characteristics**. In the same line of the appliances consumption model, the appliances characteristics should be defined following the bottom-up methodology. Nevertheless, the wide range of fuels and appliances types sometimes might require the usage of mixed models, where not only electrical energy is considered, but also hot water demand or others fuels, although in the presented case, only the electrical cooling and

heating appliances were considered. Moreover, it will also be necessary to distinguish between thermostat controlled and non-thermostat controlled equipment. This means that thermostat controlled appliances, once there are turned on, they will follow a reference temperature without the user operation by means of hysteresis cycles or variable power cycles. Opposite, the non-thermostat controlled equipment will only operate in on or off mode and the change of state will always be determined by the occupants (although it might be possible to have different power levels, which are selectable by the user).

- **Household insulation level**. The heating and cooling appliances produce internal thermal flows that are added or subtracted from the total thermals gains. However, some of those thermals gains will depend on the structural characteristics of the household. Taking into account all the parameters, geometrical structure and construction materials of the stock of houses will require a computational intensive task with a large database. Moreover, those data are not usually widely available. Thus, an alternative might be the definition of the different average levels of insulation and thermal losses, which will determine the probability of activating the heating and cooling equipment. Therefore, in a poor insulated household, the probability of turning on a device or cycling a thermostat controlled appliance is higher than in a good insulated one.

8.3.4.3 Simulation algorithm

The simulation algorithm is shown in Figure 8.6, where, as it can be seen, the same iterative procedure of the previous models is also carried out. Therefore, for each simulation interval and heating or cooling appliance, the consumption is calculated by using the above-exposed methodology and input parameters.

First, the simulation system loads all the input parameters about the dwelling characteristics and the appliances, as well as the active occupancy for each time interval and the temperature data, that must include, as it was exposed, both the daily indicators (mean, maximum and minimum temperature) and the detailed profile for the day.

Next, the degree-days are calculated for the heating (HDD) and cooling (CDD) consumption. These values will determine the overall consumption intensity for the given day, so if the HDD > 0, a heating consumption is likely to occur, whereas if the CDD > 0, the consumption will be due to the cooling appliances. If both of these values are null or negative, then the zone conditioning system will not consume electrical energy for this day.

Nevertheless, the simulation procedure continues to account the energy in a lower lever following the bottom-up methodology. Therefore, for each time interval and appliance, the active occupancy of the household is checked. If there are no active occupants, the heating and cooling appliance will be turned off, since mainly electrical heating and cooling appliances do not remain on when nobody is at home. However, as it can be observed in Figure 8.6, this condition has an asterisk in the switch off condition as it implements a percentage of exception (around 10% for the studied countries) due to the fact that a relatively significant percentage of

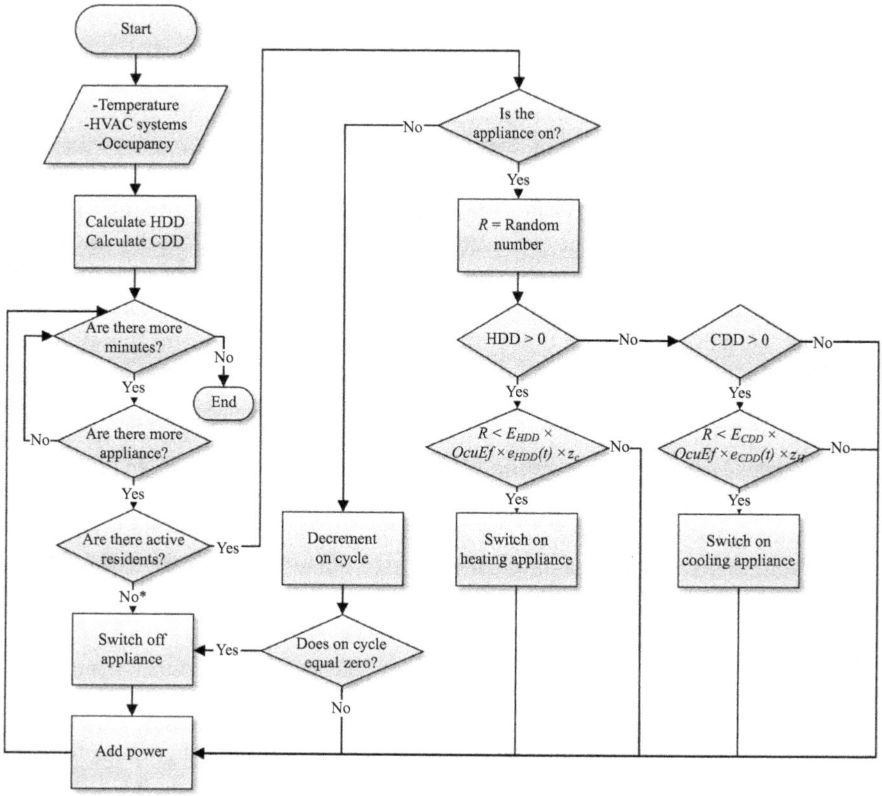

Figure 8.6 Flowchart of the cooling and heating system consumption simulation algorithm

heating and cooling appliance might be on during the night by using temporization methods and when the active occupancy is null. Thus, although the occupancy might be zero in some cases, the appliances will continue operating until their cycle is finished.

After this, if there is active occupancy, the state of the appliance is tested. If the appliance is already on, the on cycle will continue either until the end of the established average operation time or until the temperature is within the comfort limits in the case of thermostat operated systems. Opposite, when the appliance is off, the possibility of being turned on is checked. This logical and statistical test has two levels.

The first level or step is related to the energy intensity of the day. For this aim, the obtained values of the HDD and CDD are used. These indices relate the deviation from the comfort temperature with the daily energy intensity. Therefore, if any of them is not null, there is a possibility of consuming energy for this day and the logical test continues to the next stage. Otherwise, if both are zero, it means that it is no likely to use the heating or cooling devices for this day.

If the logical test of the first level is true, the system will continue to the second stage, where the possibility of turning on a heating or cooling device is now tested with a higher temporal resolution using the daily temperature profile. For this, a random number, generated in the previous steps is used and compared with the product of three probabilities.

The first one named as E_{HDD} or E_{CDD} relates the deviation from the comfort temperature with the consumption intensity. This can be calculated using the historical temperatures and the consumption figures for each day and then normalising the results as a probability distribution. The second influence factor is the effective occupancy (*OcuEf*), which as well as in the lighting consumption model takes into account the sharing of the heating and cooling appliances. Finally, the last factor is the energy intensity related to the temperature for that simulation interval ($e_{\text{HDD}}(t)$ or $e_{\text{CDD}}(t)$), obtained from the daily profile and the insulation level of the household. In addition, two calibration scalars are introduced in order to keep the results consistent with the annual consumption figures observed in the studied region (z_H or z_C).

Thus, if the result of the test is true, the appliance is turned on, setting either an operation cycle for it, in the case where no thermostat control is implemented by the device, or a comfort thermostat temperature for switching off, when the equipment has a thermostat control. Finally, the power consumption for this interval is accounted for this appliance.

Subsequently, iterating through all the time interval and defined appliance, the energy consumption due to heating and cooling equipment is simulated for a household with the given characteristics having the advantage of individually emulating the behaviour of each device in relation to the human patterns and the external climate conditions.

8.3.5 Remarks on the model

The previous sections have presented a specific model based on the bottom-up methodology and based on time of use data. Thus, it is only a concrete implementation that has been proposed by the authors due to the capabilities that it presents in relation with the integration of energy resources. Nevertheless, as it was pointed out in Section 8.2, the range of possibilities when it comes to the demand modelling techniques is broad, so in case further knowledge about the topic is desired the lecture of the references [9,10] is recommended.

Some of these capabilities of the model that will be used in the following section to show the possible applications are summarised below:

- **Simplicity of use:** The model has only a few input parameters, some of which can be defined by the user or obtained from third party sources such as the climate condition regarding irradiance and temperature. Therefore, in the most general case, the only input information required by the model will be the number of residents in each house, the number of households to be simulated, the region where they are located and the date (including month, the day of the month and the day of the week). With this information, the climate condition will be loaded from the database based on the date and location, and the household characteristics such as appliances, insulation and income level will be selected randomly on the basis of the observed statistics of the region. Nevertheless, due to the high flexibility of the model, the climate condition, the appliances sets or the household characteristics can also be defined, although deeper knowledge of the system to be modelled will be required.

- **Individual household simulation:** The results of the model are obtained with high temporal resolution (1-min resolution in the results that will be presented) for each defined household. Thus, each consumption profile will simulate the real stochastic behaviour of the residents. It is true that this will be required to perform aggregated simulation in order to obtain significant results. However, it allows us to simulate low-level phenomena such as the distributed integration of PV, micro-wind turbines or batteries at household level or the influence of each house consumption and generation in the feeder (overvoltage, overload etc.) if further details of the distribution system are known.
- **Breakdown of energy end-uses and appliance:** As well as from the distribution point of view, the proposed model has also several advantages from the end-user point of view. The division of the consumption into lighting, general appliances and heating and cooling devices allow determining not only the energy intensity of each of them but the actual daily profiles of this consumptions. In addition, the modelling of the global consumption as the aggregate of each individual appliance presents significant benefits. On the one hand, energy policies can be tested from the user point of view such as the impact that might have a technological update of lighting systems, the usage of a more efficient appliance or certain energy saving behaviours. On the other hand, the period of usage of the different appliance can be observed along the daily profile, which can be used for the development of demand response strategies that shift some of those consumptions to other period of the day depending on factors such as the variable price of the energy or the integration of non-dispatchable renewable resources.

8.4 Applications of bottom-up stochastic models

After exposing the modelling methodology, as well as the main features of the model, the usage of the system will be addressed for four tasks. First, the characteristics for demand prediction will be commented. After this, the capabilities for the assessment of energy policies and demand response strategies will be exposed. Finally, the usability of the system for energy resource integration will be introduced.

8.4.1 Demand prediction

The possibility of supplying the model with external conditions data and characteristics of the households allows using the system for the prediction of power and energy needs. Nevertheless, it has to be considered that the model methodology is based on a stochastic process, so the simulation of an individual household cannot be employed as a forecasting tool, due to the determinism of the simulation. This fact is presented in Figure 8.7 where the instantaneous power consumption with 1-min resolution has been simulated with the model for two households with three residents during a weekday of a winter month and located in Cordova, Spain.

Figure 8.7 includes the different sub-consumptions, (a) lighting, (b) general appliances, (c) heating, as well as the total consumption (d). As it can be observed, the degree of determinism of the results is extremely high as each appliance is represented at a low level. This completely tallies with the actual behaviour of the residents, whose activity is chaotic. Therefore, as the figure depicts, each time the model is run, the curves obtained are different, and subsequently, the results cannot be used for the estimation of the demand.

However, despite the stochastic methodology, the observed statistics of the population are intrinsically implemented in the model. This means that if enough simulations are performed, the results are likely to have a trend and be reproducible. In other words, a number of household with some given characteristics can be simulated, representing a community, neighbourhood or group of users connected to the same network or feeder. With this methodology, each simulated house

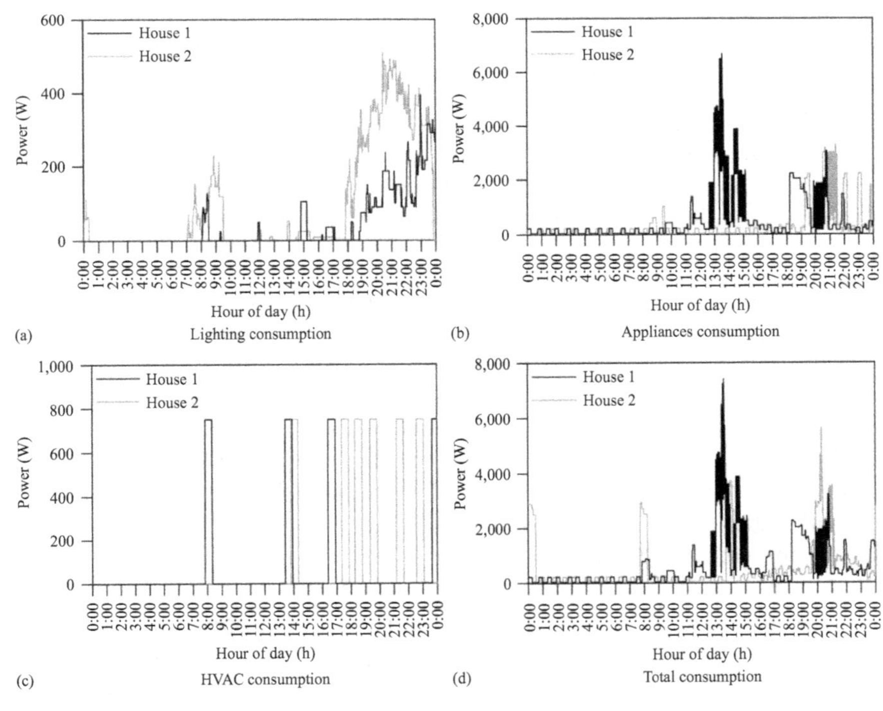

Figure 8.7 Daily simulation. Two household of three residents for a weekday in winter and located in Cordova: (a) lighting consumption; (b) appliances consumption; (c) HVAC consumption; and (d) total consumption

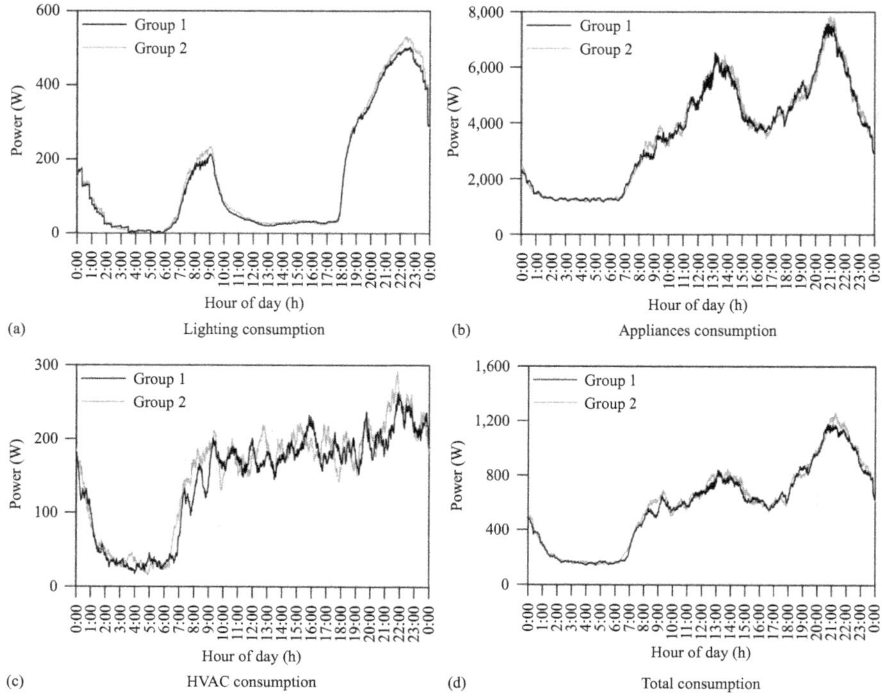

Figure 8.8 Daily simulation. Two groups of 1,000 Household of three residents for a weekday in winter and located in Cordova: (a) lighting consumption; (b) appliances consumption; (c) HVAC consumption; and (d) total consumption

will present deterministic results almost irreproducible, but the aggregated estimation will produce similar results.

This is observed in Figure 8.8, where the results for 2 groups of 1,000 household as the one previously simulated are represented. As is shown, both groups present similar daily estimations. Therefore, in the same line of the top-down models that aim to estimate the total demand for supply purpose, the aggregated data of bottom-up models can be used for the same task. However, the results have to be calibrated for each zone in order to predict consumption figures consistent with the observed energy in the area or region.

8.4.2 Energy policies and demand response strategies assessment

The previous section has shown the possibility of using the stochastic models for demand prediction. However, they present better and exclusive features in other fields such as the energy policies and demand response strategies assessment. These capabilities are a consequence of the modelling strategies, where each appliance is individually simulated with a given power consumption and operation cycle. Therefore, if the characteristics of these appliances or their relationship with the human behaviour are changed following some policies, the impact on the overall consumption can be observed.

The first element that can be simulated is the adoption of certain energy policies. Energy policies are understood as techniques that aim to address energy issues and they can be composed of legislation, incentives or guidelines. Some of those policies might be a change in the appliance technology with more efficient equipment or consumption reduction with some energy-saving strategies.

Figure 8.9 represents the impact of reducing the consumption of standby equipment in an aggregate of 10,000 households during a winter day. This reduction can be driven by the users' interaction or using specific switches that detect that consumption and deactivate the appliances. Anyhow, the consequence of this is that the consumption of the appliances, when they are off, is reduced from a few Watts to zero. This impact can be observed in the daily profile where a reduction between the base scenario (solid black line) and the action applied (solid grey line) is depicted, especially during the night period.

In energy terms, if the daily energy consumption of the both scenarios is accounted, the reduction is around 8% a day. That means that if the average consumption of a household is around 14 kWh/day, almost 1 kWh can be saved each day if the operation of a certain device in standby is avoided. Other strategies that can be tested are the implementation or update of technologies, as was shown in [8], where the implementation of the new European directive regarding the substitution of lighting technologies was assessed.

Another important field of application is the development of demand response strategies. These techniques aim to manage the different appliances' consumptions in a way that certain optimums or desired objectives are achieved. Some of these

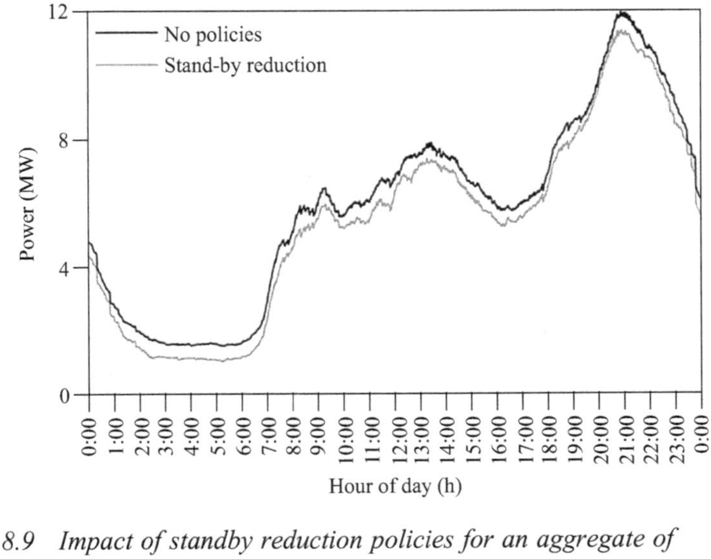

Figure 8.9 Impact of standby reduction policies for an aggregate of 10,000 households

objectives are usually the match between generation and consumption when non-dispatchable resources are used or the reduction of the monthly bill in a context where dynamic energy or electricity prices are established.

An example of the usage of the model for simulating these techniques is shown in Figure 8.10. The case of study is a set of 10,000 households with PV production whose objective is to increment the solar energy usage and reduce the night peak, which will have to be supplied either from an energy storage system or from the grid. In order to reduce the peak, a group of loads considered as shiftable during the

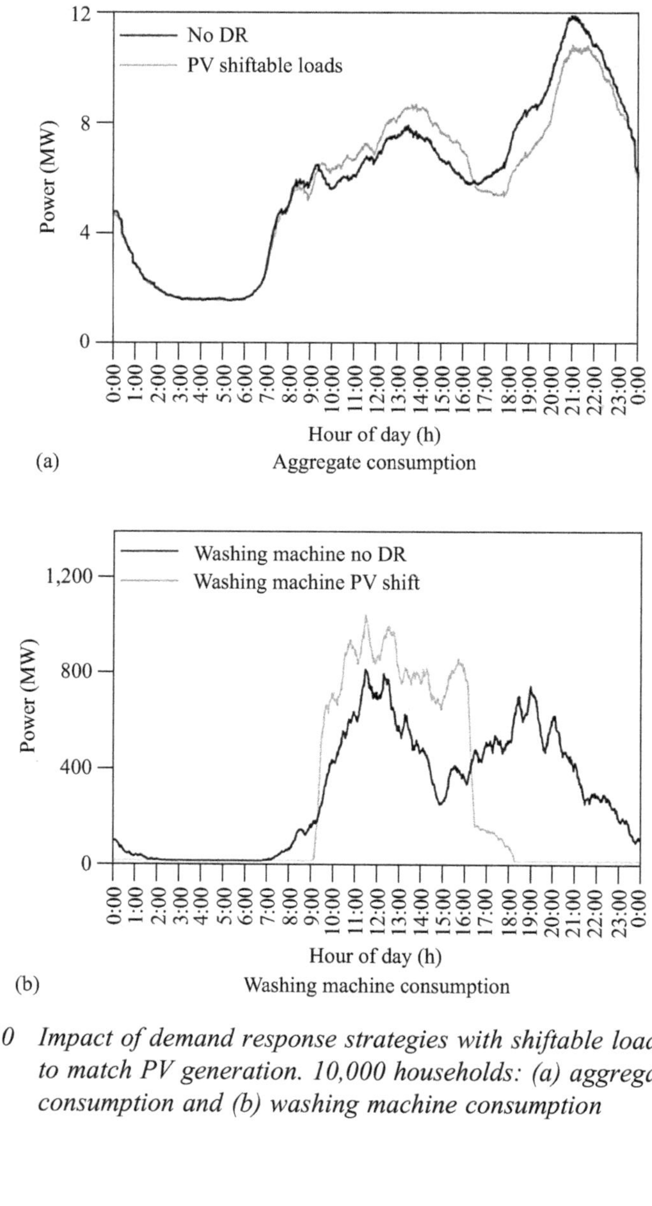

Figure 8.10 Impact of demand response strategies with shiftable loads to match PV generation. 10,000 households: (a) aggregate consumption and (b) washing machine consumption

day have been selected to be moved to the period when the irradiance is maximum. Those loads are washing machines, tumble dryer and washer dryer machines.

The global results of this strategy can be observed in Figure 8.10, where the aggregated instantaneous power consumption of the 10,000 households is represented for the base scenario (solid black line) and for the demand response scenario (solid grey line). As can be observed, the result is as expected, and the demand during the middle hours of the day has been increased, whereas the night demand peak has been reduced around 1 MW in aggregate.

Moreover, as the model is able to simulate each appliance individually the instantaneous power demand only for one of the selected appliances for demand response strategies can be observed at a low level. In this case, the aggregate demand for the washing machines is represented in Figure 8.10 for the base scenario (solid black line) and for the demand response scenario (solid grey line). It should be pointed out that although 10,000 different households are simulated; it does not mean that each of them has installed a washing machine, but they are randomly distributed according to the observed penetration rates include in the model.

The results clarify the demand response strategy. In the base scenario, the consumption due to the washing machines presents two demand peaks one around 12.00 h, and the second one around 19.00 h. The demand response strategy is to centralise these consumptions in the central hours of daylight when the PV generation is the highest, as well as avoiding using those appliances when no solar irradiation exists. Thus, in the demand response scenario the consumption is now limited from 9.00 h to around 18.00 h, increasing the utilisation of the solar resource and eliminating the demand when there is no daylight.

8.4.3 Distributed resources integration

Finally, the last field of application of these models is the evaluation of the integration in the households of distributed energy resources. As the model is able to simulate each individual house independently, the interaction with renewable resources such as PV generation, as well as with different storage systems and the energy exchange with the grid can be simulated. This is especially important when non-dispatchable resources are considered since demand and generation can be non-coincident and, therefore, although enough energy might be generated, some of this energy cannot be utilised unless a storage system is also installed [28].

Thus, at a low level, the individual behaviour can be simulated as is shown in Figure 8.11, where two individual households, located in the same community or neighbourhood, are presented. As can be seen, the consumption profiles are completely different and subsequently the interaction with the PV generation and the storage system. These types of simulations are useful for developing energy management strategies, where the grid interaction can be studied at a low level, sometimes with the possibility of including the influence on voltage and frequency of the injection of the PV excess power. In addition, peer-to-peer energy management can be also studied emulating the wide range of demand profiles and the casuistry that they might generate.

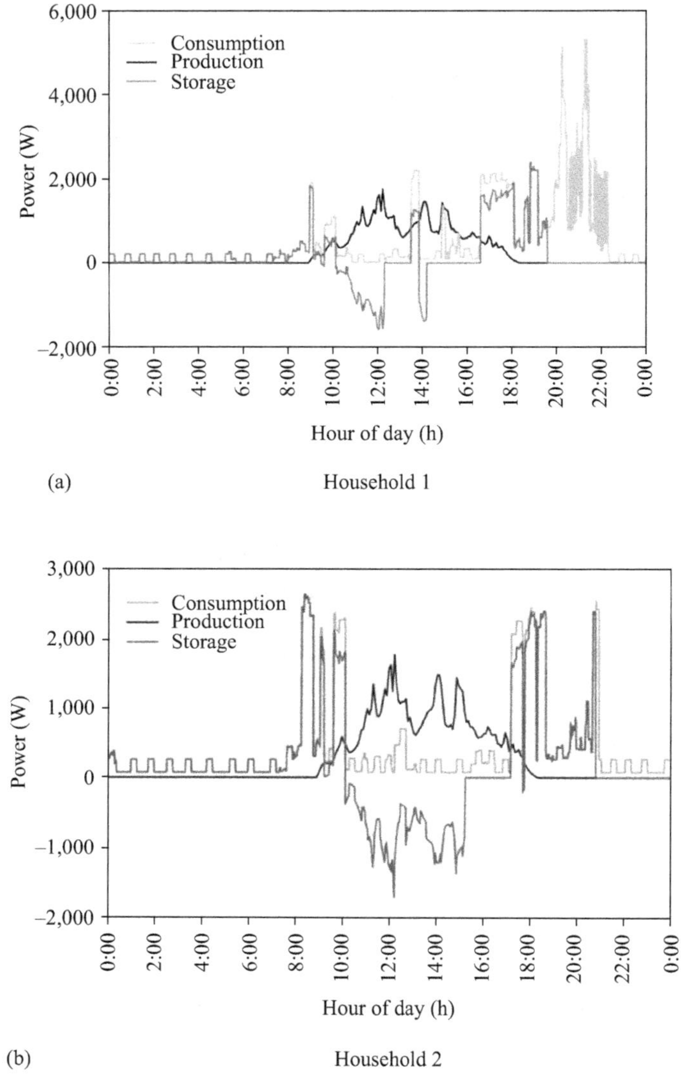

Figure 8.11 Individual simulation of two households with PV production and energy storage system: (a) household 1 and (b) household 2

Moreover, using the above-mentioned individual profiles, the aggregate curves for demand (a), production (b), grid exchanges (c) and storage (d) can be constructed for a group of households with a high temporal resolution, something that might be interesting for the planning of the distribution network when bidirectional power flows exist. That can be observed in Figure 8.12 where a contour plot has been used for representing a whole year with a 1-min resolution of a community

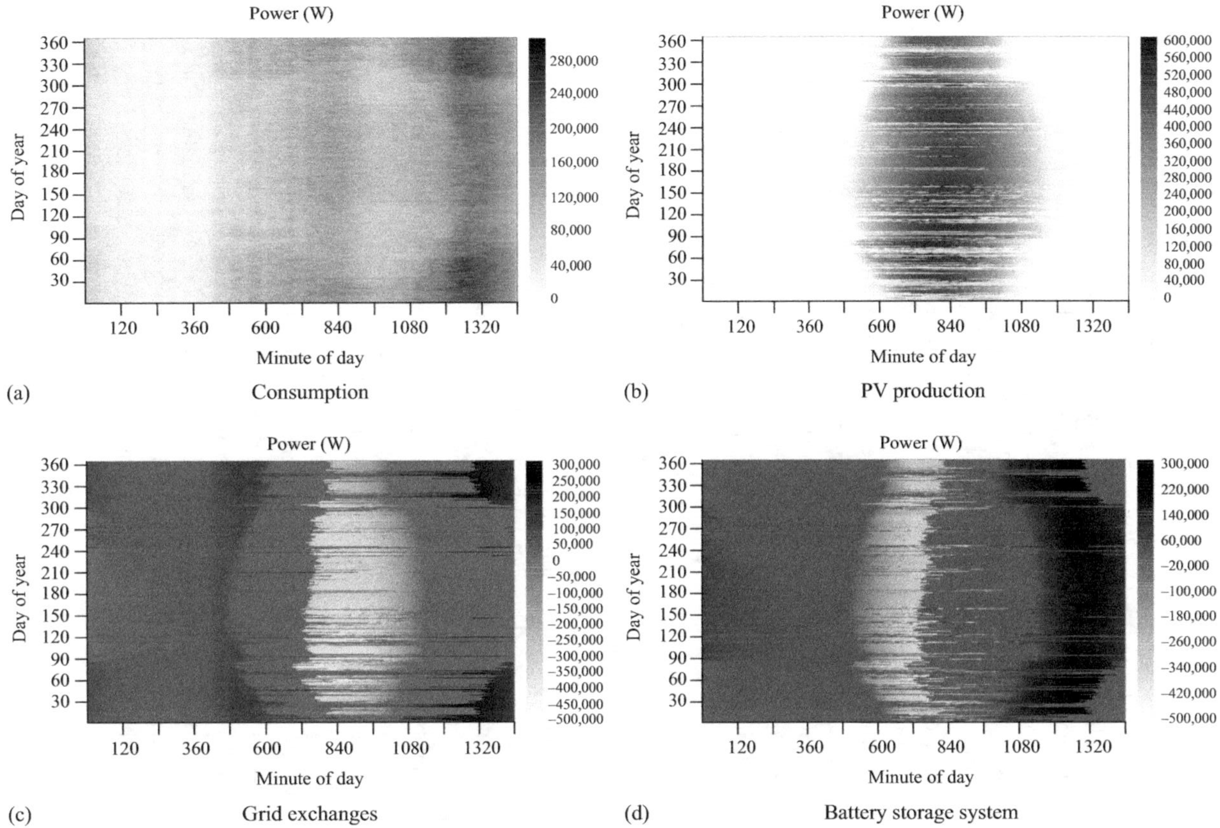

Figure 8.12 Interaction with distributed renewable resources of a 200 households community with PV production energy storage and grid exchanges: (a) consumption; (b) PV production; (c) grid exchanges; and (d) battery storage system

composed of 200 households with integrated PV production and storage located in the south of Spain. The *X*-axis represents the minute of the day, the *Y*-axis is the day of the year and the colour scale is the instantaneous power.

The complexity of the information represented in the graph is high, but many conclusions can be obtained. From the consumption plot (a), it can be seen how the power demand varies along the year, but always presenting two consumption peaks around 13.00 (780) and 22.00 h (1,320) corresponding with the lunch and dinner times in Spain. In addition, it can be observed how the demand from 20.00 (1,200) to 24.00 (1,440) increases during the winter months, whereas for the summer months the consumption is increased from 14.30 (870) to 19.00 h (1,140). In the case of the PV production (b), the increment in the hours of light from the winter to the summer months is clearly illustrated too. Moreover, the existence of many days with low irradiance during the winter period is also depicted.

These both profiles have a direct influence in the grid exchanges (positive means power imported from the grid, while negative is power injected in the grid) and the storage system (positive when the battery discharges and negative when it is charged). Therefore, in the grid exchanges (c) figure it is shown how during the winter months it is necessary to buy energy from the grid starting around 22.00 h (1,320) and sometimes early for the days with the worst weather conditions. This tallies with the behaviour of the storage system (d) that cannot be fully charged in the winter days. Opposite, during the summer time, the system can sell energy to the grid for a long time, as well as fully charge the storage system. Thus, almost no energy has to be bought from the grid.

With all this information at such a high level of resolution, it is possible to evaluate the average autonomy and energy utilisation of a given community, neighbourhood or grid area. This can be performed by means of some indicators defined in the literature named as demand cover factor (DCF) and supply cover factor (SCF). The DCF indicates the percentage of demand that can be supplied with the generation system either by direct consumption or by delayed consumption when storage systems are installed. On the other hand, the SCF accounts for the degree of utilisation of the generation system. The formal definition of these two indices can be found in [29].

Using these two indices, the proposed model and different rates of penetration of PV generation and battery storage, Figure 8.13 was constructed where the average DCF (solid lines) and the SCF (dashed lines) can be observed for a household located in the south of Spain and simulated during a year using real irradiance data. The *X*-axis represents the different PV power peaks that might be installed, whereas the lines with different colours are different capacities of the storage system.

This figure depicts that when more PV power is installed in the household the DCF increases, but with a tendency to saturate. In contrast, the SCF decreases since not all the PV capacity is used every day, but only in the worst scenarios. On the other hand, when the storage capacity is increased both indices are always improved. Nevertheless, the degree of enhancement is major for the DCF, allowing the delayed consumption of the PV generation.

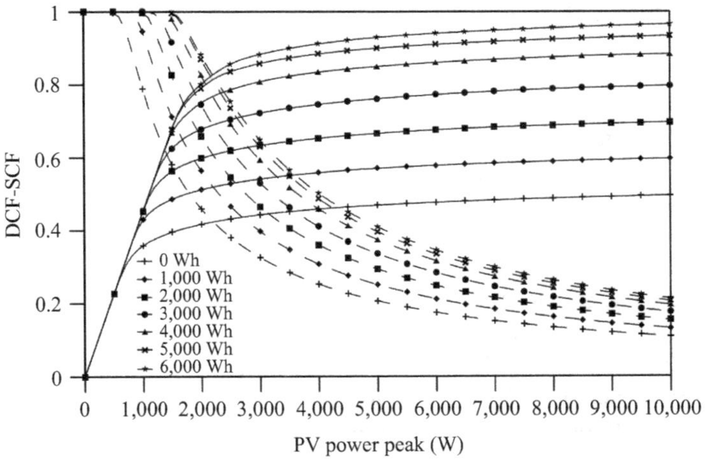

Figure 8.13 Demand cover factor (solid lines) vs supply cover factor (dashed lines)

In addition, it should be pointed out that for a given battery capacity value, both the DCF and the SCF intercept. At this point, the energy demand that cannot be covered (1-DCF) equals the portion of energy production that can be neither stored nor consumed (1-SCF). Nevertheless, if a bi-directional power flow interchange with the grid is considered, the net energy exchanged with the grid at that point will be zero and, therefore, if not additional losses at considered, the system will follow a net-zero energy philosophy. Thus, this modelling technique has several advantages as an energy planning tool.

8.5 Conclusion

The chapter has presented the current context of the demand modelling with a focus on the stochastic and bottom-up modelling techniques. In this context, this specific methodology has been contextualised between all the other strategies pointing out the reasons and arguments that make this technique the most suitable in the context of this book.

Next, the methodology for implementing this modelling technique has been shown, addressing the main parts of the system, as well as the basic and fundamentals of each of them. Subsequently, the simulation procedure that is employed for generating the results has been explained by means of detailed flow charts and exposing the main input parameters, and how they can be obtained.

Finally, the application areas have been discussed. The validity of this model for demand prediction has been exposed, although some issues derivate from the stochastic simulation methodology do not make this model the most suitable for accurate energy prediction in aggregate terms. However, they have shown a unique ability in the fields of energy policies assessment, demand response strategies development, and distribute energy resources integration and planning, all of these due to their capability of simulating the appliances at a low level and always in relation to the human behaviour. This emphasises the important role that stochastic modelling techniques can play in the energy planning and management sector.

References

[1] Suganthi L, Samuel AA. 'Energy models for demand forecasting: A review'. *Renewable Sustainable Energy Rev* 2012;16:1223–40. doi:10.1016/j.rser.2011.08.014.

[2] Pérez-Lombard L, Ortiz J, Pout C. 'A review on buildings energy consumption information'. *Energy Build* 2008;40:394–8. doi:10.1016/j.enbuild.2007.03.007.

[3] de Almeida A, Fonseca P, Schlomann B, Feilberg N. 'Characterization of the household electricity consumption in the EU, potential energy savings and specific policy recommendations'. *Energy Build* 2011;43:1884–94. doi:10.1016/j.enbuild.2011.03.027.

[4] Cao S, Sirén K. 'Impact of simulation time-resolution on the matching of PV production and household electric demand'. *Appl Energy* 2014;128:192–208. doi:10.1016/j.apenergy.2014.04.075.

[5] Widén J, Wäckelgård E, Paatero J, Lund P. 'Impacts of different data averaging times on statistical analysis of distributed domestic photovoltaic systems'. *Sol Energy* 2010;84:492–500. doi:10.1016/j.solener.2010.01.011.

[6] Torriti J. 'Demand side management for the European supergrid: Occupancy variances of European single-person households'. *Energy Policy* 2012;44:199–206. doi:10.1016/j.enpol.2012.01.039.

[7] Torriti J. 'A review of time use models of residential electricity demand'. *Renewable Sustainable Energy Rev* 2014;37:265–72. doi:10.1016/j.rser.2014.05.034.

[8] Palacios-Garcia EJ, Chen A, Santiago I, Bellido-Outeiriño FJ, Flores-Arias JM, Moreno-Munoz A. 'Stochastic model for lighting's electricity consumption in the residential sector. Impact of energy saving actions'. *Energy Build* 2015;89:245–59. doi:10.1016/j.enbuild.2014.12.028.

[9] Swan L, Ugursal V. 'Modeling of end-use energy consumption in the residential sector: A review of modeling techniques'. *Renewable Sustainable Energy Rev* 2009;13:1819–35. doi:10.1016/j.rser.2008.09.033.

[10] Grandjean A, Adnot J, Binet G. 'A review and an analysis of the residential electric load curve models'. *Renewable Sustainable Energy Rev* 2012;16:6539–65. doi:10.1016/j.rser.2012.08.013.

[11] Richardson I, Thomson M, Infield D. 'A high-resolution domestic building occupancy model for energy demand simulations'. *Energy Build* 2008;40:1560–6. doi:10.1016/j.enbuild.2008.02.006.

[12] Widén J, Wäckelgård E. 'A high-resolution stochastic model of domestic activity patterns and electricity demand'. *Appl Energy* 2010;87:1880–92. doi:10.1016/j.apenergy.2009.11.006.

[13] Lopez MA, Santiago I, Trillo-Montero D, Torriti J, Moreno-Munoz A. 'Analysis and modeling of active occupancy of the residential sector in Spain: An indicator of residential electricity consumption'. *Energy Policy* 2013;62:742–51. doi:10.1016/j.enpol.2013.07.095.

[14] Ndiaye D, Gabriel K. 'Principal component analysis of the electricity consumption in residential dwellings'. *Energy Build* 2011;43:446–53. doi:10.1016/j.enbuild.2010.10.008.

[15] Stokes M, Rylatt M, Lomas K. 'A simple model of domestic lighting demand'. *Energy Build* 2004;36:103–16. doi:10.1016/j.enbuild.2003.10.007.

[16] Richardson I, Thomson M, Infield D, Delahunty A. 'Domestic lighting: A high-resolution energy demand model'. *Energy Build* 2009;41:781–9. doi:10.1016/j.enbuild.2009.02.010.

[17] Widén J, Nilsson A, Wäckelgård E. 'A combined Markov-chain and bottom-up approach to modelling of domestic lighting demand'. *Energy Build* 2009;41:1001–12. doi:10.1016/j.enbuild.2009.05.002.

[18] Gago EJ, García JO, Estrella AE. 'Development of an energy model for the residential sector: Electricity consumption in Andalusia, Spain'. *Energy Build* 2011;43:1315–21. doi:10.1016/j.enbuild.2011.01.016.

[19] Reinhart C. 'Lightswitch-2002: A model for manual and automated control of electric lighting and blinds'. *Sol Energy* 2004;77:15–28. doi:10.1016/j.solener.2004.04.003.

[20] Hunt D. 'The use of artificial lighting in relation to daylight levels and occupancy'. *Build Environ* 1979;14:21–33. doi:10.1016/0360-1323(79)90025-8.

[21] REMODECE Project. 'Residential monitoring to decrease energy use and carbon emissions in Europe (REMODECE) database'. Institute of System and Robotics, University of Coimbra. 2015. URL: http://remodece.isr.uc.pt/

[22] ENTRANZE Project. 'Heating and cooling energy demand and loads for building types in different countries of the EU'. Energy Economics Group, Vienna University of Technology. 2014. URL: http://www.entranze.eu/

[23] Afram A, Janabi-Sharifi F. 'Review of modeling methods for HVAC systems'. *Appl Therm Eng* 2014;67:507–19. doi:10.1016/j.applthermaleng.2014.03.055.

[24] Sailor DJ. 'Relating residential and commercial sector electricity loads to climate: Evaluating state level sensitivities and vulnerabilities'. *Energy* 2001;26:645–57. doi:10.1016/S0360-5442(01)00023-8.

[25] Sarak H, Satman A. 'The degree-day method to estimate the residential heating natural gas consumption in Turkey: A case study'. *Energy* 2003;28:929–39. doi:10.1016/S0360-5442(03)00035-5.

[26] Valor E, Meneu V, Caselles V. 'Daily air temperature and electricity load in Spain'. *J Appl Meteorol* 2001;40:1413–21. doi:10.1175/1520-0450(2001)040<1413:DATAEL>2.0.CO;2.

[27] Isaac M, van Vuuren DP. 'Modeling global residential sector energy demand for heating and air conditioning in the context of climate change'. *Energy Policy* 2009;37:507–21. doi:10.1016/j.enpol.2008.09.051.

[28] Palacios-Garcia EJ, Moreno-Munoz A, Santiago I, Moreno-Garcia IM, Milanés-Montero MI. 'Smart community load matching using stochastic demand modeling and historical production data'. *2016 IEEE 16th Int. Conf. Environ. Electr. Eng.*, 2016, p. 1–6. doi:10.1109/EEEIC.2016.7555885.

[29] Munkhammar J, Grahn P, Widén J. 'Quantifying self-consumption of on-site photovoltaic power generation in households with electric vehicle home charging'. *Sol Energy* 2013;97:208–16. doi:10.1016/j.solener.2013.08.015.

Chapter 9

Distributed micro-storage systems at residential level in smart communities with high penetration of photovoltaic generation

María Isabel Milanés-Montero[1], Fermín Barrero-González[1], Eva González-Romera[1] and Enrique Romero-Cadaval[1]

One of the current problems of renewable energy is its random and intermittent character, which complicates the management of electrical networks due to its difficult prediction. Energy storage is proposed as one solution to facilitate the management of this type of energy, improving the efficiency and capabilities of these sources of generation (such as photovoltaic and wind power) and, thus, increasing their percentage in the electricity market. It is a fact that energy storage systems (ESSs) will be indispensable in the energy future of any country, contributing to the development of the smart grid, facilitating the integration of renewable energy and providing another set of ancillary services such as frequency regulation or voltage control [1].

ESS is also a crucial part of emerging energy management systems (EMSs) at homes and communities. These devices will directly influence changing consumption patterns and shifting the demand curve, allowing storing energy at low prices at times of low energy demand and selling when prices rise. Consumers can then reduce the electricity bill by reducing the contracted power [2].

Moreover, electric vehicles (EVs) and plug-in hybrid EVs (PHEVs) offer the possibility to use their batteries as distributed ESS which has great potential for expansion. These batteries can be used in combination with photovoltaic solar energy to supply homes, allowing operation modes known as vehicle-to-home, and even can inject into the grid the surplus energy stored by means of the vehicle-to-grid mode [3].

To date, most of the storage technology research has focused on the design of large storage capacity and low-cost new devices, with capacities of about 10–100 MWh [4]. In recent years, however, research has been conducted on the use of distributed micro-storage systems [5]. The term 'micro' refers to the size of the storage system compared to the generation of the grid, with a capacity of about few kWh. The development of micro-storage devices, which can be installed in homes

[1]Universidad de Extremadura, Badajoz, Spain

and in EV and PHEV, gives way to a near future where a large number of individual consumers can store small amounts of energy with the aim of taking charge of the peaks in demand and supply variability. The operation of these devices responds to the set-points generated by a central EMS located in the smart community [6].

On the other hand, most battery chargers for ESS on the market behave as non-linear loads, causing harmonic distortion, reactive consumption and in case of three-phase chargers, imbalances. Several authors have documented the impact of the penetration of massive battery chargers in the quality of the mains, demonstrating the need for research into new electronic-converter topologies and new control strategies applicable to the energy storage management system (ESMS) to ensure that these chargers are adequately integrated into the operation of the smart grid [7–9].

In this line, this chapter aims to provide an analysis of distributed micro-storage energy systems at the residential level to contribute to smart grid goals. The chapter is organized in five sections as follows:

- First an overview of micro-storage technologies, including the state of the art, is presented. It will also be shown the advantages when using different ESS unified in a hybrid energy storage system (HESS), achieving best parameters for specific applications than the ESS operating individually.
- The second section is devoted to the suitable topologies for the bi-directional electronic converter responsible for the flux between the micro-storage energy system located at home and the grid by means of the smart community energy management system (SCEMS).
- Afterwards, in the third section control strategies for the ESMS of the micro-storage device are proposed, following different compensation objectives from the set-points provided by the SCEMS. Furthermore, these strategies will ensure at all times safety and will improve the efficiency and lifetime of the ESS.
- In the fourth section, the attention is paid to the power interfaces for the integration of ESS into the grid to adapt the requirements of distributed micro-storage system in a smart community with high penetration of photovoltaic systems.
- Finally, conclusions are included at the end of the chapter.

9.1 Overview of micro-storage technologies

It has been widely discussed that the presence of ESSs will be essential in near-future power systems. Among other advantages, they are responsible for:

- Reduction of the need for transmission and distribution capacity and losses because ESSs provide power locally and because peak shaving can be performed in the demand curve.
- Better integration of renewable not-manageable energy, like wind or photovoltaic generation. By means of a storage system, inherent variability of these

energy sources can be reduced, and therefore, manageability and reliability of power system can be improved.
- Provision of ancillary services, like contribution to power-frequency regulation and voltage control in both grid-connected and isolated modes.
- Improvement of the power system reliability by providing power supply during mains outage situations. Like uninterruptible power supplies, ESSs allows to keep power supply in relatively small portions of distribution grids, during a brief or a longer period of time depending of its capacity and the presence of distributed generation.

This section deals with a summary of ESS technologies that are suitable for been used as micro-storage systems in a community EMS context, with the aim of contributing to demand management, integration of renewable energy generation systems and power flow control.

The suitability of a specific ESS to be used in different applications depends on several of their technical and economical characteristics, like rated power and energy capacity, discharge duration, response time and capital and maintenance cost.

Luo *et al.* [10] present a complete overview of ESS and their application to power systems. Both large and medium–small-scale storage systems are assessed and different technologies are discussed.

Focusing in community energy management systems, application areas of ESS in this context are related to energy management, time shifting, peak shaving and end-user electricity service reliability. Authors in [10] summarize the requirements of ESS for these and other applications and propose the most suitable technologies to assess them:

- Energy management, to decouple the timing of generation and consumption, requires large discharge duration (hours–days), but it is not highly demanding in terms of response time.
- Time shifting and peak shaving are directly related and their objective is storing electrical energy when it is cheaper to be used in high-price time periods. As a consequence of the economic strategy, power peaks are avoided in high-demand periods, with the consequent advantage for distribution grid. Both applications also require long discharge duration but they do not need fast response time.
- On the contrary, response time is important when power supply reliability is the objective, at the expense of shorter discharge duration.

Besides the traditional requirements for EMSs, nowadays the concept of microgrid has reached communities. In this new context, the whole microgrid has to be able to keep frequency and voltage stable and under boundaries in both isolated and grid-connected situations and synchronization has to be adequately performed to move between states. Since renewable energy generation systems (like photovoltaic power plants, the widest used generation technology in communities) are usually designed and controlled to inject the maximum active power with no grid control abilities, ESS are promising as responsible for the energy management and the microgrid control. These new functions require shorter time response than the previous ones.

As a conclusion, ESS technologies selected to perform energy management and microgrid control in communities must be versatile enough to encompass such opposite requirements or hybrid technologies have to be used. There are several technologies which meet the mentioned requirements, according to [10], with different maturity levels: conventional batteries, flow batteries, supercapacitors, flywheels and superconducting magnetic energy storage (SMES). The functioning principles and main characteristics of each of them are described below.

9.1.1 Conventional batteries

An electrochemical reaction in an electrolyte inside a cell bi-directionally converts chemical into electrical energy and vice versa. During discharging, two electrodes (anode and cathode) collect electricity with a specific DC voltage and current, while during charging, a DC voltage applied between electrodes restores the electrolyte to the initial conditions. The battery is composed by a series-parallel combination of cells.

Batteries are available in a wide range of sizes and energy capacities, which makes them suitable for very different applications. On the other hand, relatively low cycling times and the dependence of their lifetime on the depth of discharge are the main disadvantages of this ESS technology, as well as the recycling requirements of sometimes toxic chemical products. Several kinds of batteries are currently available in the market. They are described in the following subsections [10–12] and their main characteristics are summarized in Table 9.1.

9.1.1.1 Lead–acid batteries

Electrodes of a lead–acid battery are made of PbO_2 and Pb and the electrolyte is sulphuric acid. Its low capital cost (300–600 \$/kWh) makes it the most widely used battery. However, energy density (50–90 Wh/l) and specific energy (30–50 Wh/kg) are lower than other kinds of batteries. In other words, their size and weight are higher than other batteries with the same energy capacity. Besides, they have a short life cycle (500–1,000 cycles), limiting their use in some power operation applications.

9.1.1.2 Lithium-ion (Li-ion) batteries

These batteries have a lithium metal oxide in the cathode and a graphitic carbon anode, being the electrolyte an organic liquid with dissolved lithium salts. Among their characteristics, their small size stands out and makes them especially suitable for mobile applications, like laptops, mobile phones and EVs. However a wide range of discharge times (from seconds to weeks) makes them promising for other power systems applications despite they are currently still expensive. The main disadvantage of this kind of batteries lies on safety, since most of the metal oxide electrodes are thermally unstable and can lead to a thermal runaway at elevated temperatures. For this and other reasons, extra systems are required to avoid over-charging and over-discharging and to balance voltage between cells.

Table 9.1 Advantages and disadvantages of different batteries [10,12]

	Advantageous characteristics	Limiting characteristics
Lead–acid	Response time	Cycling time
	Daily self-discharge rate	Energy density
	Cycle efficiency	Performance at low temperature
	Capital cost	
Li-ion	Response time	Lifetime dependence on deep cycling
	Dimension/weight	Management (on board control unit required)
	Cycle efficiency	
NiCd	Robust reliability	Toxic heavy metals
	Maintenance	Memory effect
NiMH	Specific energy	Daily self-discharge rate
	Energy density	Sensitive to deep cycling
	Reduced memory effect	
	Environmental friendly	
NaS	Energy density	Operating cost
	Daily self-discharge rate	Extra system required to ensure high temperature
	Rated capacity	
	Pulse power capability	
	Recyclability	
ZEBRA	Pulse power capability	12–15 h to heat up after been frozen
	Cell maintenance	
	Daily self-discharge rate	
	Cycle life	
Metal–air	Specific energy	Commercial maturity
	Recyclability	

9.1.1.3 Nickel–cadmium (NiCd) and nickel–metal-hydride (NiMH) batteries

Together with lead–acid, nickel–cadmium (NiCd) batteries are among the most mature technologies in rechargeable batteries. Their electrodes are made of nickel hydroxide and metallic cadmium, in an alkaline electrolyte. A robust reliability and low maintenance requirements are their main advantages, besides a relatively high energy density (50–75 Wh/kg). On the other hand, they suffer from the phenomenon known as 'memory effect', which consists on a high decrease in maximum capacity if recharging after partial discharging is repeated. Another important drawback of NiCd batteries is the use of cadmium, a carcinogenic heavy metal with cumulative character [13].

Due to these issues of NiCd batteries, nickel–metal hydride ones were developed in the past 1990s initially to replace them. They have a high energy density (170–420 Wh/l) and the same advantages of NiCd, except for the rated capacity, but their memory effect is partially mitigated and the cadmium has been replaced by a hydrogen-absorbing alloy, which reduces hazards for health and environment.

9.1.1.4 Sodium–sulphur (NaS) and sodium-nickel chloride (ZEBRA) batteries

Sodium–sulphur (NaS) batteries have molten sulphur and molten sodium in electrodes, with a solid beta-alumina ceramic electrolyte. They require temperature to be kept between 300 and 350 °C, so a heat source is required that reduces the battery performance. Their main advantages come from an almost zero daily self-discharge, high-rated capacity (up to 200 MWh) and relatively high energy density (150–300 Wh/l).

A variation of high temperature batteries, similar to NaS, is the sodium nickel chloride one, better known as ZEBRA (zero emission battery research). It uses nickel chloride instead of sulphur, which leads to a slightly lower operation temperature, better safety characteristics and a higher cell voltage than NaS batteries. ZEBRA development is currently focused in the field of electric and hybrid vehicles, as well as power system applications.

9.1.1.5 Metal–air batteries

In this kind of batteries, the anode is made of a pure metal, and the cathode is connected to an air supply. The reaction produced is the oxidation of the metal, and the replacement of substances is commonly required. The theoretical specific energy of metal–air batteries is their most attractive characteristics, as it is noticeably higher than that of other electrochemical storage technologies [12]. That specific energy value excludes the weight of oxygen, because it is not stored in the battery. Several metals have been studied for the anode material, but currently, lithium and zinc are the most widely used and aluminium is becoming a new attractive option. Lithium–air batteries present the issue of a high reactivity of lithium with air and humidity, with the consequent safety risk. Zinc-air batteries have reached a certain maturity in the field of hearing aids and medical devices. Regarding aluminium-air batteries, according to [14], it is able to provide around 400 Wh/kg during a long discharging time (more than 100 h). Compared to the specific energy 100–150 Wh/kg of lithium-ion (Li-ion) batteries (the highest among the commercial technologies), it is a noticeable improvement.

9.1.2 Flow batteries

Flow batteries use two electrolytes stored in independent tanks. During the charging/discharging periods, both electrolytes are pumped into different compartments of the cell stack, separated through an ion-selective membrane. Flow batteries have an important advantage compared to conventional batteries: their rated power depends on the size and shape of the electrodes, while the energy capacity is related to the amount and concentration of electrolytes. Therefore, both parameters are independent, providing this kind of batteries with a high flexibility to be designed according to the application requirements. On the other hand, the need to pump electrolytes to the stack reduces the general performance of this kind of ESS. Among the different flow batteries proposed in literature during the last years, two of them have had special attention. They are further described in the below sections [10,12].

9.1.2.1 Vanadium redox flow battery

This is a mature kind of flow battery. The electrolytes have different oxidation states of vanadium ions (couples of V^{2+}/V^{3+} and V^{4+}/V^{5+}). Inside each of the stack compartments, vanadium is changed between the two states of one of the couples and H^+ ions are interchanged through the membrane. It has an important advantage compared to other flow batteries: as ions of the same metal are present in the whole battery, in case some of the metal ions cross the membrane some energy would be lost, but no irreversible degradation would be produced in the battery. On the other hand, production of hydrogen during battery operation constitutes a safety issue. Quick response and long cycle lifetime are their main characteristics.

9.1.2.2 Zinc–bromine (ZnBr) flow battery

Electrolytes are solutions of zinc and bromine reactive components in ZnBr batteries. During charging Zn is deposited at the carbon–plastic electrode and at discharging Zn^{2+} goes back into solution. Electrons produced as a result of the chemical reaction are interchanged through the polyolefin membrane. A deep discharge capability and a good reversibility are among the strengths of this kind of batteries, in contrast with the material corrosion and a narrow temperature band.

9.1.3 Supercapacitors

Also known as double-layer capacitors, they constitute a compromise between classical capacitors and batteries [4,10]. Composed of two porous electrodes, an electrolyte and a membrane separator, energy is stored as static charge on the electrode surfaces. Nano materials or especially porous configuration for electrodes lead to an extremely high capacitance value (thousand farads). As a result, supercapacitors are an ESS with long cycling times and very fast charges and discharges. They show a high power density but a low energy density compared to batteries. These characteristics make them suitable for applications which require frequent charge/discharge cycles but no long-term storage.

9.1.4 Superconducting magnetic energy storage

A SMES consists of a superconducting coil and a power conditioning system [15]. Operation of the coil requires temperatures below the superconducting critical temperature (up to 100 K in the most recent approaches). Therefore a cryogenic cooling system is essential for the equipment performance. The energy is stored by means of the magnetic field generated in the coil. Both response and discharge times are quick, and they present a high efficiency and a long lifetime. However, capital cost is still high and reliability strongly depends on the refrigeration system.

9.1.5 Flywheels

The kinetic energy of a flywheel rotating at a specific speed is the kind of energy stored in this ESS. The flywheel is commonly installed in a vacuum chamber to reduce losses and conversion to electric energy is performed by means of bearings, a reversible electrical motor/generator and a power electronics unit. Charging consists in increasing the rotating speed by injecting electric energy and the opposite process is performed during discharging. Advances in material research have led to high-speed flywheels, which contribute to a noticeable improvement in the specific energy at the expense of a higher cost. In this kind of ESS, the depth of discharge do not cause negative effects and they have an easy maintenance, but self-discharge rate is not negligible [16].

9.1.6 Comparison of characteristics of micro-storage system technologies

Among the different technologies available for energy storage, those previously described are suitable for community energy management, home energy management and other micro-storage applications, like, in some cases, EVs. The suitability for any of the functions desirable for ESS in this context strongly depends on their characteristics. For example, response time is essential if voltage and frequency control is necessary in a microgrid. In contrast, discharge duration is more important than response time if the purpose of the ESS is the time shift of demand between different time periods to match renewable production or to benefit from reduced energy prices. Energy and power density are vital when the space is limited and capital and maintenance cost are especially important to achieve massive sales among home consumers.

In this subsection, several characteristics of the mentioned ESS technologies are compared and discussed, with the aim of analysing their suitability for different functions (Figure 9.1).

Energy density limits the size of the ESS device, an essential parameter for mobile applications, like laptops, mobile phones and EVs. Besides metal–air batteries, which have not been included in this comparison because of their lack of commercial maturity, the highest energy density is a property of Li-ion batteries, currently the undeniable candidates for those uses.

Energy capacity measures the amount of energy that can be stored. Except for NiCd batteries and supercapacitors, most of the ESS present a wide range of this parameter, making them suitable for multiple applications. The low-energy capacity of supercapacitors reduces their interest as unique storage system for a community. However, their quick response time, high power and extraordinarily high cycling lifetime make them perfect to be combined with a slower and higher capacity ESS technology in a hybrid storage system.

The number of cycles during the lifetime is also shown in the figure. Applications related to voltage and frequency control and power quality correction require frequent changes between charging/discharging modes, in contrast to time-shifting purpose. Supercapacitors and SMES clearly outperform other technologies from this point of view. Lead–acid batteries, one of the most widely used ESS due to their cost and maturity, present the worst value of cycling time. Therefore, they are not the most adequate kind of ESS for control functions.

The efficiency during discharge is good enough when using any of the ESS technologies. However, there is an important contrast between those of supercapacitor, SMES and flywheels and those of the flow batteries. Several reasons explain these

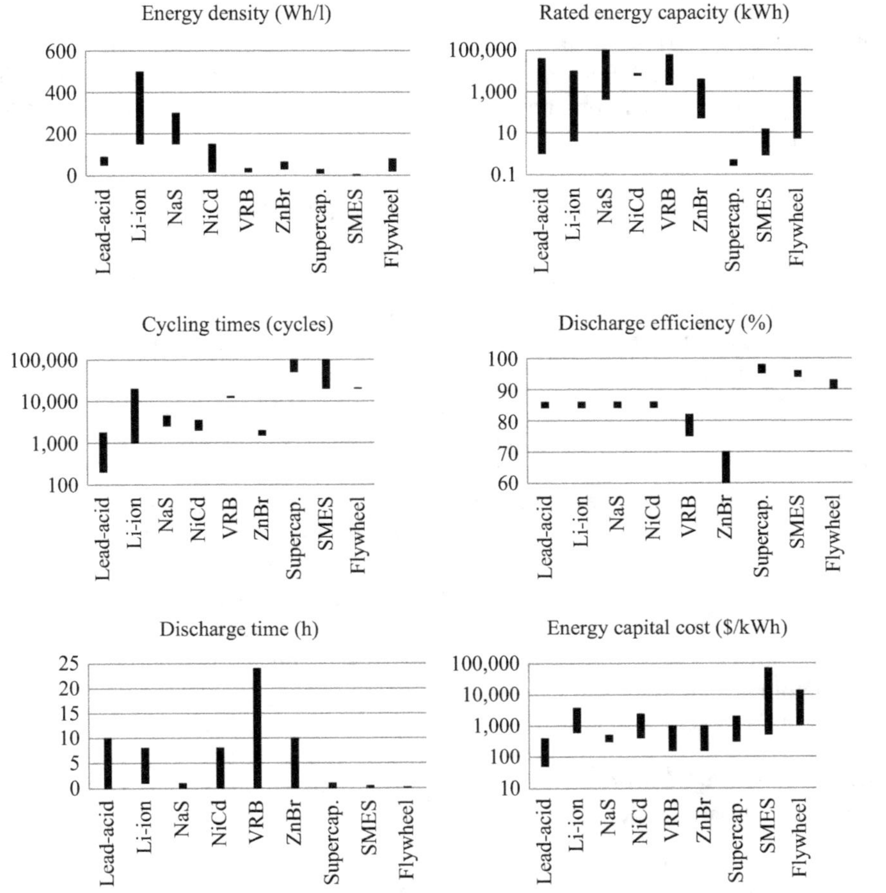

Figure 9.1 Comparison of characteristics between ESS technologies (compiled by authors based on [10])

differences. Firstly, supercapacitor and SMES directly store electric energy, and flywheels store mechanical energy. The conversion efficiency between mechanical and electric energy is usually higher than that of electrochemical energy of batteries. Besides, the need for pumping electrolytes into the stack, using the stored energy, reduces the overall performance of flow batteries, compared to conventional ones.

If the objective of the ESS is to shift demand between different periods of the day or to reserve the energy generated from renewable energy sources to be used later, a discharge time of several hours could be required. In this situation, flow batteries and some of the conventional batteries (lead–acid, Li-ion and NiCd) are the most adequate ESS. Supercapacitors, SMES, flywheels and NaS batteries would be discarded unless a hybrid ESS is planned. The discharge time of VRB flow batteries is worth mentioning, as it can reach the whole day.

The last characteristic analysed in this subsection is related to capital cost. For each kWh of energy capacity, the most cost-effective technology is lead–acid battery, but flow batteries have become competitive in this aspect. The capital cost of SMES can be extraordinarily high, although its low energy capacity also influences this parameter. A high operation cost has to be considered as well, due to the cryogenic cooling. Therefore, nowadays they are discarded as consumer devices for home and community EMSs.

9.2 Topologies for the bi-directional electronic converter

This section is devoted to the links or interfaces between ESS and the electrical network, based on power electronic converters. The idea is to give some general guidelines that permit the designer to select the most suitable topology. To do this a revision of power converter interfaces is shown. The topologies can be divided into three classes [17]: standard, multilevel and multiport. The standard topologies, also known as two-port converters, can be divided, in turn, into single stage and double stage. The single stage uses only one converter to charge and discharge the storage system and to control the power injected to the AC grid. On the other hand, the double stage configuration uses two power converters: one to manage the charge and discharge of the batteries (a DC/DC converter) and another one to interface with the AC grid (a DC/AC converter). The multi-level topologies are based on the synthesis of a desired AC voltage from several levels of DC voltages and they are particularly suitable for high-voltage application. Finally, the multi-port topologies are designed to permit the energy management from several sources, as in the case of using different types of storage systems at the same time.

9.2.1 Standard topologies

9.2.1.1 Single stage

The control of the energy flow between the energy storage and the AC grid requires a bi-directional AC/DC converter with proper charging/discharging features. The full-bridge voltage source inverter (VSI) is the simplest and most commonly used topology. Figure 9.2(a) shows the single-phase case and Figure 9.2(b) the three-phase case. In several applications, the neutral wire is required and in that case the single capacitor in the DC side of the converter must be replaced with split capacitors.

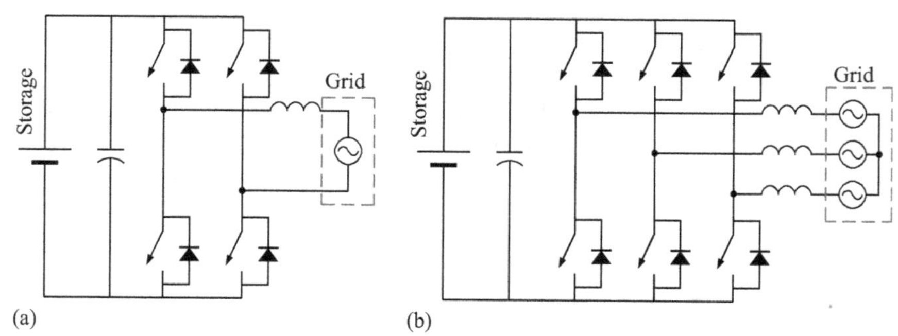

Figure 9.2 Bi-directional AC/DC converters: (a) single-phase and (b) three-phase

This type of converter requires a minimum DC voltage when connected to the grid, if a PWM sinusoidal technique is utilized. The AC output voltage cannot exceed the DC-link voltage or, in other words, the DC-link voltage has to be higher than the AC output voltage.

To overcome the limitations and problems of the traditional VSI, a lot of solutions have been proposed. The more relevance ideas can be resumed in the following: the Z-source inverter (ZSI) proposed by Peng [18], the quasi-ZSI (qZSI) [19] and the bi-directional matrix converter [20].

The advantages of the single-stage topologies are low device count, low cost and reduced losses. However, as it has said before, there is a limitation regarding the connection with low voltage storage systems. One solution for this is to connect several storage modules in series, but in this case, there is the major disadvantage of global system failure. Indeed, if one storage module fails the global system fails, the unit must be replaced or bypassed.

9.2.1.2 Double stage

The structure of a two-stage interface converter, shown in Figure 9.3, is formed with a DC/DC conversion stage and a DC/AC stage. The DC/DC stage adapts the DC voltage level to the DC-link voltage so that the inverter stage can be directly interfaced to the grid. Topologies for the DC/DC stage can be classified [21] into two main categories depending on the electrical isolation between input and output ports: non-isolated and isolated. Although many non-isolated bi-directional topologies have been studied and proposed, the research has shown that the most practical choice is the conventional bi-directional buck-boost topology.

The most widely used solution is the bi-directional half-bridge topology due to its simplicity and efficiency. Figure 9.4(a) shows this topology connected between the energy storage and the second stage DC/AC converter [22]. It consists of two transistors and one inductor. It operates in buck mode when charging the batteries and in boost mode when drawing energy from them. Its advantages are low component count and simplicity resulting in a reliable and low-cost solution. On the other hand, there are some limitations. Thus, the boost capability is limited and, as a solution, several energy storages connected in series could be needed to achieve the suitable DC-link voltage level. This topology also presents the problem that in the case of a short circuit in the DC-link, the battery is shorted as well. The double bi-directional half-bridge converter, shown in Figure 9.4(b), can solve this problem,

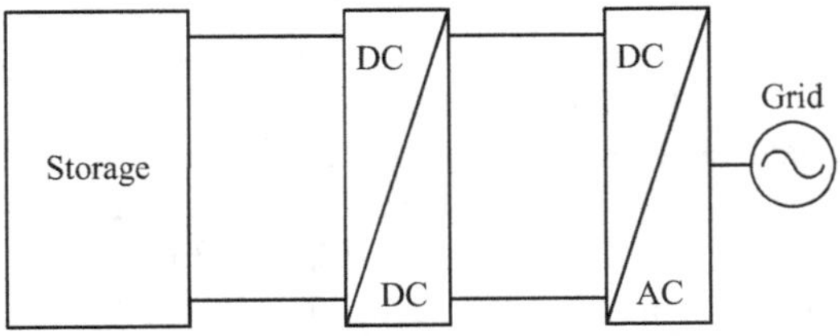

Figure 9.3 Two-stage interface converter for energy storage

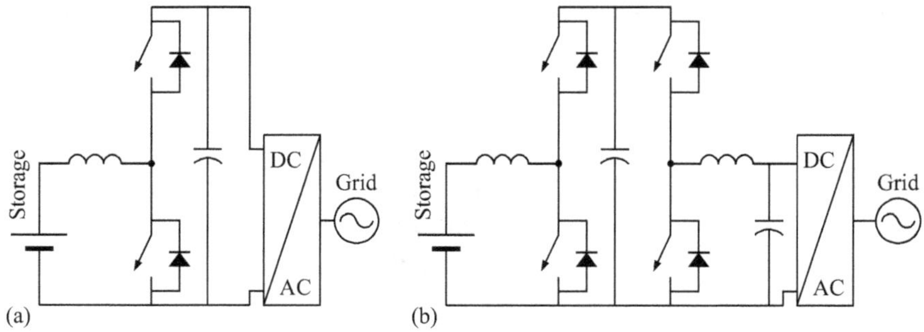

Figure 9.4 (a) Bi-directional half-bridge step up DC/DC converter. (b) Double bi-directional half-bridge DC/DC converter

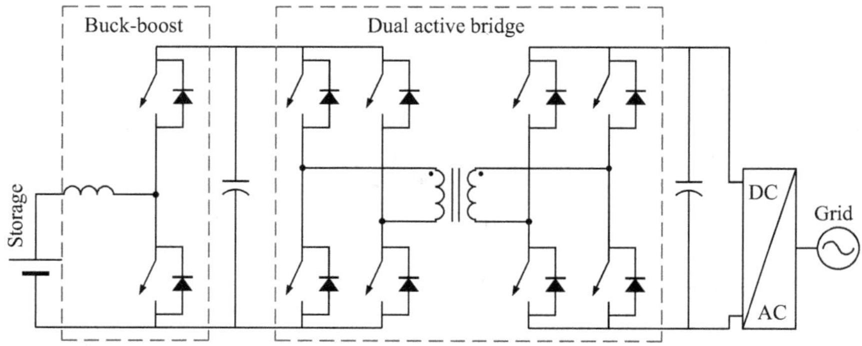

Figure 9.5 Bi-directional converter with galvanic isolation. Buck-boost and DAB

because in this case, the current can be fully controlled in both directions and faults can be isolated from both sides: energy storage side and inverter side.

Galvanic isolation between energy storage and the grid is a stipulation mandated by many standards. Personnel safety, noise reduction and correct operation of protection systems are the main reasons for this requirement.

The isolation can be done through a line (50 or 60 Hz) transformer. This solution makes the equipment heavy and bulky. This problem can be solved by using a high-frequency transformer as the isolation stage. The best location to implement such a high-frequency transformer is in the DC/DC conversion stage. Furthermore, with the appropriate control, the converter could be used not only as a conventional transformer but also for active functions for smart grid such as reactive power regulation and harmonic elimination [23].

The bi-directional buck-boost converter combined with a dual active bridge (DAB) is the most frequently used solution (Figure 9.5). The bi-directional buck-boost regulates the input voltage of DAB and keeps it constant, allowing DAB to work in one operating point and with a maximum efficiency. The transformer provides additional boost and, therefore, this solution is useful in the cases where higher boost ratio is required. The main disadvantage is the relatively high component count, which increases losses.

9.2.2 Multi-level topologies

A multi-level voltage source converter synthesizes the desired AC voltage from several levels of DC voltages. These topologies are well established for medium and high voltage applications; however, the latest progress in research and development of low-voltage power semiconductors, for example, silicon carbide or gallium arsenide, leads to the idea of use multi-level inverters for low-voltage applications [24].

Many multi-level converter structures have been proposed during the last years, however, the most common topologies are the cascaded H-bridge converter (CHB) structure and the neutral point clamped multi-level inverter (NPC).

The CHB structure [25] is based on the series connection of H-bridges with separate DC sources. The H-bridges can be replaced for other type of converter obtaining several variants of the same structure: without isolation by means of a double-stage scheme or with isolation by means of a DAB. In any case, this structure overcomes the disadvantage of series storage systems: if one of the elements fails, it must be replaced or bypassed.

The NPC multi-level inverter uses a series string of capacitors to divide the DC side voltage into several levels. Figure 9.6 shows a single-stage grid connected structure of the three-level NPC and Figure 9.7 shows a three-level NPC based on a two-stage converter structure.

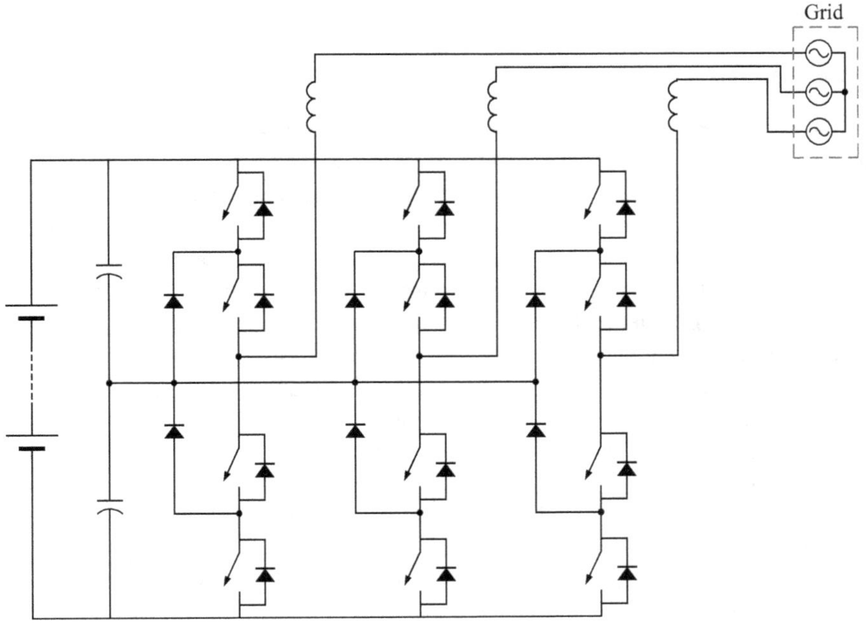

Figure 9.6 Single stage three-phase three-level NPC converter structure

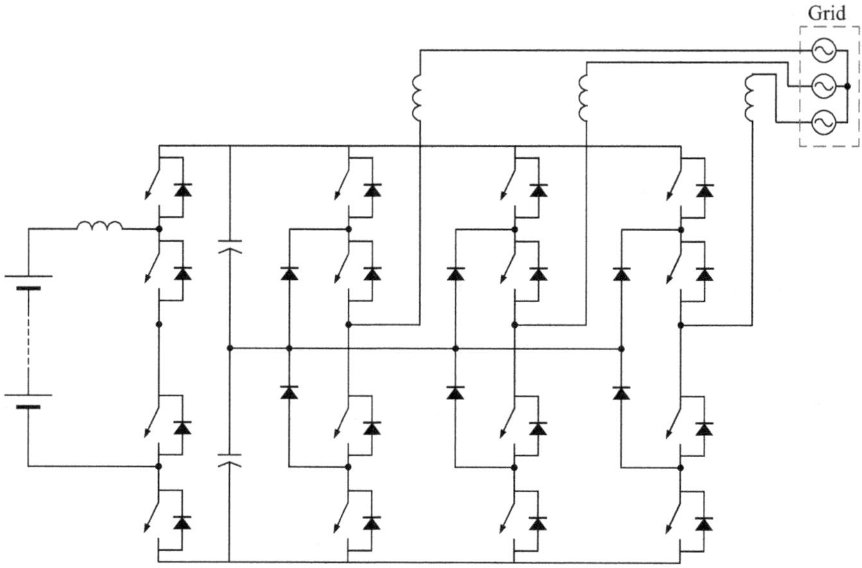

Figure 9.7 Two stage three-phase three-level NPC converter structure

As an alternative solution, an interesting topology from the family of single-stage buck-boost multi-level inverters is presented and analysed in [26]. It combines a qZSI topology (before mentioned in Section 9.2.1.1) and a three-level NPC topology.

9.2.3 Multi-port topologies

The power converters described above are usually known as two-port converters, due to the fact that they have an input port that connects to the energy source and an output port that connects to the grid. However, there are systems with multiple energy sources. In this case, the energy management can be done by mean of two types of power converter structures. The first option consists of an intermediate power bus where the multiple sources and the output are connected by two-port converters, independently controlled. The second option has a single power-processing stage with multiple interfacing ports. This structure, known as multi-port topology, has several advantages over the two-port topology, such as high power efficiency and power density, lower cost and control simplicity. The multi-port converter is interesting in systems where energy is collected from different energy sources in combination with energy storage or in systems where different types of storage systems are used. This last case can be posed at home with several micro-storage energy systems.

The following description refers to the first stage of a double stage multi-port topology, that is to say, the output of the multi-port converter is directly connected to the DC-link that is the input to the second stage converter, the DC–AC converter.

Several multi-port topologies with and without galvanic isolation have been proposed [27,28]. Figure 9.8 shows a topology without a galvanic isolation: two of

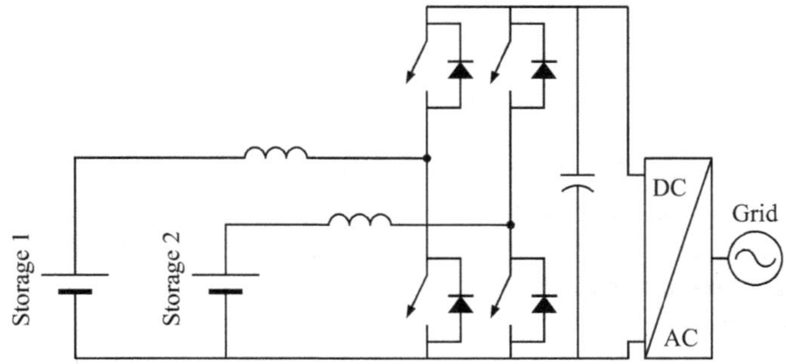

Figure 9.8 Three-port converter with a DC-link

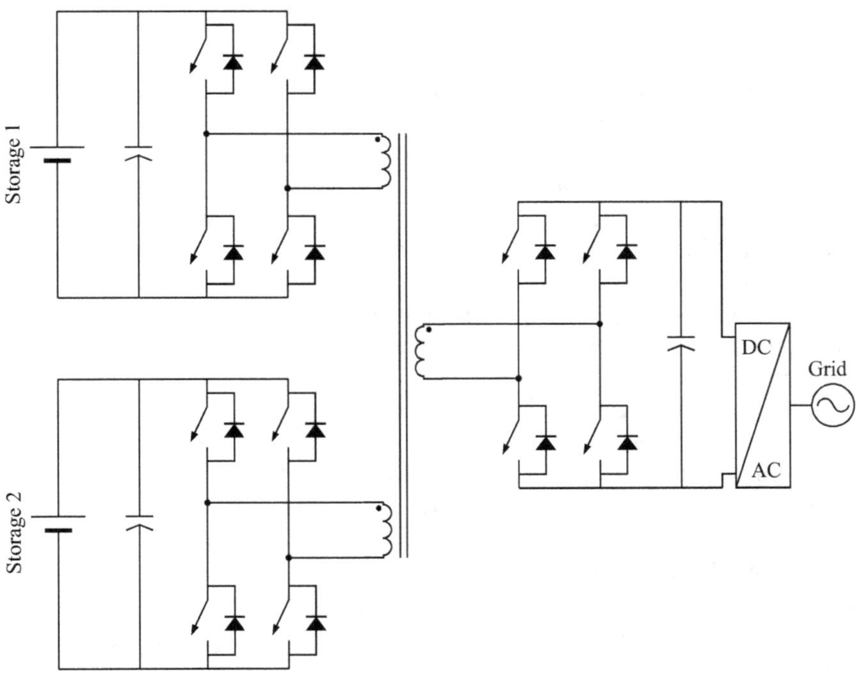

Figure 9.9 Three-port triple-active-bridge converter

the ports connect the storage systems, and the third port connects the DC–AC converter by a DC-link. Figure 9.9 presents a magnetically coupled three-port converter consisting of three full-bridge converter coupled by a high-frequency three-winding transformer. This converter has the additional advantage of easily matching different port voltage levels. This can be done by choosing the appropriate numbers of turns for the transformer windings.

The comparison of the three-port DC–DC converters proposed by different researchers has been carried out as a guide for the appropriate selection of converters for utilization in practical applications in [29].

9.3 Control strategies for the ESMS of the storage device

In this section, control strategies for the ESMS of the micro-storage device located at a home of a smart community are presented. These strategies aim to fulfil the set-points provided by the SCEMS, regarding active power for charging or discharging the ESS according to the expected demand and production [30,31].

Besides, common residential appliances such as microwaves, dish-washers, washing machines, air conditioners and fridges, consume reactive power for which the residential customers do not pay, but the utility is responsible to deliver. By using storage devices at homes of the smart community, capable of supplying this reactive power locally, energy losses due to reactive power consumption could be reduced, and so the global efficiency of the community would increase [32]. Some control strategies for reactive control have been proposed in the technical literature concerning the use of the EV batteries [33,34]. In the context of a smart community, the SCEMS is responsible for sending reactive power set-points to compensate the reactive power of the community increasing the global power factor or to contribute to voltage control.

Finally, bi-directional chargers employed for the ESS operation could also be used to improve the power quality locally [35], compensating the typical harmonic currents demanded by a residential customer, so that the harmonic distortion of these currents could be reduced. And also, in case of three-phase systems, control strategies could be configured to reduce imbalance ratios in the consumption of the customer. These additional functions could reduce the losses in the smart community due to harmonic and unbalanced current components, increasing the efficiency.

Active, reactive and power quality and imbalance control strategies are exposed in this section, considering a home, part of a smart community, connected to a three-phase four-wire system, to take into account the least favourable situation regarding triplen harmonic and homopolar or zero-order components, which produce the circulation of current in the neutral conductor. In case of active and reactive control, the set-points are sent by the SCEMS, but in case of power quality and imbalance control, the reference current is calculated according to the measurement of the demanded current and ensuring that the charger does not exceed its nominal parameters.

The active and reactive control strategies are based on the generalized theory of instantaneous reactive power for three-phase four-wire power systems [36], which is a modification of the original p–q theory proposed by Akagi *et al.* [37].

For a three-phase four-wire power system, instantaneous voltages, u_a, u_b and u_c, and instantaneous currents, i_a, i_b and i_c, are expressed as instantaneous space vectors, **u** and **i** in a–b–c coordinates:

$$\mathbf{u}_{(a,b,c)} = \begin{bmatrix} u_a \\ u_b \\ u_c \end{bmatrix}, \quad \mathbf{i}_{(a,b,c)} = \begin{bmatrix} i_a \\ i_b \\ i_c \end{bmatrix}. \tag{9.1}$$

By using the Park transformation [38], it is possible to transform these space vectors from the three-phase stationary coordinate system *a–b–c* to the 0–*d*–*q* rotating coordinate system:

$$\mathbf{u}_{(0,d,q)} = \begin{bmatrix} u_0 \\ u_d \\ u_q \end{bmatrix} = \mathbf{P} \begin{bmatrix} u_a \\ u_b \\ u_c \end{bmatrix}, \tag{9.2}$$

where $\mathbf{P}$ is the so called transformation matrix, satisfying $\|\mathbf{P}\| = 1$, $\mathbf{P}^{-1} = \mathbf{P}^T$:

$$\mathbf{P} = \sqrt{\frac{2}{3}} \begin{bmatrix} \frac{1}{\sqrt{2}} & \frac{1}{\sqrt{2}} & \frac{1}{\sqrt{2}} \\ \cos\theta & \cos\left(\theta - \frac{2\pi}{3}\right) & \cos\left(\theta + \frac{2\pi}{3}\right) \\ -\sin\theta & -\sin\left(\theta - \frac{2\pi}{3}\right) & -\sin\left(\theta + \frac{2\pi}{3}\right) \end{bmatrix} \tag{9.3}$$

In this matrix, θ is the angle formed by the *d*-axis in the rotating reference frame with the *a*-axis in the stationary reference frame.

The inverse Park transformation allows to transform the space vector from the 0–*d*–*q* to *a*–*b*–*c* coordinates:

$$\begin{bmatrix} u_a \\ u_b \\ u_c \end{bmatrix} = \mathbf{P}^{-1} \begin{bmatrix} u_0 \\ u_d \\ u_q \end{bmatrix}. \tag{9.4}$$

9.3.1 Active power control strategies

The SCEMS will manage the local ESS at homes, by sending an active power set-point, P_{ref}, to each individual ESS, according to previous market negotiations. Two modes of operation can be distinguished: Grid to ESS mode (G2ESS-P mode) when the ESS is extracting active power to charge the storage system, and ESS to grid mode (ESS2G-P mode), when the ESS is injecting active power from the storage device into the grid. The sign criterion for the set-point is positive in the ESS2G-P mode ($P_{\text{ref}} > 0$), and negative in the G2ESS-P mode ($P_{\text{ref}} < 0$).

A sinusoidal current (SC) control strategy is proposed [39]. This strategy aims that the current demanded or injected into the grid will be in phase with the positive-sequence fundamental source voltage. It guarantees, on the one hand, that the ESS will operate with unity displacement power factor, *dPF*, and on the other hand, that the charger current will have neither harmonic content nor imbalance.

The reference charger current will be given by

$$\mathbf{i}_{ch,\text{ref}} = K \cdot \mathbf{u}_1^+, \tag{9.5}$$

where $\mathbf{u}_1^+$ is the positive-sequence fundamental source voltage space vector. If the *d*-axis employed in the Park transformation is in phase with the positive-sequence fundamental source voltage space vector, then, this space vector in 0–*d*–*q* coordinates is expressed as

$$\mathbf{u}_1^+ = \begin{bmatrix} 0 \\ u_{d1}^+ \\ 0 \end{bmatrix}. \tag{9.6}$$

The instantaneous active power delivered by the ESS, by using the generalized theory of instantaneous reactive power for three-phase four-wire power systems [36] is

$$p_{ch} = \mathbf{u}\cdot\mathbf{i}_{ch,\mathrm{ref}} = \mathbf{u}\cdot K\cdot\mathbf{u}_1^+ = K\left(\mathbf{u}\cdot\mathbf{u}_1^+\right) = K\left(u_d u_{d1}^+\right). \tag{9.7}$$

The constant K will be determined with the condition that the active power P_{ch} (mean value of the instantaneous active power) delivered by the ESS equals the set-point, P_{ref}, provided by the SCEMS, so

$$K = \frac{P_{ch}}{\left(u_d u_{d1}^+\right)_{\mathrm{DC}}} = \frac{P_{\mathrm{ref}}}{u_{d1}^{+2}}. \tag{9.8}$$

By substituting (9.6) and (9.8) into (9.5), finally, the reference charger current in 0–*d*–*q* coordinates will be given by

$$\begin{bmatrix} i_{ch,\mathrm{ref}0} \\ i_{ch,\mathrm{ref}d} \\ i_{ch,\mathrm{ref}q} \end{bmatrix} = K\begin{bmatrix} 0 \\ u_{d1}^+ \\ 0 \end{bmatrix} = \frac{P_{\mathrm{ref}}}{u_{d1}^{+2}}\begin{bmatrix} 0 \\ u_{d1}^+ \\ 0 \end{bmatrix}. \tag{9.9}$$

To extract the positive-sequence fundamental component of the source voltage it is necessary to use a phase-locked-loop system [40], such as the autoadjustable synchronous reference frame (ASRF) proposed in [41].

The block diagram of the control strategy is displayed in Figure 9.10. The inputs are the measured phase-to-neutral source voltage in *a*–*b*–*c* coordinates, $\mathbf{u}_{(a,b,c)}$ and the set-point P_{ref} sent by the SCEMS. The output is the reference charger current for active power *P*-control in *a*–*b*–*c* coordinates.

9.3.2 Reactive power control strategies

Local ESS at homes of smart communities can be used as distributed storage sources to provide voltage control or to improve globally the power factor of the

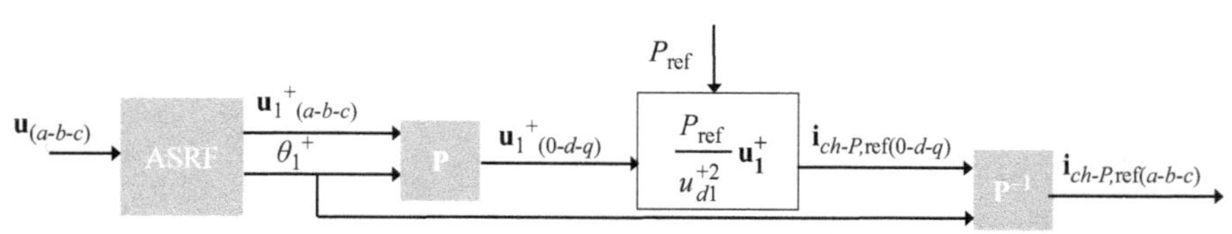

Figure 9.10 Block diagram of the active power control strategy

community. Thereby, the SCEMS is responsible for calculating a fundamental reactive power set-point, $Q_{1\text{ref}}$, for each individual ESS. The modes of operation are: G2ESS-Q mode, when the ESS is absorbing fundamental reactive power from the grid, so $Q_{1\text{ref}} < 0$, and ESS2G-Q mode, when $Q_{1\text{ref}} > 0$, since the storage system is injecting fundamental reactive power to the grid.

A quadrature SC control strategy is proposed in this case, equivalent to the previous SC strategy, but taking over the reactive power, instead of the active power set-point. This strategy assures that the current demanded or injected into the grid will be sinusoidal, balanced and in quadrature with the positive-sequence fundamental source voltage, achieving a null *dPF*.

The reference charger current will be given by

$$\mathbf{i}_{ch,\text{ref}} = K \cdot \mathbf{u}_{1\perp}^{+}, \tag{9.10}$$

where $\mathbf{u}_{1\perp}^{+}$ is the positive-sequence fundamental source voltage vector being $+90°$ phase shifted. Since the *d*-axis is in phase with the positive-sequence fundamental source voltage space vector, $\mathbf{u}_{1\perp}^{+}$ in the 0–*d*–*q* reference frame is

$$\mathbf{u}_{1\perp}^{+} = \begin{bmatrix} 0 \\ 0 \\ u_{d1}^{+} \end{bmatrix}, \tag{9.11}$$

and the reference charger current vector is

$$\mathbf{i}_{ch,\text{ref}} = \begin{bmatrix} 0 \\ 0 \\ Ku_{d1}^{+} \end{bmatrix}. \tag{9.12}$$

By using the generalized *p*–*q* theory [36], the reactive power delivered by the ESS, q_{ch}, is the norm of the instantaneous reactive power $\mathbf{q}_{ch}$, calculated as

$$\begin{aligned} \mathbf{q}_{ch} &= \mathbf{u} \wedge \mathbf{i}_{ch,\text{ref}}, \\ q_{ch} &= \|\mathbf{q}_{ch}\| = \overline{q}_{ch} + \tilde{q}_{ch}, \end{aligned} \tag{9.13}$$

where $\overline{q}_{ch}$ is the mean or constant value of the charger reactive power, and $\tilde{q}_{ch}$ is the oscillating component. Taking into account that only the components of current and voltage of the same harmonic order and sequence gives an average value,

$$\overline{q}_{ch} = \left\| \mathbf{u}_{1}^{+} \wedge \mathbf{i}_{ch,\text{ref}} \right\| = Ku_{d1}^{+2}. \tag{9.14}$$

The constant K will be determined with the condition that the constant reactive power delivered by the ESS has to be equal to the set-point, $Q_{1\text{ref}}$, provided by the SCEMS, so

$$K = \frac{\overline{q}_{ch}}{u_{d1}^{+2}} = \frac{Q_{1\text{ref}}}{u_{d1}^{+2}}. \tag{9.15}$$

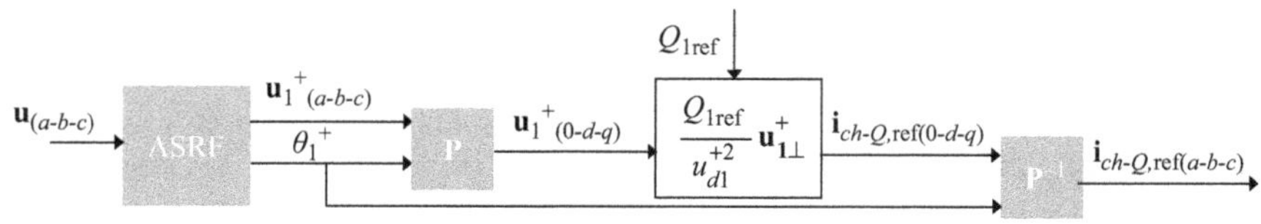

Figure 9.11 Block diagram of the reactive power control strategy

Finally, the reference charger current will be given by

$$\begin{bmatrix} i_{ch,\text{ref}0} \\ i_{ch,\text{ref}d} \\ i_{ch,\text{ref}q} \end{bmatrix} = K \begin{bmatrix} 0 \\ 0 \\ u_{d1}^{+} \end{bmatrix} = \frac{Q_{1\text{ref}}}{u_{d1}^{+2}} \begin{bmatrix} 0 \\ 0 \\ u_{d1}^{+} \end{bmatrix}. \tag{9.16}$$

The block diagram of the control algorithm for the reactive power control is shown in Figure 9.11. Again it is necessary to employ an ASRF to extract the positive sequence fundamental source voltage component and to create the space vector $\mathbf{u}_{1\perp}^{+}$. In this case, the inputs are the measured phase-to-neutral source voltage in *a–b–c* coordinates, $\mathbf{u}_{(a,b,c)}$ and the set-point $Q_{1\text{ref}}$ provided by the SCEMS. The output is the reference charger current for reactive power *Q*-control in *a–b–c* coordinates.

9.3.3 Power quality and imbalance reduction control strategies

If the ESS is below 100% capacity, it can be in charge of an additional function concerning the power quality improvement of the current demanded by each home and so, contributing to the global power quality consumption of the community.

Besides, it is also possible to use the remaining capacity to compensate imbalance, due to negative-sequence or zero-sequence components in the fundamental current demanded by the home. It contributes to reduce imbalance ratios in the consumption locally and, therefore, to decrease the imbalance ratios in the voltage at the point of coupling of the smart community.

This harmonic and imbalance control function has only a unidirectional operation mode, from the charger to the grid: ESS to grid mode (ESS2G-H-I mode).

A total harmonic and imbalance compensation (THIC) control strategy is proposed. It aims to provide as reference charger current the whole harmonic spectrum of the current demanded by the house, neglecting the fundamental frequency, and the fundamental unbalanced components (negative- and zero-sequence terms).

If the harmonic and imbalance reference current added to the current for *P* and *Q* control exceeds the nominal current of the charger, $I_{ch,n}$, the reference current has to be saturated to prevent overload. The maximum RMS harmonic and imbalance charger current is obtained as

$$I_{ch-H-I,\max} = \sqrt{I_{ch,n}^2 - I_{ch-P}^2 - I_{ch-Q}^2}, \tag{9.17}$$

where $I_{ch\text{–}P}$ and $I_{ch\text{–}Q}$ are the RMS charger current components responsible for the *P* and Q control, respectively.

The reference charger current can be calculated as

$$\mathbf{i}_{ch-H-I,\text{ref}} = \mathbf{i}_L - \mathbf{i}_{L1}^{+}, \tag{9.18}$$

where $\mathbf{i}_L$ is the load current vector demanded by the house and $\mathbf{i}_{L1}^{+}$ is the positive-sequence fundamental component load current vector, obtained from the ASRF. By using this control strategy, all the harmonic currents $\mathbf{i}_{Lh}$ and the unbalanced fundamental components, that is, the negative-sequence fundamental load current $\mathbf{i}_{L1}^{-}$ and the zero-sequence fundamental load current $\mathbf{i}_{L1}^{0}$, will be eliminated or reduced in the source current demanded to the grid.

This reference charger current hast to be limited to the maximum RMS harmonic and imbalance charger current, $I_{ch\text{–}H\text{–}I,\max}$, resulting a reference charger current after saturation:

$$\begin{aligned} \mathbf{i}_{ch-H-I,\text{ref}}^{sat} &= \mathbf{i}_{ch-H-I,\text{ref}} && \text{if } I_{ch-H-I,\text{ref}} \leq I_{ch-H-I,\max} \\ \mathbf{i}_{ch-H-I,\text{ref}}^{sat} &= \frac{\mathbf{i}_{ch-H-I,\text{ref}}}{I_{ch-H-I,\text{ref}}} I_{ch-H-I,\max} && \text{if } I_{ch-H-I,\text{ref}} > I_{ch-H-I,\max} \end{aligned} \tag{9.19}$$

where, $I_{ch\text{–}H\text{–}I,\text{ref}}$ is the RMS value of $i_{ch\text{–}H\text{–}I,\text{ref}}$.

The block diagram of the THIC strategy is displayed in Figure 9.12 in the most general situation, when the charger current for the *P*- and *Q*-control are not null. An ASRF to extract the positive sequence fundamental load current component is employed. The inputs are the measured phase-to-neutral source voltage in *a–b–c* coordinates, $\mathbf{u}_{(a,b,c)}$ the measured load current in *a–b–c* coordinates, $\mathbf{i}_{\mathrm{L}(a,b,c)}$ and the nominal charger current, $I_{ch,n}$. The reference charger current for the *P*- and *Q*-control are also inputs in this algorithm to calculate (9.17). The output of this control strategy is the saturated reference charger current in *a-b-c* coordinates, $\mathbf{i}_{ch-H-I,\text{ref}}^{sat}$, when applying the harmonic and imbalance reduction control strategy. Since this figure shows the global algorithm when the *P*-, *Q*- and *H-I*-control are operating at the same time, finally the output is the total reference charger current in *a–b–c* coordinates, $\mathbf{i}_{ch,\text{ref}(a,b,c)}$, calculated by adding the reference current obtained with each independent control algorithm.

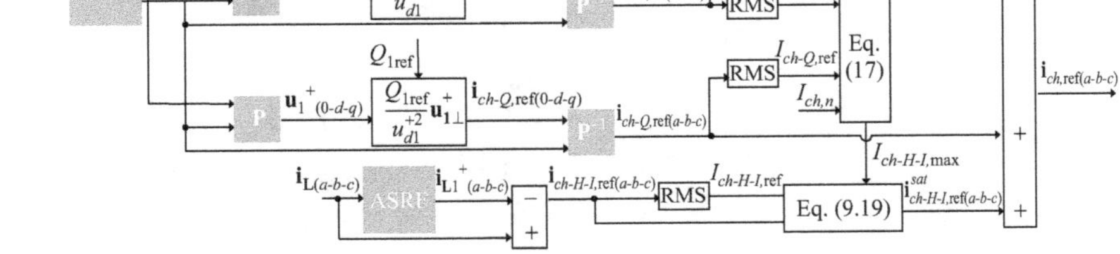

Figure 9.12 Block diagram of the active, reactive and harmonic-imbalance control strategy

9.4 Power interfaces

In this section, power interfaces for the integration of ESMS of the micro-storage device located at a home of a smart community are discussed. The main idea is to discuss how the low voltage used typically in this kind of ESS is adapted to be able to connect to the grid.

As it has been discussed before there are several drawbacks in the technically available ESSs and a perfect system does not exist. One possibility to improve ESSs, and achieve required operation specifications is to build a hybrid system, combining kind of technologies. Hybrid systems improve the indexes that can be achieved if a single technology is considered. Some works have compared different ESS alternatives, and have classified them according to different criteria as availability, energy storage density, price and so on. Supercapacitors seem to be a good alternative, but they are nowadays away from being commercially competitive.

The nominal grid voltage is a main parameter to be taken into account when designing an EMS for controlling the ESS charge and discharge. Usually an AC/DC converter is needed, and a minimum voltage at the DC side is required for a correct operation [42–49].

There are different chargers topologies, but there are based in two main ways for matching the ESS with the grid voltage, schematically represented in Figure 9.13 as follows:

- to adapt the DC voltage by re-designing the ESS or by including a DC/DC converter (with or without high frequency transformer), or
- to adapt the AC voltage, by including a conventional low frequency (50 or 60 Hz) transformer.

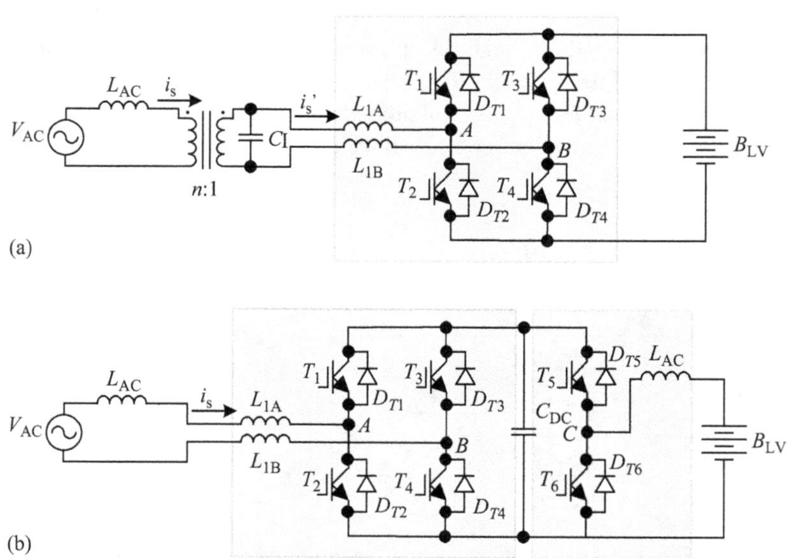

Figure 9.13 Typical connection of a low-voltage ESS: (a) with conventional transformer, (b) with additional DC/DC converter

9.4.1 Analysis of typical solution

There are many solutions for interfacing ESS with the electrical grid [42–49]. Typical solutions for storing energy in low-voltage single-phase applications are shown in Figure 9.13. Alternative shown in Figure 9.13(b) is considered in this section.

As the ESS operates in DC, a bi-directional AC/DC converter is required. This converter requires a minimum DC voltage when connected to the grid, if a PWM sinusoidal technique is applied and operation in the lineal region are required:

$$V_{\mathrm{DC,min}} = \frac{\sqrt{2}V_{\mathrm{AC}}}{n}. \tag{9.20}$$

If LV-ESS is implemented with Li-ion batteries associated in series and conforming a pack of 48 V (that can be assumed as the minimum operation DC voltage), and the ESS system is connected to the a grid with a nominal voltage of 230 V 50 Hz, the transformer should have a conversion relation, n, of

$$n \geq \frac{\sqrt{2}V_{\mathrm{AC}}}{V_{\mathrm{LV,min}}} = \frac{\sqrt{2}\cdot 230}{48} \approx 6.78. \tag{9.21}$$

The problem associated with matching the AC and DC voltages is even worse if supercapacitors are used in the ESS, as the supercapacitor DC voltage depends highly on the stored energy, according to the next well-known equation

$$E = \frac{1}{2}C_{\mathrm{SC,ESS}}V_{\mathrm{DC}}^2. \tag{9.22}$$

ESS storage system based in supercapacitors requires a minimum DC voltage that is usually adopted as the 50% of the nominal supercapacitor voltage. For example, if a 48 V 195 F supercapacitor ESS is considered, the minimum DC voltage is 48 V·50% = 24 V, requiring a transformer with a conversion ratio given by

$$n \geq \frac{\sqrt{2}V_{\mathrm{AC}}}{V_{\mathrm{LV,min}}} = \frac{\sqrt{2}\cdot 230}{24} \approx 13.56, \tag{9.23}$$

that is much bigger that in the case of using batteries. In addition, establishing this DC voltage as the minimum operation value, produces that the supercapacitors never discharge totally, having a residual stored energy equal to

$$\begin{aligned} E_{\mathrm{res}} &= \frac{1}{2}C_{\mathrm{SC,ESS}}V_{\mathrm{DC,min}}^2 = \frac{1}{2}C_{\mathrm{SC,ESS}}(0.5V_N)^2 \\ &= 0.25\left(\frac{1}{2}C_{\mathrm{SC,ESS}}V_N^2\right) = 25\% E_N. \end{aligned} \tag{9.24}$$

Due to this residual only the 75% of the nominal capacity of supercapacitors ESS can be used in the energy management strategy, and, as consequence, the ESS size has to be increased by 133%.

Table 9.2 Main parameters values used in simulation of typical solution

Symbol	Description (units)	Values
SC_{ESS}	ESS supercapacitor (F)	0.1
L_{ESS}	ESS Filter inductance (mH)	2
R_{ESS}	Resistance of ESS filter inductance (ohm)	0.01
f_{BH}	Frequency of the synchronous band (kHz)	20
rt	Transformer ratio (V/V)	230/16
V_s	Grid voltage: voltage (V), frequency (Hz)	230/50

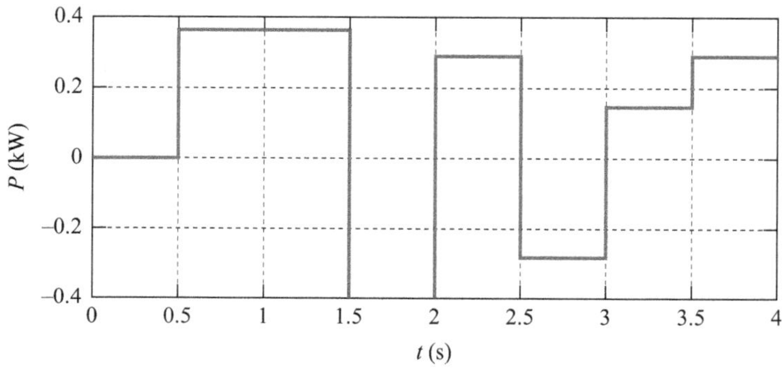

Figure 9.14 Profile of power used for ESS charge/discharge in simulation tests

The performance of the typical solution when using supercapacitors is tested by simulation using MATLAB®/Simulink™ with the main values shown in Table 9.2.

Among different high-level operation strategies that have been introduced previously, in this section it is supposed that the ESS power interface receives the power reference for charging-discharging the ESS shown in Figure 9.14, which has been selected with the main objective of showing the operation of the system without requiring large simulation times.

In the considered solution, an adaptation transformer is required, with a conversion ratio close to 14. So as it can be seen the ESS can be used, while the supercapacitor voltage is in the operative range, being 24 V the minimum operative voltage and 48 V the maximum voltage, which is the nominal voltage of supercapacitors.

This is clearly shown in the case shown in Figure 9.15, where no minimum operative voltage is applied and the system is not able to operate when voltage reaches the minimum value ($t = 1.65$ s).

This fact is easily seen if waveforms of current injected into the grid (Figure 9.16) are analysed. It can be observed that grid currents that are sinusoidal

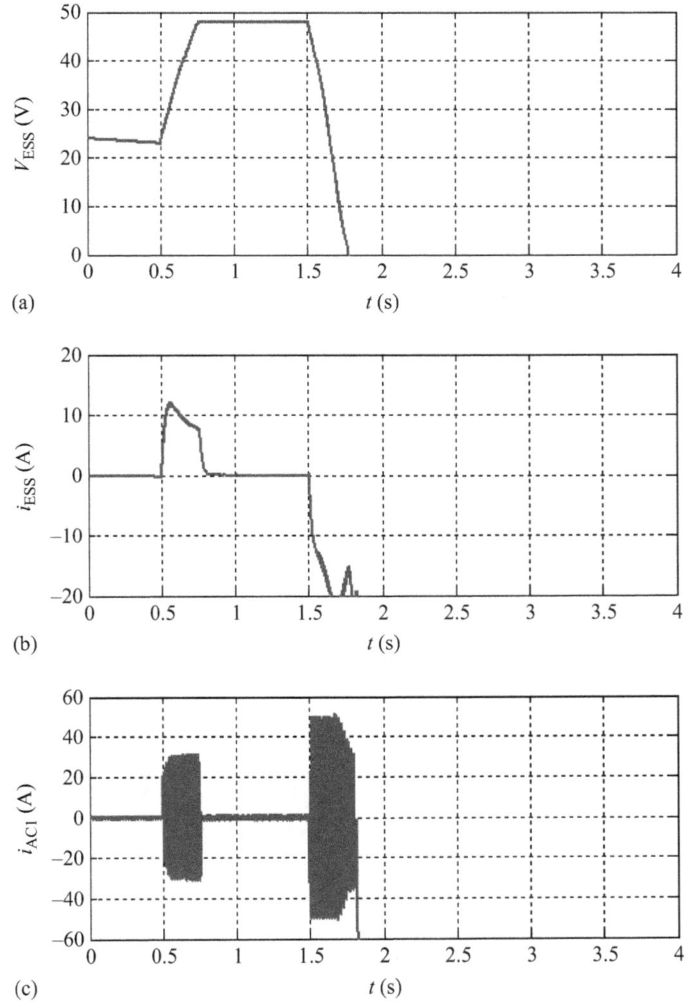

Figure 9.15 Representative waveforms of ESS: (a) ESS voltage, (b) ESS DC current and (c) ESS AC current

in normal operation when the ESS voltage is greater than the minimum operative value (for example, in $t = 1.55$ s) are no longer sinusoidal when the ESS voltage is lower than the minimum required value (as when $t = 1.65$ s). Finally, when the ESS voltage is very low the system is unable to operate and collapse.

To solve this situation, the system should stop its operation before to collapse. In the case of Figure 9.17, the minimum voltage is applied and a correct operation is achieved. Nevertheless this way of operate produces that the power profile cannot be followed and that a residual energy keep stored as a base energy in the ESS bases in supercapacitors.

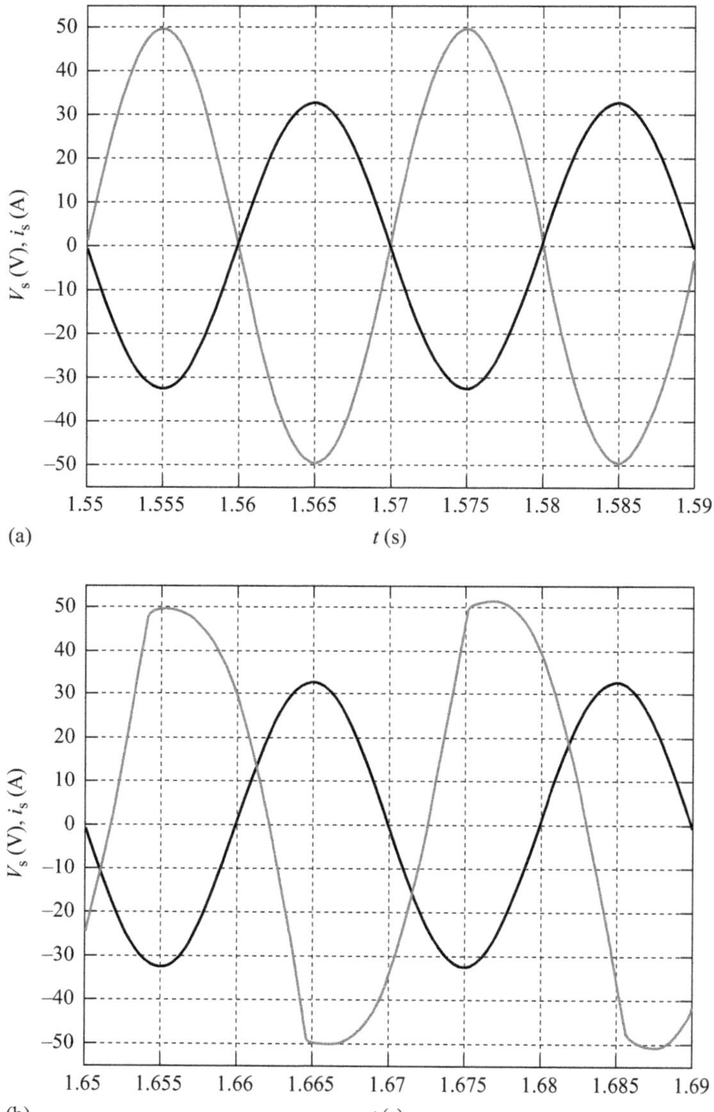

Figure 9.16 Grid voltage and current when discharging: (a) when the ESS voltage is greater than the minimum operative voltage, (b) when is lower than the minimum operative voltage

9.4.2 An improved solution based on cooperative converters

There could be used an alternative solution based in two cooperative converters. The idea of cooperative converters was initially presented and explained by examples in [50] and validated for photovoltaic inverter applications in [51–53].

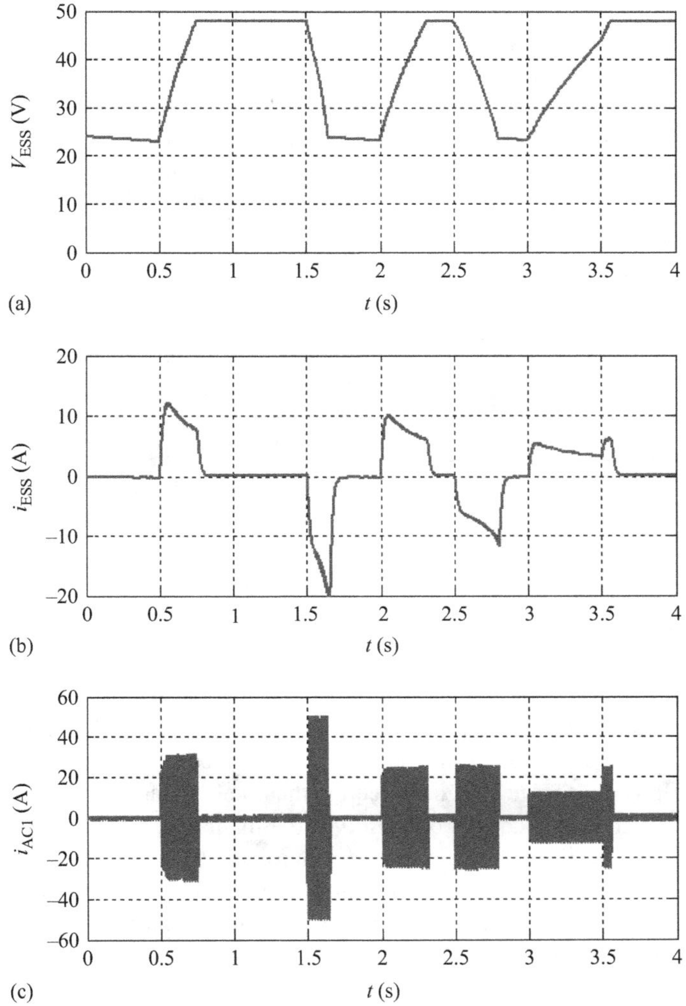

Figure 9.17 Representative waveforms of ESS: (a) ESS voltage, (b) ESS DC current and (c) ESS AC current

It consists on systems that use more than one converter that performs optimally the associated specific functions that are planned for optimize the whole converter operation. For this case, both converters are based on full bridge inverters, and the associate specific functions are: to charge and discharge the ESS for the main one and to maintain a minimum DC voltage for the second auxiliary one as proposed in [54].

Whit this improved solution, the minimum DC voltage is determined to guarantee a good operation of the EMS when connected to the grid, avoiding the requirements of maintaining a minimum ESS voltage and having a transformer. Main converter (on the top of Figure 9.18) is controlled using a power reference to

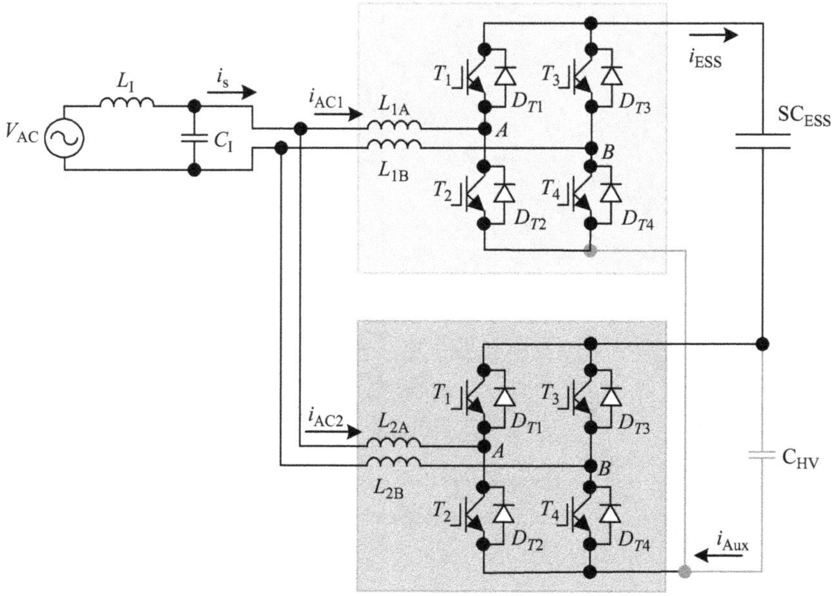

Figure 9.18 Alternative solution based on cooperative converters

charge/discharge the ESS, establishing the current RMS value of this inverter and taking into account that a unitary power factor has to be achieved and the operative voltage limits of the ESS, meanwhile the auxiliary converter (on the bottom of Figure 9.10) is in charge of guaranteeing the total voltage is greater than the minimum that ensures a correct operation of the main one.

The results obtained for the proposed solution are shown in Figure 9.19, where it can be seen how the alternative solution is able to operate properly till fully charge or discharge the ESS, without any minimum operation voltage.

This alternative solution is specially suitable for the case of using super-capacitors as ESS, as other solutions do not allow to fully discharge it, so a residual energy is always stored in the system (being usually equal to the 25% of the rated storing capacity) increasing the ESS size and cost. This could be easily followed from Figure 9.20, where the stored energy for conventional and cooperative solutions are shown.

9.5 Conclusions

This chapter delivers an analysis of micro-storage energy systems located at homes of a smart community with the aim of contributing to demand management, integration of renewable energy generation systems and power flow control.

A summary of ESS technologies that are suitable for been used as micro-storage systems in a community EMS context has been presented. It has been concluded that ESS technologies selected to perform energy management and microgrid control in

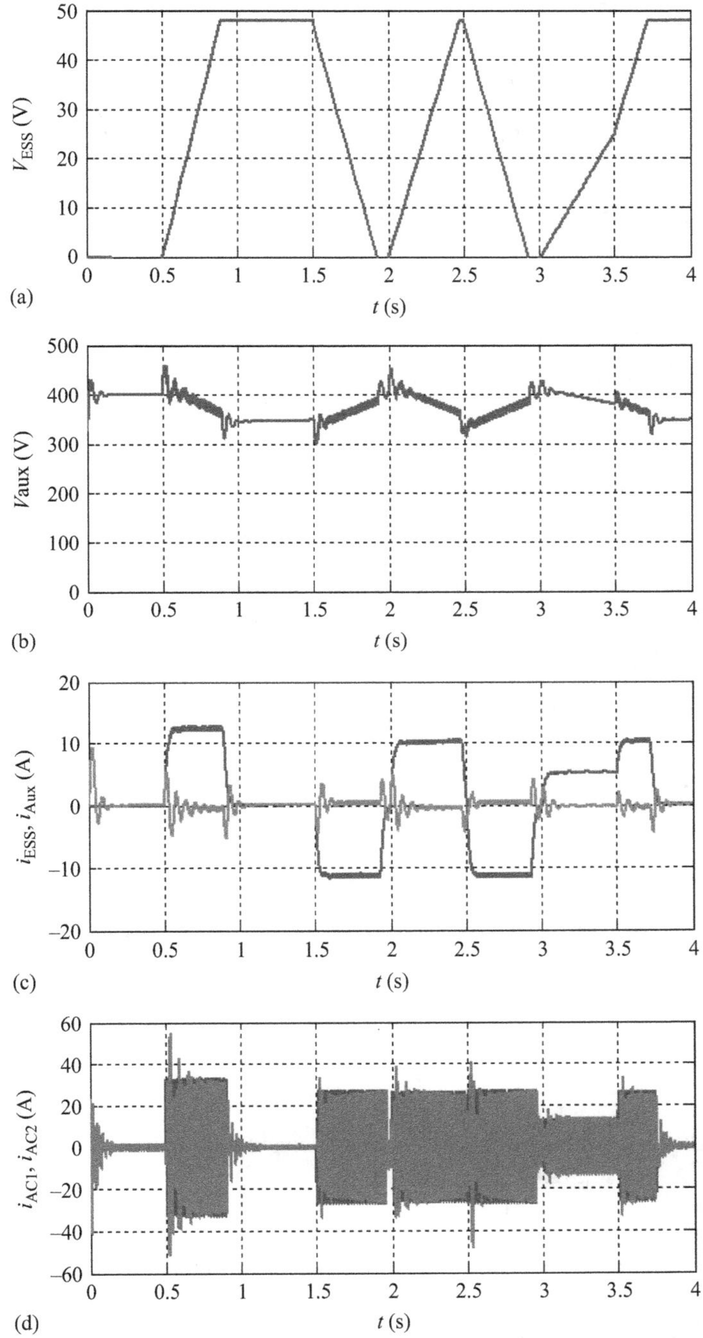

Figure 9.19 Representative waveforms of alternative ESS cooperative operation: (a) ESS voltage, (b) Auxiliary voltage, (c) ESS and auxiliary filtered DC currents and (d) ESS and auxiliary AC currents

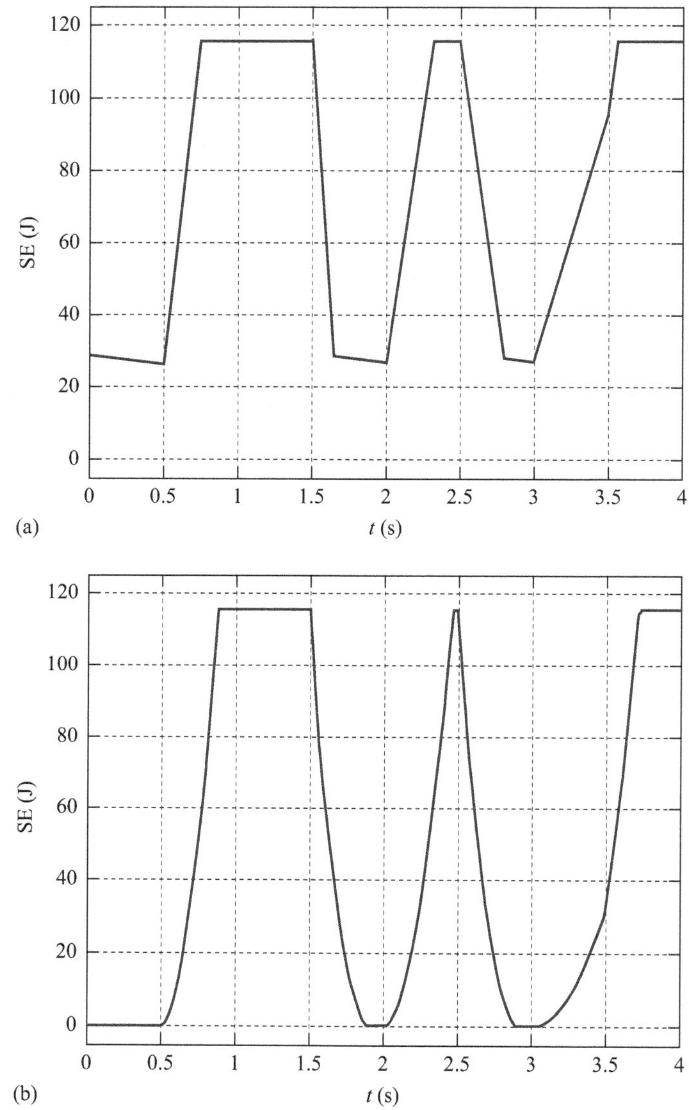

Figure 9.20 Instantaneous stored energy: conventional (a) and cooperative (b) solutions

communities must be versatile enough to encompass such opposite requirements or hybrid technologies have to be used. There are several technologies which meet the mentioned requirements, with different maturity levels: conventional batteries, flow batteries, supercapacitors, flywheels and SMES. The functioning principles and main characteristics of each of these ESS have been described. Finally, these ESS technologies have been compared and discussed, with the aim of analysing their suitability for different functions, according to these parameters: energy density, rated energy capacity, number of cycles during the lifetime, efficiency during discharge, discharge time and energy capital cost.

A review of the topologies based on power-electronic converters for the bi-directional flux between the local ESS of the smart community and the grid is provided. Standard, multi-level and multi-port topologies are proposed, explaining the characteristics and advantages of each configuration, as guidelines to select the most suitable topology.

Control strategies for local energy micro-storage systems in homes of smart communities regarding active and reactive power control have been proposed. The aim of these strategies is to fulfil the set-points received from the SCEMS and, at the same time, to ensure that the charger current will have neither harmonic content nor imbalance. In addition, a power quality and imbalance reduction control strategy is presented to decrease the harmonic and unbalanced currents demanded by the individual home where the micro-storage system is located. This control only comes into operation once the ESMS verifies that the ESS has available capacity and, if necessary, saturates the harmonic and imbalance load current compensation to ensure that the charger does not exceed its nominal parameters. These strategies improve the ESS performance and the power quality and imbalance ratios of the charger current, and therefore, of the global community.

The power interface with the grid has to be able to fully implement the control strategies, injecting or demanding the desired current to/from the grid guaranteeing its sinusoidal waveform and the displacement angle. At the same time it has to fully fit with the ESS requirements, mainly do not overpass its nominal current, to operate in the required voltage margin and to detect fully charge or discharge. It is mainly important to guarantee the minimum voltage required to properly operate the DC/AC when connected to the grid, being critical if supercapacitor based ESS are used.

An alternative solution for the power interfaces is to use cooperative converters to improve the performance of typical solutions and to reduce the residual energy that is always stored in the system (being usually equal to the 25% of the rated storing capacity in the case of supercapacitors).

References

[1] *The future role and challenges of Energy Storage*. European Commission. 2013. http://ec.europa.eu/energy/infrastructure/doc/energy-storage/2013/energy_storage.pdf.

[2] Aman, S., Simmhan, Y., Prasanna, V. K. 'Energy management systems: state of the art and emerging trends', *IEEE Communications Magazine*. 2013; **51**: 114–119.

[3] Ota, Y., Taniguchi, H., Nakajima, T., Liyanage, K. M., Baba, J., Yokoyama, A. 'Autonomous distributed V2G (vehicle-to-grid) satisfying scheduled charging', *IEEE Transactions on Smart Grid*. 2012; **3**(1): 559–564.

[4] Guerrero, M. A., Romero, E., Barrero, F., Milanes, M. I., Gonzalez, E. 'Overview of medium scale energy storage systems'. *Compatibility and Power Electronics, 2009. CPE'09*; Badajoz, Spain. 2009, pp. 93–100.

[5] Vytelingum, P., Voice, T. D., Ramchurn, S. D., Rogers, A., Jennings, N. R. 'Theoretical and practical foundations of large-scale agent-based micro-storage in the smart grid'. *Journal of Artificial Intelligence Research*. 2011; **42**: 765–813.

[6] Pérez-Romero, M., Gallardo-Lozano, J., Romero-Cadaval, E., Lozano-Tello, A. 'Local energy management unit for residential applications'. *Electronics and Electrical Engineering*. Kaunas, Lithuania: Kaunas University of Technology. ISSN 1392-1215. 2013; **19**(7): 61–64.

[7] Onar, O. C., Khaligh, A. 'Grid interactions and stability analysis of distribution power network with high penetration of plug-in hybrid electric vehicles'. *Twenty-Fifth Annual IEEE Applied Power Electronics Conference and Exposition (APEC)*. 2010, pp. 1755–1762.

[8] Moses, P. S., Deilami, S., Masoum, A. S., Masoum, M. A. S. 'Power quality of smart grids with plug-in electric vehicles considering battery charging profile'. *Innovative Smart Grid Technologies Conference Europe (ISGT Europe)*. 2010, pp. 1–7.

[9] Gallardo-Lozano, J., Milanes-Montero, M. I., Guerrero-Martinez, M. A., Romero-Cadaval, E. 'Electric vehicle battery charger for smart grids'. *Electric Power Systems Research*. 2012; **90**: 18–29.

[10] Luo, X., Wang, J., Dooner, M., Clarke, J. 'Overview of current development in electrical energy storage technologies and the application potential in power system operation'. *Applied Energy*. 2015; **137**: 511–536.

[11] Chen, H., Cong, T. N., Yang, W., Tan, C., Li, Y., Ding, Y. 'Progress in electrical energy storage system: a critical review'. *Progress in Natural Science*. 2009; **19**(3): 291–312.

[12] *Electrical energy storage: white paper. Technical report. Prepared by electrical energy storage project team*. International Electrotechnical Commission (IEC), 2011. http://www.iec.ch/whitepaper/pdf/iecWP-energy storage-LR-en.pdf.

[13] Lacerda, V. G., Mageste, A. B., Santos, I. J. B., da Silva, L. H. M., da Silva, M. do C. H. 'Separation of Cd and Ni from Ni–Cd batteries by an environmentally safe methodology employing aqueous two-phase systems'. *Journal of Power Sources*. 2009; **193**(2): 908–913.

[14] Chacón, J., Soler, P. R. 'Electrically rechargeable Aluminum–air batteries to power Smart Cities'. *2013 International Conference on New Concepts in Smart Cities: Fostering Public and Private Alliances (SmartMILE)*. 2013, pp. 1–4.

[15] Ali, M. H., Dougal, R. A. 'An overview of SMES applications in power and energy Systems'. *IEEE Transactions on Sustainable Energy*. 2010; **1**: 38–47.

[16] Pena-Alzola, R., Sebastian, R., Quesada, J., Colmenar, A. 'Review of flywheel based energy storage systems'. *2011 International Conference on Power Engineering, Energy and Electrical Drives*, Malaga, 2011, pp. 1–6.

[17] Fernão Pires, V., Romero-Cadaval, E., Vinnikov, D., Roasto, I., Martins, J. F. 'Power converter interfaces for electrochemical energy storage systems—a review'. *Energy Conversion and Management*. 2014; **86**: 453–475.
[18] Peng, F. Z. 'Z-source inverter'. *IEEE Transactions on Industry Applications*. 2003; **39**: 504–510.
[19] Anderson, J., Peng, F. Z. 'Four quasi-Z-source inverters'. *Power Electron Specialists Conference*. 2008, pp. 2743–2749.
[20] Ishii, S., Hara, H., Higuchi, T., *et al.* 'Bidirectional dc–ac conversion topology using matrix converter technique'. *Power Electronics Conference (IPEC)*. 2010, pp. 2768–2773.
[21] Kouro, S., Leon, J. I., Vinnikov, D., and Franquelo, L. G. 'Grid-connected photovoltaic systems: an overview of recent research and emerging PV converter technology'. *IEEE Industrial Electronics Magazine*. 2015; **9**(1): 47–61.
[22] Dong, D., Cvetkovic, I., Boroyevich, D., Zhang, W., Wang, R., Mattavelli, P. 'Grid-interface bidirectional converter for residential DC distribution systems—part one: high-density two-stage topology'. *IEEE Transactions on Power Electronics*. 2013; **28**(4): 1655–1666.
[23] Miñambres-Marcos, V., Roasto, I., Romero-Cadaval, E., Strzelecki, R., Barrero-González, F. 'Single-phase power electronics transformer with active functions for smart grid'. *Ninth International Conference on Compatibility and Power Electronics (CPE)*. 2015, pp. 528–533.
[24] Giuliani, F., Barater, D., Concari, C., *et al.* 'Modular photovoltaic inverter with high-frequency DC/DC stage based on low-voltage FETs'. *IEEE Energy Conversion Congress and Exposition (ECCE)*. 2014, pp. 39–46.
[25] Malinowski, M., Gopakumar, K., Rodriguez, J., Perez, M. A. 'A survey on cascaded multilevel inverters'. *IEEE Transactions on Industrial Electronics*, 2010; **57**(7): 2197–2206.
[26] Husev, O., Roncero-Clemente, C., Romero-Cadaval, E., Vinnikov, D., Stepenko, S. 'Single phase three-level neutral-point-clamped quasi-Z-source inverter'. *IET Power Electronics*. 2015; **8**(1): 1–10.
[27] Khosrogorji, S., Torkaman, H., Karimi, F., 'A short review on multi-input DC/DC converters topologies,' *Power Electronics, Drives Systems & Technologies Conference (PEDSTC), 2015 6th*, Tehran, 2015, pp. 650–654.
[28] Tao, H., Kotsopoulos, A., Duarte, J. L., Hendrix, M. A. M. 'Family of multiport bidirectional DC–DC converters'. *IEE Proceedings Electric Power Applications*. 2006; **153**: 451–458.
[29] Zhang, N., Sutanto, D., Muttaqi, K. M. 'A review of topologies of three-port DC–DC converters for the integration of renewable energy and energy storage system'. *Renewable and Sustainable Energy Reviews*, 2016; **56**: 388–401.
[30] Mets, K., Verschueren, T., De Turck, F., Develder, C. 'Exploiting V2G to optimize residential energy consumption with electrical vehicle (dis)charging'. *Smart Grid Modeling and Simulation (SGMS), 2011 IEEE First International Workshop on*, Brussels. 2011, pp. 7–12.

[31] Onar, O. C., Starke, M., Andrews, G. P., Jackson, R. 'Modeling, controls, and applications of community energy storage systems with used EV/PHEV batteries'. *2012 IEEE Transportation Electrification Conference and Expo (ITEC)*. 2012, pp. 1–6.

[32] Kisacikoglu, M. C., Ozpineci, B., Tolbert, L. M. 'EV/PHEV bidirectional charger assessment for V2G reactive power operation'. *IEEE Transactions on Power Electronics*. 2013; **28**(12): 5717–5727.

[33] Pinto, J. G., Monteiro, V., Gonçalves, H., Afonso, J. L. 'Onboard reconfigurable battery charger for electric vehicles with traction-to-auxiliary mode'. *IEEE Transactions on Vehicular Technology*. 2014; **63**(3): 1104–1116.

[34] Kisacikoglu, M. C., Kesler, M., Tolbert, L. M. 'Single-phase on-board bidirectional PEV charger for V2G reactive power operation'. *IEEE Transactions on Smart Grid*. 2015; **6**(2): 767–775.

[35] Tanaka, T., Sekiya, T., Tanaka, H., Hiraki, E., Okamoto, M. 'Smart charger for electric vehicles with power quality compensator on single-phase three-wire distribution feeders'. *2012 IEEE Energy Conversion Congress and Exposition (ECCE)*. 2012, pp. 3075–3081.

[36] Peng, F. Z., Lai, J. S. 'Generalized instantaneous reactive power theory for three-phase power systems'. *IEEE Transactions on Instrumentation and Measurement*. 1996; **45**(1): 293–297.

[37] Akagi, H., Kanazawa, Y., Nabae, A. 'Instantaneous reactive power compensators comprising switching devices without energy storage components'. *IEEE Transactions on Industry Applications*. 1984; **20**(3): 625–630.

[38] Park, R. H. 'Two-reaction theory of synchronous machines generalized method of analysis—part I'. *Transactions of the American Institute of Electrical Engineers*. 1929; **48**(3): 716–727.

[39] Montero, M. I. M., Cadaval, E. R., Gonzalez, F. B. 'Comparison of control strategies for shunt active power filters in three-phase four-wire systems'. *IEEE Transactions on Power Electronics*. 2007; **22**(1): 229–236.

[40] Rodriguez, P., Pou, J., Bergas, J., Candela, J. I., Burgos, R. P., Boroyevich, D. 'Decoupled double synchronous reference frame PLL for power converters control'. *IEEE Transactions on Power Electronics*. 2007; **22**(2): 584–592.

[41] Milanes-Montero, M. I., Romero-Cadaval, E. R., de Marcos, A., Minambres-Marcos, V. M., Barrero-Gonzalez, F. 'Novel method for synchronization to disturbed three-phase and single-phase systems' *2007 IEEE International Symposium on Industrial Electronics*. 2007, pp. 860–865.

[42] Xue, Y., Chang, L., Kjaer, S. B., Bordonau, J., Shimizu, T. 'Topologies of single-phase inverters for small distributed power generators: an overview', *IEEE Transactions on Power Electronics*. 2004; **19**(5): 1305–1314.

[43] Khaligh, A., Zhihao, L. 'Battery, ultracapacitor, fuel cell, and hybrid energy storage systems for electric, hybrid electric, fuel cell, and plug-in hybrid electric vehicles: state of the art', *IEEE Transactions on Vehicular Technology*. 2010; **59**(6): 2806–2814.

[44] Zhou, H., Bhattacharya, T., Tran, D., Khambadkone, A. M., Siew, T. S. T. 'Composite energy storage system involving battery and ultracapacitor with dynamic energy management in microgrid applications'. *IEEE Transactions on Power Electronics*. 2011; **26**(3): 923–930.

[45] Choi, M., Seo, S., Kim, S. 'Energy management optimization in a battery/supercapacitor hybrid energy storage system'. *IEEE Transactions on Smart Grid*. 2012; **3**(1): 463–472.

[46] Hwang, T.-S., Tarca, M. J., Park, S.-Y. 'Dynamic response analysis of DC–DC converter with supercapacitor for direct borohydride fuel cell power conditioning system'. *IEEE Transactions on Power Electronics*. 2012; **27** (8): 3605–3615.

[47] Wang, G., Ciobotaru, M., Agelidis, V. G. 'Power smoothing of large solar PV plant using hybrid energy storage'. *IEEE Transactions on Sustainable Energy*. 2014; **5**(3): 834–842.

[48] Kollimalla, S. K., Mishra, M. K., Narasamma, N. L. 'Design and analysis of novel control strategy for battery and supercapacitor storage system'. *IEEE Transactions on Sustainable Energy*. 2014; **5**(4): 1137–1144.

[49] Hu, S., Liang, Z., He, X. 'Ultracapacitor-battery hybrid energy storage system based on the asymmetric bi-directional Z-source topology for EV'. *IEEE Transactions on Power Electronics*. 2016; **31**(11): 7489–7498.

[50] Romero-Cadaval, E., Milanes-Montero, M. I. 'Cooperative converters in power electronic systems'. *2010 12th Biennial Baltic Electronics Conference*. 2010, pp. 51–60.

[51] Romero-Cadaval, E., Milanes-Montero, M. I., Gonzalez-Romera, E., Barrero-González, F. 'Power injection system for grid-connected photovoltaic generation systems based on two collaborative voltage source inverters'. *IEEE Transactions on Industrial Electronics*. 2009; **56**(11): 4389–4398.

[52] Miñambres-Marcos, V., Romero-Cadaval, E., Guerrero-Martínez, M. A., Milanés-Montero, M. I. 'Cooperative operation of inverters for grid-connected photovoltaic generation systems'. *Electric Power Systems Research*. 2013; **96**: 47–55.

[53] Miñambres-Marcos, V., Romero-Cadaval, E., Guerrero-Martínez, M. A., Milanés-Montero, M. I. 'Cooperative converter for improving the performance of grid-connected photovoltaic power plants'. *IET Renewable Power Generation*. 2013; **7**(2): 110–117.

[54] Romero-Cadaval, E., Guerrero-Martínez, M. A., Milanes-Montero, M. I., Barrero-González, F., González-Romera, E. 'A cooperative converter for managing low-voltage energy storage systems'. *2016 10th International Conference on Compatibility, Power Electronics and Power Engineering (CPE-POWERENG)*. 2016, pp. 162–167.

Index

www.ingramcontent.com/pod-product-compliance
Lightning Source LLC
LaVergne TN
LVHW011912200926
848484LV00002B/605